Association Descartes - The *Association Descartes* was founded in 1989 under the sponsorship of the "Ministère de l'Enseignement Supérieur et de la Recherche (France)". Its activity centres on a dual mediation: firstly, to improve the intelligence of relations that are developing between the growing body of scientific and technical knowledge and our social systems; secondly, to promote encounters between the world of research and the world of action.

EUROPEAN DIRECTORY OF BIOETHICS

1993 - 1994

Under the direction of
Gérard HUBER

published with the support of
the Ministère de l'Enseignement Supérieur et de la Recherche (France)
and the cooperation of
the European Association of Centers of Medical Ethics

John Libbey Eurotext / Association Descartes

British Library Cataloguing in Publication Data
A catalogue for this book is available from the British Library.

ISBN 2-7420-0043-7

Édition John Libbey Eurotext
6, rue Blanche, 92120 Montrouge, France. Tél. : (1) 47.35.85.52

John Libbey & Company Ltd
13, Smith Yard, Summerley Street, London SW 18 4HR, England.
Tel. : (1) 947.27.77

Jonh Libbey CIC
Via L. Spallanzani, 11,00161 Rome, Italy. Tel. : (06) 862.289

A French version of this book was published by the "Association Descartes" in 1993.
A complete updating in bilingual version (french and english) will come out in 1995.

In accordance with usual practices, we have published information relating to people's activities only in cases where explicit permission was given in the response to our questionnaire.

Furthermore, we have naturally taken into account explicit requests not to be included in the Directory. These rules, which correspond to legal and statutory obligations, explain any omissions which the user might notice.

To make this Directory as complete and useful an instrument as possible, however, certain omissions -which would have appeared as gaps, given the importance of the omitted individuals or organisations- have been filled in as recommended by DG XII of the Commission of European Communities. These individuals and organisations appear in the Complementary List.

The Publisher asks those persons and institutions whom we were unable to identify, and who therefore were not contacted, to kindly excuse us. We should be grateful if they could make themselves known, so that they may be included in the next edition of the Directory.

Table of contents

Preface

This book is much more than a directory. Besides meeting the urgent need for a list of names and addresses (as attested by the many excellent answers to our questionnaire), it provides a silhouette, faint as yet, of a dawning field of awareness. Bioethics is not built like a discipline, even less like a science buttressed by a solid body of knowledge. It is a whole aggregate of questions that we seek to answer, each by his or her own means. In the ledger that the other sciences have opened, this new field can thus do no more than add notes in the margin.

Yet this book is, in itself, a source of learning. Not only through its list of addresses, but through our respondents' list of "centres of interest", described with remarkable precision. This family of themes (sometimes subtly linked), this sum of new relations between such diverse disciplines, this gamut of shared concerns, together provide the best, perhaps even the only definition of bioethics. There is knowledge and wisdom that can only take root after passing through the winnows of language. This is the role of the lists presented here: lists of names compile practices; lists of theme words compile written texts.

Our list has not been drawn up by chance, nor just to meet our objectives. It is based on the INSERM list, inspired by the BIOETHICS data base, which in turn is connected to the MeSH base of the National Library of Medicine in the United States. It is not a predefined nomenclature, cast in iron, although it was used as such to meet the needs of our first survey. This list uses the language of indexes, a language that grows with each new word a scientist contributes, quick to seize on new concepts, recording new relations and quantifying recurrent themes. This language captures, weighs, joins, and compares objects that have even once appeared relevant. It is a tool for keeping what can be called an "epistemological watch".

In any other discipline, such an instrument would be nothing more than a bare list. Here, it enables us to apprehend an unstable ensemble with uncertain limits and far off horizons, but ever so cautiously: our perception must remain open and relative.

Open, because the list of sciences concerned with bioethics will never be closed. Indeed, the thesaurus must expand beyond the medical context of its origins, and already psychoanalysts, philosophers, lawyers, and historians are adding their names, introducing their own perspectives.

Relative, because each concept must be interpreted in terms of its history, the stakes involved, the surrounding culture. Here the thesaurus, unlike others of its kind, must not seek to unify terminology too quickly. Dual or parallel meanings are preferable to standardisation.

This directory foreshadows an edifice that is yet to be built. To compile and then review these questions together would require more than a classical documentation centre, ill equipped to deal with such a wide range of issues in such a sensitive manner. The new sciences of our times are to be found at the crossroads of the old disciplines (I am thinking of psycholinguistics and of all that makes up the "cognitive sciences", which involve information science as much as grammar, psychology, and semantics). For these sciences, it is necessary to invent a new mode of documentation management. In such a system, each document would remain inseparable from its source and context, and the system would not rely on its own collection but could call, as needed, upon the various libraries of the world. Such an observatory would not even possess its own collection of documents, but rather its own methods for pinpointing questions and highlighting how they are approached.

This directory is as different from a catalogue as knowledge is from information. For it to thrive, this is how it must be understood and critically analysed.

Michel Melot
Vice Chairman of the
Conseil supérieur des bibliothèques*

*Higher Council of Libraries

Foreword

In Europe as a Community, biomedical research generates multifarious thought, opinions, recommendations, codes of conduct and legal texts, and an equal number of answers to the problems that the very rapid progress in biological knowledge and its medical applications, both preventive and therapeutic, imposes on society and on the individual.

Legal, executive, and advisory bodies devote great amounts of time to this issue, with special concern for protecting the dignity of the individual without holding back research that is expected to conquer the major diseases.

It this context, European ethics changed over the last ten years as national ethics committees were created in Belgium, Denmark, France, Italy, Luxembourg, and Portugal, and discussions on the matter were started in Spain, and the United Kingdom. Ireland and Germany have opted for bodies adapted to specific fields. Recently, under the auspices of the Council of Europe, Madrid hosted a meeting of interested national committees and similar bodies.

This movement has also made its mark on the Commission of European Communities, which formed an advisory group on ethics in biotechnology; DG XII, notably, included in its BIOMED 1 programme (Area IV) a careful examination of ethical questions, in accordance with decisions of the European Council and Parliament.

Since the major research programmes are joint-European endeavours, biomedical ethics should be reflected at the European level. This was the position espoused at major conferences such as The European Bioethics Conference — Human Embryos and Research (Mainz, 7-9 November 1988) organised by DG XII and BMFT (the German Ministry of Research and Technology in Bonn), and the Colloque, Patrimoine Génétique et Droits de l'Humanité (Symposium on Genetic Heritage and the Rights of Humanity — Paris, 25-28 October, 1989) organised by Association Descartes.

This thought and discussion indicate above all that the field of ethics is not limited to biomedical ethics per se. That is why a new concept has been forged, the concept of bioethics. It means that beside the welfare of the patient - which is still the main objective — biologists, physicians, sociologists, philosophers, psychoanalysts, lawyers, economists, and parliamentarians of all countries, in particular those from the European Community are, from their particular vantage points, wondering about the effects of biomedicine-driven change on human mores, the conceptual evolution that may result therefrom, and the new approaches that they can suggest to practitioners and decision-makers. We can expect this movement to gradually affect public opinion at large.

As groundwork for better understanding, the Association Descartes took the initiative to produce and publish this European Directory of Bioethics. It is essentially composed of a list of committees, centres, organisations, institutions, and resource people in the Community who are known for their work on bioethics.

This directory will be updated regularly. We hope that the next edition will include all the information relative to the non-members countries.

This book is the result of much joint effort. Our gratitude is due to many authorities and colleagues, starting with the Steering Committee and the Board of Trustees of the Association Descartes who supported the project, and then to the Reading Committee which helped us formulate a most relevant questionnaire. We should also like to express our appreciation to the Comité consultatif national d'éthique (the national advisory committee on ethics) and the Centre de Documentation et d'Information en Ethique (at INSERM - the Centre for documentation and information on ethics) for their valuable assistance. We owe special gratitude of the Ministère de L'Enseignement Supérieur et de la Recherche (France) whose financial support made the project possible.

Last, special tribute is due to the efficiency of Miss Sonia Le Bris, and the dynamism of Rev. Edouard Boné, Drs. Louis and Nicole Lery, Emilio Mordini, Catherine Perrotin and Maurice De Wachter, members of the European Association of Centers of Medical Ethics.

Gérard Huber
Head of the "Sciences et Technologies du Vivant" Department, Association Descartes

European Community

Problems of bioethics, at the European Community level, are being viewed with growing and active interest. As the European area becomes more unified, a minimum of harmonisation seems necessary between the often dissimilar regulatory provisions obtaining in the member States. In the spirit that characterises the construction of Europe, with its own specific features, the objective is to prepare a foundation of fundamental principles, accepted by all, and without jeopardy to national and cultural identities in which the various perceptions of the ethical aspects of science are rooted.

Community activities in the field of bioethics are carried out at various levels. Generally many acts adopted by the Community in its traditional domains (health, agriculture, environment, intellectual property, etc.) include bioethics. The provisions that have been made show the interlinkage between the economic, scientific and ethical dimensions in the way that the Community treats the issues at stake. Economic logic is tempered by thoughts on respect for the human person, respect for diversity and for the integrity of the living person, etc. One field that is especially concerned is biotechnology. Two directives devoted, respectively, to voluntary dissemination in the environment and to the limited use of genetically modified organisms were adopted in 1990. A directive on the legal protection of biotechnological inventions is currently being discussed.

The European Parliament has long been deeply concerned with questions relating to bioethics. In 1988 it adopted a resolution on European harmonisation in the field of medical ethics. In 1989 two resolutions were adopted on the ethical and legal problems of genetic engineering and on artificial fertilisation in vitro and in vivo. The Parliament also introduced several amendments to proposals for Community research programmes in the fields of human genome analysis and medical research. It is largely thanks to a Parliament initiative that greater importance is being assigned to questions of ethics in these two programmes.

In 1988 the Commission of the Communities, at the initiative of the German Government, organised two conferences. One was held in Mainz, on the ethical problems of research on human embryos and the other in Berlin, on the effects of disseminating genetically modified organisms in the environment. It also participated in conferences on bioethical issues organised under the Group of the Seven Most Industrialised Countries (G7). In this context, it prepared in 1989 the Brussels conference on the ethics of the environment.

In 1990, the European ministers of research held an informal meeting in Kronberg (Germany). The goal was to identify the ethical principals of biomedical research on which the Member States agreed, to examine discordant positions in the two fields of embryo research and human genome analysis, and to discuss possibilities of bringing themselves closer to a consensus. Going beyond certain differences in problem assessment, the participants were able to come to agreement on a certain number of issues. In the field of human embryo research, certain practices that were deemed unjustifiable were rejected, e.g. cloning by substitution of nuclei, research on embryos that could be carried out on animals, etc. In the field of human genome analysis, the principles of the individual's right to free decision and of confidentiality in data processing were accepted, gene therapy of somatic cells was authorised. Two working groups were established as a result of the Kronberg meeting.

The first is responsible for monitoring the legal and practical aspects of research on human embryos in the Member States and identifying sectors where a consensus could be reached. (WG-HER, Working Group on Human Embryos and Research). In 1992, a first report was published on research on embryos before implantation. In an annex, the state of legislation on embryo research in the Member States was reviewed. Three groups of countries were identified: countries that have not yet passed legislation, countries where all research on embryos is forbidden, and countries where research on embryos is allowed under certain conditions, i.e. no research on embryos over 14 days of age, prior authorisation from an ethics committee, etc.

Activities of the second group concern the ethical, social, and legal aspects of human genome analysis. This group's mission is to stimulate public discussion and make recommendations to the Commission in preparation of future initiatives, including legal initiatives related to applications of human genome analysis (WG-ESLA, Working Group on Ethical, Social, and Legal Aspects of Human Genome Analysis). Here again, a first report was published in 1992. The group expressed its commitment to the principle of free and informed consent for diagnosis and selection, and reasserted the general value of the confidentiality principle applied to individual genetic information. It expressed opposition to all forms of patentability of the human genome as such (but not, however, of products and procedures stemming from knowledge acquired on the genome). It felt the principle of somatic gene therapy to be acceptable, but deemed unacceptable, as the issue now stands, any modification of human germ cells.

As part of the Community's research programme in the field of human genome analysis (1990-1992), a discussion has been organised on the ethical, social and legal aspects of this type of research. Eighteen studies have already been started on various themes, e.g. genetic information and life insurance, the ethical aspects of experimental genetic therapy, ethical standards in genetic counselling in Europe, etc. These studies will be completed by the end of 1992 and then published by the Commission.

One whole phase of the Community's research programme on biomedicine and health (1990-1994) gives special attention to research on biomedical ethics. The focus is on problems related to three other fields of research covered by the programme, viz., prevention, care and health systems; diseases with a strong socio-economic impact (AIDS, cancer, cardio-vascular diseases, mental illnesses and neurological disorders, ageing); human genome analysis. A certain number of general ethical effects of research on human beings are also taken into account.

The biotechnology research programme (1990-1994) also includes research on the ethical aspects of activities conducted in this field.

Especially in the field of biotechnology, in order to shed light on the way these questions are broached in the various member States and to initiate discussion between representatives of the various disciplines, the Commission has created an ethics advisory group responsible for determining the potential impact of the Community's activities in the field of biotechnology on society and on the individual person. It can take the initiative of preparing reports for the Commission or may be consulted by the Commission on specific questions.

Other Community programmes relate to bioethics in the broad sense of the term. Part of the environmental research programme (1990-1994), for instance, bears on the economic and social aspects of environmental problems. On the "Ecology and Biology of Populations" theme, the biotechnology research programme also works on both the ecological implications of biotechnology and the conservation of genetic resources.

Activities of the European Community in the field of bioethics are carried out in close cooperation with parallel activities in the Member States and in other European organisations such as the Council of Europe. The Commission, in particular, is a member of the Bioethics Steering Committee of the Council of Europe together with which the European Parliament organised a joint meeting in 1992 and studied the possibility of creating a European committee on bioethics.

NB: The names of the members and/or managers of the institutions and groups mentioned are compiled in the *European Institutions* section (HER, ESLA, Advisory group on ethics of biotechnologies, European Parliament).

Director General of DG XII: Paolo M Fasella.
Head of Unit on Ethical and Legal Aspects of Life Sciences: José Elizalde.
200, rue de a Loi — 1049 Brussels
✆ (322) 295 72 87 Fax: (322) 296 20 07

The Council of Europe,

a centre for reflection and action in the field of biomedical ethics

The Council of Europe brings together all the democratic countries of Europe, i.e. 32 member states[1] as well as many countries of Central and Eastern Europe, which enjoy special guest status with the Parliamentary Assembly for a wide-ranging task: European unity, to be attained by political and technical co-operation. It is active in many fields, ranging from law, human rights and health, to education, culture, environment and mass media. Bioethics is one of these activities. The Council acts as an international clearing house for reflection and action regarding the social, legal, human rights, ethical and moral consequences of the development of the biomedical sciences.

The results of the Council of Europe's action may take the form of multilateral treaties, i.e. binding, or of recommendations setting out guidelines for the policy and legislation of the member states. In addition to the two principal organs, the Parliamentary Assembly and the Committee of Ministers, there are several conferences of specialised ministers. One of these Conferences, that of the European Ministers of Justice, closely follows and supports the work on bioethics.

The ad hoc Committee of experts on bioethics (CAHBI/CDBI)[2]

In March 1985, the European Ministerial Conference on Human Rights adopted, in Vienna, Resolution N°3, asking that the Council of Europe become the focal point and clearing house for information, opinions and, where appropriate, joint international action with regard to biomedicine. This Resolution was based on a report, presented by the French Minister of Justice, on the "protection of human beings and their physical and intellectual integrity in the context of the progress being made in the fields of biology, medicine and biochemistry". The report drew attention to the fact that the life sciences are creating a new situation, which upsets the traditional legal order. In his report, the French Minister called on the law-makers of Europe to proclaim and guarantee the inviolability of the human being and appealed to his colleagues to undertake the analysis of these problems in an international context.

This Resolution of the Human Rights Ministers was followed by an initiative of the European Ministers of Justice who held in June 1985 in Edinburgh an informal meeting on the basis of a note of the Spanish Minister entitled "Human fertilisation and embryology". The Ministers of Justice noted that any exclusively national regulation in the biomedical field would run the risk of remaining ineffective in practice because it would always be possible to use certain techniques in another country where the regulation was different, hence the importance of European harmonisation.

In response to these two Ministerial Conferences, the Committee of Ministers decided to intensify and enhance the work which was already going on in the Council of Europe on the impact of the biomedical sciences. The ad hoc Committee of experts on bioethics (CAHBI) was instructed to carry out the activities of the Council of Europe in this field. It is a multidisciplinary body — lawyers, biologists, medical experts, ethicists and human rights experts — which counts, apart from experts of the member States, participants from several non-member States: Australia, Bulgaria, Canada, Japan, Holy See, the United States of America and Russia as well as the Parliamentary Assembly, the European Community, the World Health Organisation, UNESCO, OECD, the International Commission on Civil Status, the European Science Foundation and the Steering Committees of the Council of Europe in the fields of health (CDSP) and human rights (CDDH).

The main object of the CAHBI is to fill the political and legal gaps that may result from the rapid development of the biomedical sciences. The principle of the Rule of Law, to which all member States of the Council of Europe are committed, implies not only their duty to apply and respect existing law but also to create where necessary new law, adapted to the future.

The CAHBI is aware of the difficulty for the member States to arrive at a consensus on these extremely delicate problems. Prior to the setting up of the CAHBI it has happened on two occasions that draft standard-setting texts on which there was a wide measure of agreement remained a dead letter for lack of unanimity. The CAHBI therefore actively explores every means to promote constructive dialogue between the member States and to avoid deadlocks. It encourages them to carry out in a harmonised manner law reform and policy-making in the field of bioethics. The texts of the CAHBI reaffirm the major principles and values which must guide any regulations on bioethics and also indicate which limits must at all cost be respected. The CAHBI has made special efforts in order to identify the fundamental points on which the Member States are unanimous, such as respect for human dignity, recognition of the inviolability of the human body, avoidance of commercial profiteering in the field of biomedicine, as well as the principle of self-determination of the individual.

Results of the work of the Council of Europe in the field of bioethics

The Council of Europe's work in bioethics deals with the life sciences and their applications. It focuses on the beginning of life, or access to life, and on the quality of life. In this context, it is important to acquire better insight into the genetic aspects of the human organism. The Council of Europe therefore follows with keen interest the progress made in the biomedical sciences. This is true both for fundamental research -e.g. the Human Genome Project- and for its practical applications that make it possible to detect particular disorders in persons and families. The Council examines in this context the legitimate applications of biomedicine and the limits to be set with regard to the use of research findings. Since 1978, when it began dealing with problems of biomedicine, the Council of Europe has adopted or published the following texts:

Human substances

1978 Resolution (78) 29 on the harmonisation of legislation of member States relating to removal, grafting and transplantation of human substances

1979 Recommendation R (79) 5 on international exchange and transportation of human substances

Mental disorders

1983 Recommendation R (83) 2 on legal protection of persons suffering from mental disorders and placed as involuntary patients

Recombinant DNA

1984 Recommendation R (84) 16 concerning notification of work involving recombinant deoxyribonucleic acid (DNA)

Euthanasia

1987 Opinion given by the CAHBI at the request of the Government of the Netherlands on the question of "terminating the life of a patient at his express request" (this document remains confidential)

Human artificial procreation

1989 Report on human artificial procreation

Medical research on human beings

1990 Recommendation R (90) 3 on medical research on human beings

Genetic screening and testing for medical purposes

1990 Recommendation R (90) 13 on prenatal genetic screening, prenatal genetic diagnosis and associated genetic counselling

1992 Recommendation R (92) 1 on the use of analysis of deoxyribonucleic acid (DNA) within the framework of the criminal justice system

1992 Recommendation R (92) 3 on genetic testing and screening for health care purposes

Protection of the human person in the context of the development of biomedical sciences

Since 1991, preparation of a bioethics Convention and Protocols relating to (i) organ transplants and (ii) medical research on human beings

The Council of Europe, a bioethics forum

In order to make the Council of Europe's work in the field of bioethics better known to interested circles (universities, research and training centres, ethics committees, politicians, churches, the business world, international and non-government organisations etc.) and to promote exchange of information and ideas, the CAHBI organised in Strasbourg in December 1989 the first Symposium on Bioethics. Three hundred persons from all over the world took part in this Symposium.

The principal themes of the Symposium were: (i) training and teaching (ii) research in bioethics (iii) information and documentation (iv) ethics committees. The Symposium enabled the Council of Europe to register the participants' opinions about new ideas and proposals launched during the Symposium, in particular those which were submitted by the Portuguese Minister of Justice, the French Deputy Minister of Foreign Affairs and the Secretary General of the Council of Europe.

The proceedings of the Symposium have been published by the Council of Europe under the title "Europe and Bioethics". During the Symposium the film "Lifelines", produced specially for the occasion, was shown[3]. The Second Symposium will take place at the end of 1993. Its principal theme will be "Genetics".

Teaching, exchange of information and documentation

The Council of Europe actively supports co-operation and exchanges between research and documentation centres in the field of bioethics and the publication of journals and newsletters. It is represented on the Steering Committee of the International Association "Law, Ethics and Health". In March 1991 Oxford University organised, jointly with the Council of Europe, an international Symposium on "Ethics, Law and Communication Skills". Following this Symposium, the Council of Europe's Health Division has started to build up a network of bioethics teachers.

Ethics committees

The Council of Europe will organise regular meetings of chairmen of national ethics committees and similar bodies. The first Round Table of Ethics Committees was held in Madrid on 24 and 25 March 1992. Its aim was to enable these eminent figures to discuss problems of common concern and to offer Member States considering setting up an ethics committee an opportunity to learn about the experience of existing committees. This meeting's report was drafted by Mrs. Sonia Le Bris, Montréal.

Towards an International Convention

In September 1991 the Committee of Ministers instructed the CAHBI to prepare a framework Convention, open to non member-States, setting out common general standards for the protection of the human person in the context of the development of the biomedical sciences. By this decision, the Ministers have given a positive reply to Resolution N°3 of the 17th Conference of European Ministers of Justice (Istanbul, June 1990) which underlined that the most fundamental rights of human beings are likely to be affected by the development of biomedical sciences and that it is therefore desirable to promote as much as possible the harmonisation of national laws in this field. The universality of the rights of the human person implies the necessity for States to protect these rights in their international context.

A preliminary draft for this Convention and protocols dealing with (a) organ transplants and (b) medical research on human beings is at present being negotiated. The final text of this Convention could be proposed for signing at the beginning of 1994.

Liaison with the Parliamentary Assembly

In the field of bioethics there is a permanent dialogue between the intergovernmental and parliamentary organs of the Council of Europe. The Assembly has inspired and encouraged this work in particular by the following Recommendations.

Recommendation 779 (1976) on the rights of the sick and dying
Recommendation 818 (1977) on the situation of the mentally ill
Recommendation 934 (1982) on genetic engineering
Recommendation 1046 (1986) on the use of human embryos and foetuses for diagnostic, therapeutic, scientific, industrial and commercial purposes
Recommendation 1100 (1989) on the use of human embryos and foetuses in scientific research
Recommendation 1160 (1991) on the preparation of a Convention on bioethics

On 7 February 1992 a joint meeting took place in Strasbourg of members of the Parliamentary Assembly and of the European Parliament interested in bioethics.

For any further information apply to
Division I, Directorate of Legal Affairs
Council of Europe, BP 431 R 6
67 006 Strasbourg Cedex
Contact: Carlos de Sola
Tel: (33) 88 41 20 04 - Fax: (33) 88 41 27 94

1. Austria, Belgium, Bulgaria, Cyprus, Czech Republic, Denmark, Estoniya, Finland, France, Germany, Greece, Hungary, Iceland, Ireland, Italy, Liechtenstein, Lithuania, Luxembourg, Malta, Netherlands, Norway, Poland, Portugal, Romania, San Marino, Slovakia, Slovenija, Spain, Sweden, Switzerland, Turkey, United Kingdom.

2. Since April 1992, the name of this committee has been changed to Comité Directeur pour la Bioéthique CDBI (Steering Committee for Bioethics). The current president is Mr. Octavi Quintana-Trias, Spain.

3. The film was produced by Pontes Film and sponsored by Boehringer Manheim GmbH. It is distributed by Engel Verlag, D-7640 Kehl/Rhein, Fax: (49) 7851-4234

Notice

The European Directory of Bioethics is the result of a survey carried out across Europe among people and institutions known to be active in the field of bioethics.

These people and institutions have been identified as the result of a two-stage procedure: firstly, use was made of the lists drawn up by the European Association of the lists drawn up by the European Association of Centres of Medical Ethics at the time of a first European survey (carried out at the request of the Association Descartes by Doctor Nicole Léry and Professor Edouard Boné); then contact was made with the people mentioned by those who, in answering the questionnaire sent to them, were kind enough to indicate the names and addresses of the people and institutions they thought should be included in the Directory. The bioethics community itself has thus supplied the names listed in this edition of the European Directory of Bioethics.

In accordance with usual practices, we have published information relating to people's activities only in cases where explicit permission was given in the response to our questionnaire. Furthermore, we have naturally taken into account explicit requests not to be included in the Directory. These rules, which correspond to legal and statutory obligations, explain any omissions which the user might notice. To make this Directory as complete and useful an instrument as possible, however, certain omissions -which would have appeared as gaps, given the importance of the omitted individuals or organisations- have been filled in as recommended by DG XII of the Commission of European Communities. These people and organisations appear in the Complementary List.

Structure of the Directory

Fourteen sections make up the body of this European Directory of Bioethics: the first is devoted to European institutions; the next twelve each cover one country of the European Community; the fourteenth contains the Complementary List. Each geographical section is itself divided into two chapters: Organisations and Individuals. Regarding the content of the latter sections, the following comments may be useful to the reader.

Where no entry was made under a heading, the heading does not appear in the Directory. Where an entry was made but no details were supplied, the word "yes" appears under the heading. The various headings are listed below in the order in which they appear in the Directory.When the sections have not been filled in, the corresponding section heading does not appear in the Directory. When the sections have been filled in, but without details, they are given with the word "yes" after them. The order of the sections presented below is the order in which they appear in the Directory when they have all been filled in.

Directory of organisations

Reference number. The reference number in the left-hand margin begins with an O, followed by the usual "number plate" code of the country (F for France, DK for Denmark, etc.) and a serial number. Its purpose is to enable rapid identification of entries in the book, notably when the reader wishes to use the indexes.

Affiliation. Indicates the organisation to which the respondent body belongs (in the case of a university department, for example).

Other fields of activities. Indicates other fields of activity when bioethics is not the organisation's sole field of activity.

Date of foundation.

Executive board. This section comprises a maximum of ten people's names (and possibly their positions within the body). Where the list supplied contained more than ten names, we have retained either the first ten or, where possible, the names of the principal directors: for instance, the chairman, the secretary general, and the first eight members listed).

Officially regulated organisation. Indicates the legal texts or regulations that may govern the body's mission or its operation (as is generally the case for national ethics committees, for instance). The responding organisations have sometimes referred to non-specific texts in their entries (such as the law of 1901 governing the operation of associations in France).

Teaching. Indicates teaching given (in principle, specifically bioethics teaching).

Publications.

Documentation centre. Indicates the existence of a documentation centre and specifies whether it is open to the public or reserved for internal use.

Colloquiums, symposiums.

Working on, Interested in. These sections contain lists of keywords indicated by the respondent and corresponding to the reference nomenclature supplied with the questionnaire (see Thesaurus section at end of Directory). "Working on" refers to actual fields of research, teaching, or publication. "Interested in" refers to fields on which the respondent would like to exchange or receive information, without these necessarily being current fields of work. (A thematic index at the end of this Directory gives an alphabetical list of keywords with, under each one, the persons or organisations who mentioned the keyword in their response).

Directory of individuals

Reference number. This obeys the same principle as the reference number of the bodies listed, except that it does not begin with a letter O.

Positions. The name of the person, cited is followed by an indication of the position(s) held — generally as filled in on the questionnaire, or, where necessary, translated.

Affiliations. Indicates the name and the country of the body or bodies (maximum three) to which the respondent declares himself to be affiliated, with an indication, where applicable, of the position held within the organisation (member, chairman, etc.).

Teaching. (Same principle as the corresponding heading in the "Organisations" chapter.)

Publications. When the section has been filled in, the words "Publication:

Publications. When the section has been filled in, the words "Publication: yes" appear. It has been decided - for reasons of consistency with the main purpose of the Directory - not to give more details in this section, and to reserve the references cited for use in a systematic bibliography of the field.

Working on. Interested in. (Same principle as the corresponding sections in the "Organisations" chapter.

Translation

The European Directory of Bioethics is published in two separate editions, one in French and the other in English.

It has been judged necessary to go beyond simply translating the introductory pages, and to supply the public with two separate versions which are as "monolingual" as possible, especially since many answers (concerning positions and teaching in particular) were given in all the other languages of the European Community, and not just in English and French.

Naturally, all the problems inherent in this type of undertaking (translation of job titles, in particular) were encountered, and in some cases it has been decided to favour understanding on the part of the reader rather than give exact but unintelligible translations of certain terms.

Indexes

The book has two indexes: one is a thematic index which can be used, on the basis of a keyword, (in an alphabetical list) to find the people and organisations who have described their work by this keyword; the other is an index of the names cited.

Each person or body cited in these indexes is designated by a name followed (in square brackets) by the reference number within the Directory. (For example, the reference number [O F 12] refers to the twelfth organisation in the "Organisations" list in the France section; reference number [DK 3] refers to the third name in the "Individuals" list in the Denmark section).

Thesaurus

This contains the list of keywords as presented in the survey questionnaire, i.e. in hierarchical form).

This thesaurus was developed by the Centre de Documentation et d'Information en Ethique CDIE (Ethics Documentation and Information Centre) of the INSERM (National Institute for Health and Medical Research) within the scope of a project sponsored by the Ministère de l'Enseignement Supérieur et de la Recherche. It comprises the key-words of the Bioethics Thesaurus published by the Kennedy Institute of ethics (USA).

The thesaurus makes it possible to consider "clusters" of related concepts. This sometimes enabled us, where too many concepts were listed, to group a set of concepts under a single inclusive entry.

Directory

European Institutions

Members of the Working Group on Ethical, Social, and Legal Aspects of Human Genome Analysis (ESLA)

Belgium

BONÉ (Père) E. **[L 2]**
18, rue des Trévires
1040 Brussels (Belgium)
✆ (32) 2 735 48 04 Fax: (32) 2 230 05 56

VAN DEN ENDEN (Prof. Dr) H.
Faculteit van den Letteren en Wijsgegeeret
Dienst voor Moralfilosofie
RUG – Blandijnberhg
9600 Ghent (Belgium)
✆ (32) 91 64 39 74 Fax: (32) 91 64 41 45

Denmark

MIKKELSEN (Dr) M. **[DK 33]**
Department of Medical Genetics
J.F. Kennedy Institute
7GL Landevej
2600 Glostrup (Denmark)
✆ (45) 42 45 22 28 Fax: (45) 43 43 11 30

TYBKJAER (Mrs) H.
Landesforeningen Til Bekaempelse Af Cystik Fibrose
Hyrdebakken 246
8800 Uttborg (Denmark)
✆ (45) 86 67 44 22 Fax: (45) 86 67 66 66

France

GROS François **[F 99]**
Comité consultatif national d'éthique (CCNE)
INSERM
101, rue de Tolbiac
75013 Paris (France)

MOULIN (Dr) J.
Secrétariat général du Conseil national de l'Ordre des médecins
60, boulevard Latour-Maubourg
73340 Paris Cedex 07 (France)
✆ (33) (1) 47 53 24 00 Fax: (33) (1) 45 51 75 96

QUESTIAUX Nicole
13, avenue de Bretteville
92200 Neuilly-sur-Seine (France)

Germany

PROPPING Peter **[D 39]**
University of Bonn
Institute of Human Genetics
Wihelmstr. 31
5300 Bonn (Germany)
✆ (49) 228 287 23 46 Fax: (49) 228 287 23 80

DEUTSCH (Prof.) E.
Institute of Pharmaceutical and Medical Law
University of Göttingen
3400 Göttingen (Germany)
✆ (49) 551 39 73 96 Fax: (49) 551 39 48 72

HÜBNER Jürgen **[D 23]**
Foschungsstätte des Evangelischen
Institute for Interdisciplinary Research (FEST)
Schmeilweg 5
6900 Heidelberg (Germany)
✆ (49) 6221 18 40 61 Fax: (49) 6221 16 72 57

TIEMANN (Dr) S.
Secrétariat européen des professions libérales, intellectuelles et sociales
Im Meisengrund 4A
5000 Köln 50 (Germany)
✆ (49) 32 25 11 44 39 Fax: (49) 32 25 11 01 24

Greece

MANIATIS George **[GR 10]**
Faculty of Medicine
Department of Biology
261 10 Patras (Greece)
✆ (30) 61 997 621 Fax: (30) 61 997 689

Ireland

O'DWYER (Dr) Timothy V. **[IRL 6]**
Department of Health
Hawkins House
Hawkins Street
Dublin 2 (Irlande)
✆ (353) 1 71 47 11 Fax: (353) 1 71 45 08

Italy

SINISCALCO (Prof.) M.
Imperial Cancer Research Fund
PO box 123

44 Lincoln's Inn Fields
WC2A 3PX London (Royaume-Uni)
✆ (44) 71 269 30 53 Fax: (44) 71 269 34 79
or CNR Institute of Molecular Genetics
Viale S. Pietro 43bis
07100 Sassari (Italy)

TERRAGNI (Dr) F.
Gruppo di Attenzione sulle Biotechnologia
Via Iglesias 33
20128 Milano (Italy)
✆ (39) 2 27 00 11 35 Fax: (39) 2 25 52 281

GERIN (Prof.) Guido **[I 26]**
Institut international d'études des droits de l'homme
Centre international d'études de bioéthique
Via Cantù n°10
34127 Triste (Italy)
✆ (39) 40 52 121 Fax: (39) 40 37 07 77

Luxembourg

BETZ (Dr) A.
Centre de la recherche publique Santé
120, route d'Arlon
1150 Luxembourg (Luxembourg)
✆ (352) 45 32 13 Fax: (352) 45 32 19

Netherlands

NIERMEIJER (Prof. Dr) M. F.
Department of Clinical Genetics
University Hospital Dykzigt
P.O. box 1738
3000 DR Rotterdam (Netherlands)
✆ (31) 10 463 43 07 Fax: (31) 10 408 72 13

Portugal

ARCHER Luis Jorge Peixeto **[P 2]**
Faculdade de Ciências e Tecnologia
Quinta da Torre
2825 Monté Da Caparica (Portugal)
✆ (351) 1- 295- 4464 Fax: (351) 1- 295 44 61

Spain

PALACIOS Marcelo **[E 27]**
Congreso de los Diputados
Calle Floridablanca, 1
28071 Madrid (Spain)
✆ (34) 1 429 51 93 Fax: (34) 1 429 96 27

QUINTANA-TRIAS (Dr) O. **[E 27b]**
Ministerio de Sanidad y Consumo
Paso del Prado 18-20
28071 Madrid (Spain)
✆ (34) 1 420 23 75 or 1 338 03 18
Fax: (34) 1 420 10 42

United Kingdom

BRENNER (Dr) S.
MRC Molecular Genetics Unit
Hills Road
CB2 2QH Cambridge (United Kingdom)
✆ (44) 223 24 80 11 or 223 40 24 06
Fax: (44) 223 21 01 36

DUNSTAN (Prof.) G.R. **[GB 14]**
9, Maryfield Ave.
Pennsylvania
Exeter EX4 6JN Devon (United Kingdom)
✆ (44) 392 21 46 91

METTERS (Dr) J.S. (President)
Department of Health, Richmond House
79 Whitehall
SW1A 2NS London (United Kingdom)
✆ (44) 71 210 55 91 Fax: (44) 71 210 55 72

MODELL (Dr) B.
Perinatal Centre
UCM Obstetrics Unit
86-96 Chenies Mews
WC1E 6HX London (United Kingdom)
✆ (44) 71 387 93 00 Fax: (44) 71 380 98 64

Observers :

HONDIUS (Dr) F.W.
Council of Europe
1, avenue de l'Europe
67000 Strasbourg Cedex (France)
✆ (33) 88 41 22 03 Fax: (33) 88 41 27 94

MUFFAT-JEANDET Danièle
Council of Europe, DAJ
BP 431 R6
67006 Strasbourg Cedex (France)
✆ (33) 88 41 22 22 Fax: (33) 88 41 27 94

Secretary :

M. A. KLEPSCH
Commission of the European Communities
DG XII, 200, rue de la Loi
1049 Brussels
✆ (32) 2-295 32 10 Fax: (32) 2-295 53 65

Members of the Working Group on Human Embryos and Research (HER)

Belgium

VAN DEN BERGHE (Dr) Herman **[B 48]**
Center for Human Genetics
Campus Gasthuisberg O en N
Herestraat 49
3000 Leuven (Belgium)
✆ (32) 16 21 58 78 Fax: (32) 16 21 59 92

Denmark

BOLUND (Prof.) Lars **[DK 3]** (President)
University of Aarhus
Institute of Human Genetics
Bartholinbygningen
Wilhem Meyers Alle
8000 Aarhus C (Denmark)
✆ (45) 86 12 31 73 Fax: (45) 86 19 12 77

France

JOUANNET (Prof.) P.
Laboratoire de biologie de la reproduction et du développement
Centre hospitalier universitaire Bicêtre
78, avenue du Général-Leclerc
94270 Kremlin-Bicêtre (France)
✆ (33-1) 45 21 28 03 Fax: (33-1) 45 21 26 58

Germany

ESER (Prof. Dr) Albin **[D 10]**
Max-Planck Institut für Ausländisches und internationales Strafrecht
Güntherstalstr. 73
7800 Freiburg i. Br. (Germany)
✆ (49) 761 70 811 Fax: (49) 16 21 59 92

Greece

DALLA-VORGIA (Dr) Panaglota **[GR 3]**
School of Medicine
University of Athens
Department of Hygiene and Epidemiology
Mikras Asias
115 27 Athens
✆ (30) 1 777 11 65 Fax: (30) 1 770 42 25

Ireland

O'DWYER (Dr) Timothy V. **[IRL 6]**
Department of Health
Hawkins House
Hawkins Street
Dublin 2 (Ireland)
✆ (353) 1 71 47 11 Fax: (353) 1 71 45 08

Italy

GERACI (Dr) Domenico **[I 25]**
Istituto di Biologia dello Sviluppo
Consiglio Nazionale delle Ricerche
Via Archirafi, 20
90123 Palermo (Italy)
✆ (39) 91 616 18 92 Fax: (39) 91 616 56 65

Luxembourg

BETZ (Dr) A.
National Committee of Ethics
4, rue Jos. Trekert
2620 Luxembourg (Luxembourg)
✆ (352) 45 32 13 Fax: (352) 45 32 19

Netherlands

DE BEAUFORT (Prof. Dr) Inez **[NL 5]**
Javaweg
2202 Ba Noorwijk (Netherlands)
✆ (31) 17 192 06 89

Portugal

OSSWALD (Prof. Dr) W.
Serviço de Farmacologia e Terapeutica Geral
Faculdade de Medicina da Unicersidade do Porto
Alam. Prof. Hernani Monteiro
4200 Porto (Portugal)
✆ (351) 2 49 56 94 Fax: (351) 2 49 81 19

Spain

MONTOYA (Prof. Dr) Eladio **[E 25]**
Comision interministeral de Cienca y Tecnologia CICYT
Det? técnico de Calidad de Vida y Recursos Naturales
c/Rosario Pino 14-16
28020 Madrid (Espagne)
✆ (34) 1 279 83 83 Fax: (34) 1 571 57 81

United Kingdom

MCLAREN (Dr) Anne
The Royal Society
6 Carlton House Terrace
SW1Y 5 AG London (United Kingdom)
✆ (44) 71 389 55 61 Fax: (44) 71 930 21 70

Secretary

LODER (DR) B.
Commission of the European Communities
DG XII
200, rue de la Loi
B-1049 Bruxelles (Belgium)
✆ (32) 2-296 31 93 Fax: (32) 2-295 53 65

Advisory group for biotechnology ethics

LENOIR Noëlle **[F 127]**
Conseil constitutionnel
2, rue Montpensier
75001 Paris (France)
✆ (33-1) 40 15 30 08 Fax: (33-1) 42 60 17 41

SINISCALCO (Prof.) M.
Imperial Cancer Research Fund,
PO box 123
44 Lincoln's Inn Fields
WC2A 3PX London (United Kingdom)
✆ (44) 71 269 30 53 Fax: (44) 71 269 34 79

MIKKELSEN (Prof. Dr) Margareta **[DK 33]**
Department of Medical Genetics
J.F. Kennedy Institute,
7GL Landevej
2600 Glostrup (Denmark)
✆ (45) 42 45 22 28 Fax: (45) 43 43 11 30

WARNOCK (Lady)
Brick House
Axford
Nr. Marlborough
SN8 2EX Wiltshire (United Kingdom)
✆ (44) 672 514 686

ZACHER (Prof.) Hans F.
Max Planck Institut
Leoploldstrasse 24-26
8000 Munich (Germany)
✆ (49) 89 210 82 12 Fax: (49) 89 22 98 50

President :

OREJA AGUIRRE Macelino
Committee on institutional affairs/ European Parliament/ Genova 13
28004 Madrid (Spain)
✆ (34) 1 410 11 43 Fax: (34) 1 308 15 12

Secretary :

F. D NAPOLI
Commission of the European Communities
Secrétariat général, 200, rue de la Loi
1049 Brussels (Belgium)
✆ (322) 295 55 01 Fax: (322) 295 06 15

European Parliament

Committee on legal affairs and citizen rights (secretary) :
PASSOS Ricardo
Parlement européen
Bâtiment Rema – 712
Rue Béliard, 97/113
1049 Brussels (Belgium)
✆ (32) 2 284 24 81 Fax: (32) 2 231 12 57

Committee on the environment, public health and consumer protection
(secrétariat) :
VAN DE PERRE Lieven
Parlement européen
Bâtiment Rema – 712
Rue Béliard, 97/113
1049 Brussels
✆ (32) 2 284 25 58 Fax: (32) 2 231 12 57

Committee on energy, research and technology
(secretary) :
SPENCE James
Parlement européen
Plateau de Kirchberg
Bâtiment Schuman – 5/58
2929 Luxembourg
✆ (352) 4300 2935 Fax: (352) 43 53 59

Programme STOA, Scientific and Technological Options Assessment (secretary) :
HOLDSWORTH Dick
Parlement européen
Plateau de Kirchberg
Bâtiment Schuman – 5/58
2929 Luxembourg
✆ (352) 4300 2511 Fax: (352) 43 53 59

The STOA programme of the European Parliament has just published its rapport: *«Bioethics in Europe»*, Luxembourg PE 158 453, DG for research 1992, directed by F. Terragni

Belgium

Area: 30 513 km²
Population: 9.9 m
Gross domestic product: BFr 7.4trn; US$ 211bn
Gross domestic product per head: $21,820
Gross domestic product growth: 1991: 2,3%
1992: 2,5%

Organisations

O B 1 **ASSOCIATION EUROPÉENNE DES CENTRES D'ÉTHIQUE MÉDICALE**
Centrum voor Bio-Medische Ethiek en Recht
Kapucijnonvoer 35
3000 Leuven (Belgique)
✆ (32) 16 33 69 51 Fax : (32) 16 33 69 52

O B 2 **BELGISCHE VERENIGING VOOR MEDISCHE MORAAL EN ETHIEK**
Dr. R. Lenaerts
Lindenlei, 63
2640 Mortsel (Belgium)
✆ (32) 3 449 86

■ OTHER FIELDS OF ACTIVITIES: organises public debates on medical ethics and bioethics.
■ OFFICIALLY REGULATED ORGANISATION: statutes, French and Dutch rules —
■ DATE OF FOUNDATION: 5/05/1973.
■ EXECUTIVE BOARD: Dr. Jacques Achslogh (Chairperson *(French branch)*), Me Yves Thiry (Executive Secretary *(French branch)*), Dr. Jean Bacquelaine (Vice-Chairperson *(French branch)*), Professor Paul Cosyns (Chairperson *(Dutch branch)*), Dr. Raymond Lenaerts (Secretary *(Dutch branch)*), Yves Thiry (*(Magistrate)* Vice Chairperson *(Dutch branch)*), Dr. Léopold Van der Borght (Vice-Chairperson).
■ PUBLICATIONS: annual publication of board proceedings (discussions and information): 13 volumes published.
■ COLLOQUIUMS, SYMPOSIUMS: Seminars in Antwerp, Brussels, Liège and Ostende: annual — Annual public debate in Brussels.
■ WORKING ON: fundamental rights of the individuals [B2], common good [B45], bioethics [B89], professional ethics [C51], ethics and humanities [C54], science [E2], life sciences [E4], research [E14], health policy [G13], family practice [G60], health occupations [G61], diagnosis [I2], biological specimens procurement, blood transfusion, organ transplantation [I8], treatment [I30], patients' rights [J2], confidentiality [J20], physicians and researchers accountability [J23], legal personality [K2], legislation and law [K34], childhood difficulties, diseases, protection [L21], aged — problems related with aging [L45], sexuality and procreation [M2], reproductive technologies [M26], pregnancy and childbirth [M39], abortion [M47], fetal therapy [M57], common symptomatology [N2], cancer [N17], cardiovascular diseases [N19], congenital defects [N23], communicable diseases [N27], nervous system diseases [N40], determination of death [O2], resuscitation [O21], behavioral and mental disorders [P2], outpatient commitment [P18], psychiatric technics [P24], genetic defects and hereditary diseases [Q2], biotechnology — genetic engineering [Q20], medical genetics [Q35].

O B 3 **CENTRE DE RECHERCHES INTERDISCIPLINAIRES EN BIOÉTHIQUE** (CRIB)
Université de Bruxelles
Avenue Adolphe Buyl 145 (CP 188)
1050 Bruxelles (Belgium)
✆ (32) 2 - 650 26 28 Fax: (32) 2 - 650 36 47

■ AFFILIATION: Interdisciplinary Research Centre affiliated to the Institute for Philosophy, Free University of Brussels.
■ DATE OF FOUNDATION: 1986.
■ EXECUTIVE BOARD: Professor Gilbert Hottois (*Director*), Professor Charles Susanne (*Co-director*), Professor Fernand Leroy (*Member of the Board*), Professor Yvon Kenis (*Member of the Board*), Professor Yvon Englert (*Member of the Board*), Professor Yvon Englert (*Member of the Board*), Dr. Jean-Marie Missa (*Secretary*), Colette Maton (*Administrative Secretary*).
■ TEACHING: Introduction course on bioethics.
■ PUBLICATIONS: Editing of the "sciences-éthiques-sociétés" review published by De Boeck (Brussels)and Erpi (Montreal).
■ DOCUMENTATION CENTRE: internal.
■ COLLOQUIUMS, SYMPOSIUMS: December 91 "Aux fondements d'une éthique contemporaine: H. Jonas et HT Engelhardt" International Colloquium .

O B 4 **CENTRE DE RÉFLEXION DE BIOÉTHIQUE DE LA FACULTÉ DE MÉDECINE**
Centre Hospitalier Universitaire
Sart Tilman
4000 Liège (Belgium)
✆ (32) 41 - 66 70 05

O B 5 CENTRE INTERFACULTAIRE DE DROIT, ÉTHIQUE ET SCIENCES DE LA SANTÉ (CIDES)

Facultés Universitaires N.D. de la Paix
Rue de Bruxelles 61
5000 Namur (Belgium)
✆ (32) 81 - 72 41 30 or 81 - 72 47 89

■ AFFILIATION: Interdisciplinary Research Centre affiliated to the Medical School, the Faculty of Law, and the Faculty of Science.
■ DATE OF FOUNDATION: 1986.
■ EXECUTIVE BOARD: Professor Joseph Duchène (*Co-ordinator*), Professor Xavier Dijon (*Delegate of the Faculty of Law*), Professor Dominique Lambert (*Delegate of the Faculty of Science*), Jean Burton (*Member*), Marie-Luce Delfosse (*Assistant*).
■ TEACHING: inter-faculty seminars on bioethics for students in medical sciences, pharmacology, law, philosophy, economics and biology.

O B 6 CONSEIL NATIONAL DE L'ORDRE DES MÉDECINS

Place de Jambline de Mieux, 32
1040 Bruxelles (Belgium)

O B 7 EUROPEAN SOCIETY OF HUMAN REPRODUCTION AND EMBRYOLOGY

AZ-Vrise Universiteit Brussel
Laarbeeklaan 101
1090 Bruxelles (Belgium)

O B 8 INSTITUT SUPÉRIEUR DE PHILOSOPHIE

Collège Thomas More
Chemin d'Aristote 1
1348 Louvain-la-Neuve (Belgium)
✆ (32) 2 - 764 43 30

O B 9 SOCIÉTÉ BELGE D'ETHIQUE ET DE MORALE MÉDICALE

38, rue Froissart
1040 Bruxelles (Belgium)
✆ (32) 2 - 287 59 88

O B 10 UNITÉ DE PHILOSOPHIE DES SCIENCES BIOMÉDICALES

Boîte UCL 43/ 4534
Promenade de l'Alma, 51
1200 Bruxelles (Belgium)
✆ (32) 2 - 764 43 30 Fax: (32) 2 - 764 53 22

■ AFFILIATION: School of Medicine. — Catholic University of Louvain (U.C.L.).
■ OFFICIALLY REGULATED ORGANISATION: yes: internal rule of the University.
■ DATE OF FOUNDATION: 1982.
■ EXECUTIVE BOARD: Professor Jean-François Malherbe (*Unit Director*), Dominique Jacquemin (*Co-director 92-93*).
■ TEACHING: School of Bioethics: 6 courses of 40 hours, plus dissertation. — Philosophy and bioethics at the School of Medicine. — Up-dating courses for nurses.
■ PUBLICATIONS: "Bioethica" (magazine), "Catalyse" (review)
■ DOCUMENTATION CENTRE: open to the public.
■ COLLOQUIUMS, SYMPOSIUMS: Colloquium "Quand l'autre meurt"(When the other dies). — Colloquium "Vivre-mourir-rester" (Living, dying, or staying). — Training seminars on helpful actions.
■ DECISIONS, ADVICE: CADE Help group for medical decision. — Ethics Committees. — Internal seminars (FIV (in-vitro fertilization)/AIDS/ liver transplants from living donors).
■ WORKING ON: bibliography [A4], documentation [A8], publications [A9], fundamental rights of the individuals [B2], human person [B12], traffic of organs [B15], ethics committees [B22], patient advocacy [B42], ethical rules and principles [B47], deontology [B59], code of deontology [B60], biomedical ethics education [B78], nursing education [B79], medical education [B80], universities [B84], bioethics [B89], ethicists [B92], history of biomedical ethics [B94], bioethical issues [B96], specific approaches to ethics [C1], religion and religious ethics [C19], Christian ethics [C22], Roman catholic ethics [C23], professional ethics [C51], social control [D3], legislation [D13], sex offenses [D92], research [E14], biomedical technologies [E28], quality of environment [E41], ecology [E43], biomedical research [F2], animal experimentation [F10], human experimentation [F11], research subjects [F12], public health [G5], artificial organs [G24], nurses [G39], physicians [G40], medical staff [G41], resource allocation [H4], health care costs [H9], tissue donation [I13], donors [I14], organ

donors [I15], organ transplantation [I24], patients' rights [J2], consent to treatment [J8], consent forms [J14], referral and consultation [J24], law, legislation and jurisprudence [K1], medical law [K51], conscience clause [K63], child and family [L31], aged — problems related with aging [L45], health care services and aged [L48], sexuality and procreation — unborn child [M1], coma [N41], persistent vegetative state [N42].

■ INTERESTED IN: ethical rules and principles [B47], deontology [B59], code of deontology [B60], biomedical ethics education [B78], nursing education [B79], medical education [B80], universities [B84], bioethics [B89], European Convention of Bioethics [B97], specific approaches to ethics [C1], Christian ethics [C22], professional ethics [C51], social control [D3], legislation [D13], family members [D24], EC — European Communities [D62], sex offenses [D92], biomedical technologies [E28], quality of environment [E41], ecology [E43], biomedical research [F2], human genome project [F8], public health [G5], artificial organs [G24], resource allocation [H4], donors [I14], organ transplantation [I24], fetal tissue transplantation [I27], patients' rights [J2], law, legislation and jurisprudence [K1], medical law [K51], conscience clause [K63], child and family [L31], child diseases and problems [L37], health care services and aged [L48], sexuality and procreation — unborn child [M1], coma [N41], persistent vegetative state [N42].

O B 11 **VAKGROOP GEZONDHEIDSETHIEK EN WIJSBEGEERTE (RL- MAASTRICHT)**

Centrum voor biomedische ethiek en recht - KUL
Kapucijnonvoer 35
3000 Leuven (Belgium)
✆ (32) 16 - 21 69 51 Fax: (32) 16 - 21 69 52

Individuals

B 1 **ACHSLOGH** Jacques

Head of the Neurosurgery Department, Institut neurochirurgical de Bruxelles. — Member of the ethics committee of the Fonds National de la Recherche Scientique (FNRS).

Sociéte belge d'éthique et de morale médicale
Rue Froissart 38
1040 Bruxelles (Belgium)
✆ (32) 2-287 59 88

■ PUBLICATIONS: yes.
■ WORKING ON: specific approaches to ethics [C1], professional ethics [C51].

B 2 **ANDRÉ** Armand

Professor Emeritus, University of Liège. — Teacher. — Former Chairperson of the Ecole Liégeoise de Criminologie Jean-Constant. — Delegate of the University of the Belgian Medical Association. — Vice-Chairperson of the Académie Royale de Médecine de Belgique.

Quai de Rome 48
bte 102
4000 Liège (Belgium)
✆ (32) 41 - 52 61 82 Fax: (32) 41 - 41 69 69

■ AFFILIATIONS: Conseil National de l'ordre des Médecins, Bruxelles, Belgium (*President*).
■ TEACHING: Lectures on the main guidelines of bioethics and the organisation of the ethics committee.
■ PUBLICATIONS: yes.
■ WORKING ON: review [A2], respect of human dignity and human rights [B1], privacy [B11], human person [B12], human body commercialization [B13], traffic of organs [B15], ethics committees [B22], CCNE — Comité Consultatif National d'Ethique (France) [B23], Council of Europe [B25], professional deontology organs [B26], College of Physicians [B27], CCPPRB — French committees for the protection of human subjects of biomedical research [B32], patient advocacy [B42], ethical rules and principles [B47], biomedical ethics education [B78], nursing education [B79], Bioethics bill, 1992 (France) [B98], specific approaches to ethics [C1], medical ethics [C52], nursing ethics [C53], state responsibility [D75], peer review [E63], biomedical research [F2], human experimentation [F11], research subjects [F12], healthy volunteers [F13], animal testing alternatives [F14], methodology of

research and experimentation [F15], organization of health care; facilities, manpower and services; health occupations [G1], health facilities [G22], organization of health care [G35], health care [G45], medicine [G63], biological specimens procurement, blood transfusion, organ transplantation [I8], biological substances contamination [I9], health care professionals, researchers and patient relationship [J1], patients' rights [J2], confidentiality [J20], physicians and researchers accountability [J23], inter-personal relationship [J27], wrongful death [K12], terminal care [O7], attitudes to death [O20].

■ INTERESTED IN: *idem.*

B 3 **BERTRAND** Yves

Medical Director, Clinique Saint-Jean. — Physician.

Clinique Saint-Jean
44, rue du Marais
1000 Bruxelles (Belgium)
✆ (32) 2 - 221 91 11

■ TEACHING: Ethics committee of the clinic.
■ WORKING ON: determination of death [O2], euthanasia [O15], resuscitation [O21], patients' rights [J2], confidentiality [J20], abortion [M47].
■ INTERESTED IN: life sciences [E4], public health [G5], reproductive technologies [M26], abortion [M47], determination of death [O2], terminal care [O7], euthanasia [O15], resuscitation [O21].

B 4 **BINAMÉ** Georges

Parliamentary Agent.

Maison des parlementaires — Chambre des Représentants
21, rue de Louvain
1200 Bruxelles (Belgium)
✆ (32) 2 - 519 85 08 Fax: (32) 2 - 519 87 96

■ AFFILIATIONS: Association Internationale Droit, Ethique et Science, Strasbourg, France (*Founding Member, Administrator, Treasurer*).
■ PUBLICATIONS: yes.
■ WORKING ON: human body commercialization [B13], traffic of organs [B15], ethics committees [B22], drug industry [E37], biological specimens procurement, blood transfusion, organ transplantation [I8], biological substances contamination [I9], required request [I10], sperm banks [I11], tissue banking [I12], tissue donation [I13], donors [I14], organ donors [I15], donor cards [I16], anonymous donation [I17], directed donation [I18], blood transfusions [I19], blood donation [I20], blood substitutes [I21], transplantation [I22], transplant recipients [I23], organ transplantation [I24], organ donation [I25], organ procurement [I26], fetal tissue transplantation [I27], intracerebral fetal tissue transplantation [I28], bone marrow transplantation [I29], consent to treatment [J8], informed consent [J9], presumed consent [J10], third party consent [J11], spousal consent [J12], parental consent [J13], consent forms [J14], coma [N41], persistent vegetative state [N42], biotechnology — genetic engineering [Q20], biotechnology [Q21], genetic intervention [Q22], genome mapping [Q23], cloning [Q24], clones [Q25], sex determination [Q26], hybrids [Q27], genetically modified organisms [Q28], transgenic animals [Q29], sex preselection [Q30], recombinant DNA research [Q31], DNA fingerprinting [Q32], DNA linkage [Q33], behavioral genetics [Q34], medical genetics [Q35], genetic counseling [Q36], genetic screening [Q37], genetic screening [Q37], eugenics [Q39], negative eugenics [Q40], positive eugenics [Q41], gene therapy [Q42], population genetics [Q43], genetic identity [Q44], gene pool [Q45], carriers [Q46], human genome [Q47].
■ INTERESTED IN: CCNE — Comité Consultatif National d'Ethique (France) [B23], Council for International Organization of Medical Sciences [B24], Council of Europe [B25], drug industry [E37], experimentation [F9], animal experimentation [F10], human experimentation [F11], research subjects [F12], healthy volunteers [F13], animal testing alternatives [F14], WHO — World Health Organization [G36], developing countries [H21], biological specimens procurement, blood transfusion, organ transplantation [I8], biological substances contamination [I9], required request [I10], sperm banks [I11], tissue banking [I12], tissue donation [I13], donors [I14], organ donors [I15], donor cards [I16], anonymous donation [I17], directed donation [I18], blood transfusions [I19], blood donation [I20], blood substitutes [I21], transplantation [I22], transplant recipients [I23], organ transplantation [I24], organ donation [I25], organ procurement [I26], fetal tissue transplantation [I27], intracerebral fetal tissue transplantation [I28], bone marrow transplantation [I29], consent to treatment [J8], informed consent [J9], presumed consent [J10], third party consent [J11], spousal consent [J12], parental consent [J13], consent forms [J14], anatomy, physiology, development [L2], human body [L3], body parts and fluids [L4], hearts [L5], brain [L6], livers [L7], bone marrow [L8], kidneys [L9], animal organs [L18], death and resuscitation [O1], determination of death [O2], autopsies [O3], cadavers [O4], death [O5], brain death [O6], terminal care [O7], palliative care [O8], life-sustaining treatment

[O9], withholding treatment [O10], allowing to die [O11], terminally ill [O12], prolongation of life [O13], quality adjusted life years [O14], biotechnology — genetic engineering [Q20], biotechnology [Q21], genetic intervention [Q22], genome mapping [Q23], cloning [Q24], clones [Q25], sex determination [Q26], hybrids [Q27], genetically modified organisms [Q28], transgenic animals [Q29], sex preselection [Q30], recombinant DNA research [Q31], DNA fingerprinting [Q32], DNA linkage [Q33], behavioral genetics [Q34], medical genetics [Q35], genetic counseling [Q36], genetic screening [Q37], preimplantation genetic diagnosis [Q38], eugenics [Q39], negative eugenics [Q40], positive eugenics [Q41], gene therapy [Q42], population genetics [Q43], genetic identity [Q44], gene pool [Q45], carriers [Q46], human genome [Q47].

B 5 **BURNY** Arsène

Professor of Molecular Biology .

Faculté des sciences agronomiques
Chaussée de Namur 65
5030 Gembloux (Belgium)
✆ (32) 81 - 68 21 52 Fax: (32) 81 - 61 38 88

■ AFFILIATIONS: Société belge d'éthique et de morale médicales, Bruxelles, Belgium (*Member*).
■ WORKING ON: cancer [N17], acquired immunodeficiency syndrome [N35], gene therapy [Q42].
■ INTERESTED IN: *idem.*

B 6 **BURTON** Jean

Scientific Attaché, Centre Interfacultaire Droit, Éthique et Sciences de la santé (CIDES) of the Facultés Notre-Dame-de-la-Paix (Namur). — Priest, Society of Jesus.

Centre Interfacultaire Droit, éthique et sciences de la santé (CIDES)
Facultés universitaires
61, rue de Bruxelles
5000 Namur (Belgium)
✆ (32) 81-72 41 11 Fax: (32) 81 - 23 03 91

■ AFFILIATIONS: Centre Interfacultaire Droit, Éthique et Sciences de la santé — CIDES, Namur, Belgium (*Member*).
■ TEACHING: Superviser of undergraduates (medicine, law, biology). — Seminar for academics and professionals.
■ WORKING ON: Christian ethics [C22], Roman Catholic ethics [C23], primates [C50], biology [E5], behavioral research [E21], animal experimentation [F10].
■ INTERESTED IN: information sources, data bases [A1], review [A2], bibliography [A4], book review [A6], literature [A7], documentation [A8], specific approaches to ethics [C1], Christian ethics [C22], Roman Catholic ethics [C23], philosophy of biology [C42], biology [E5], animal experimentation [F10].

B 7 **CASPAR** Philippe

Physician. — Philosopher. — Writer. — Scientific Consultant, Dispositif-CARAT.

Avenue Bon-Air, 17
1310 La Hulpe (Belgium)
✆ (32) 2 - 653 41 27

■ AFFILIATIONS: Société Française de Réflexion Bioéthique — SFRB, Paris, France (*Member*).
■ TEACHING: Institut Robert Schumann.
■ PUBLICATIONS: yes.
■ WORKING ON: human person [B12], deontology [B59], mass media [B72], theology [C41], history of science [E3], abortion [M47], fetal development [M54], twinning [M55], mentally handicapped [P7].
■ INTERESTED IN: deontology [B59], history of science [E3], twinning [M55], mentally handicapped [P7].

B 8 **CASSIERS** Léon

Dean of the Medical School, Catholic University of Leuven.

Université Catholique de Louvain
Off bte UCL 50 20
Avenue Emmanuel-Mounier 50
1200 Bruxelles (Belgium)
✆ (32) 2 - 764 50 30 Fax: (32) 2 - 764 50 35

■ AFFILIATIONS: Unité de recherche en philosophie de la médecine et de bioéthique, Bruxelles, Belgium (*Administrator*).
■ TEACHING: Ethics in psychiatry and psychotherapy. (this course is part of the advanced curriculum organised by the research department of medical philosophy and bioethics)
■ PUBLICATIONS: yes.
■ WORKING ON: mental institutions [G29], psychiatry [G69], placebos [I36], informed consent [J9], spousal consent [J12], mind [L20], children's rights [L25], adoption [L32], suicide [O19], behavioral and mental disorders [P2], outpatient commitment [P18], psychiatric technics [P24].
■ INTERESTED IN: human person [B12], respect [B63], biomedical ethics education [B78], curriculum [B83], medical ethics [C52], psychology [C68], counseling [C87], couple [D26], parents [D32], disadvantaged [D46], stigmatization [D54], research design [E19], progress [E26], computers [E33], records [E34], computerized file [E35], peer review [E63], social control of science [E64], cognitive research [F7], medical informatics [F21], mental institutions [G29], convenience medicine [G57], family practice [G60], costs and benefits [H10], risks and benefices [H11], organ donors [I15], transplant recipients [I23], organ donation [I25], intracerebral fetal tissue transplantation [I28], privileged communication [J21], nurse patient relationship [J30], physician nurse relationship [J31], physician patient relationship [J32], professional patient relationship [J33], parent child relationship [J34], brain [L6], ill-treat children [L39], wish of children [M16], FIV centre [M30], excess embryos [M34], mother fetus relationship [M43], fetal therapy [M57], HIV seropositivity [N36], suicide [O19], behavioral and mental disorders [P2], outpatient commitment [P18], psychiatric technics [P24], genetic counseling [Q36], negative eugenics [Q40], positive eugenics [Q41], gene therapy [Q42].

B 9 **CHAPELLE** Albert
Professor of Moral and Fundamental Theology, Institut d'Études Théologiques (IET), Brussels.

Institut d'Etudes Théologiques (IET)
60, rue du Collège-Saint-Michel
1150 Bruxelles (Belgium)
✆ (32) 2 - 73 93 446 Fax: (32) 2 - 73 66 369

■ TEACHING: Teaches in the Brussels Faculty.
■ PUBLICATIONS: yes.
■ WORKING ON: respect of human dignity and human rights [B1], ethical rules and principles [B47], philosophical ethics [C2], death and resuscitation [O1].
■ INTERESTED IN: protection of rights — involved institutions [B16], bioethics [B89], religion and religious ethics [C19], history before the 20th century [C58], philosophy [C61], international aspects [D59], war [D87], sexuality and procreation — unborn child [M1], death and resuscitation [O1].

B 10 **COBBAUT** Jean-Philippe
Assistant Philosophy teacher at the Medical School of the Catholic University of Leuven. — Barrister-at-law, Brussels.

Centre d'études bioéthiques
bte. UCL 43 / 4534
Promenade de l'Alma 51
1200 Bruxelles (Belgium)
✆ (32) 2 - 764 43 30

■ TEACHING: Philosophy and bioethics — Seminars in ethics and medical philosophy (Post graduates).
■ WORKING ON: fundamental rights of the individuals [B2], institutions involved in the protection of rights [B17], ethics committees [B22], patient advocacy [B42], humanity heritage [B46], ethical rules and principles [B47], philosophical ethics [C2], social and political issues [D1], social issues [D2], social control [D3], informal social control [D4], coercion [D5], social dominance [D6], public opinion [D7], punishment [D8], exclusion [D9], formal social control [D10], model legislation [D14], socioeconomic factors [D18], political systems [D79], organization of health care; facilities, manpower and services; health occupations [G1], health care professionals, researchers and patient relationship [J1], law, legislation and jurisprudence [K1], preconception injuries [K27], economic value of life [K28], accountability [K29], legal liability [K30], ghost surgery [K31], therapeutic risk [K32], wrongful life [K33], aged — problems related with aging [L45], death and resuscitation [O1].

B 11 **COSYNS** Paul
Professor of psychiatry, medical psychology and deontology, University of Antwerp. — Head of the Psychiatry Department, University Hospital of Antwerp.

Universitair Ziekenhuis Antwerpen
Wilrijkstraat 10

2650 Edegem, Antwerpen (Belgium)
✆ (32) 3 - 839 11 11 Fax: (32) 3 - 829 05 20

■ AFFILIATIONS: Belgische Vereniging voor Medische Moraal en Ethiek, Mortsel, Belgium (*Chairperson*).
■ TEACHING: Medical deontology, University of Antwerp.
■ PUBLICATIONS: yes.
■ WORKING ON: ethics committees [B22], biomedical ethics education [B78], medical ethics [C52], prisoners [D12], informed consent [J9], third party consent [J11], expert testimony [K18], suicide [O19], dangerousness [P15], behavior control [P25], psychotherapy [P30], psychosurgery [P38].
■ INTERESTED IN: World Medical Assembly [B20], professional deontology organs [B26], College of Physicians [B27], codes of ethics [B49], biomedical ethics education [B78], normative ethics [C5], situational ethics [C9], value of life [C18], scientific misconduct [E25], human genome project [F8], compensation [H12], intracerebral fetal tissue transplantation [I28], physical restraint [I40], professional deontology [J25], legally incompetent person [K4].

B 12 **DEBRY** Jean-Michel

Biologist, in charge of an artificial procreation laboratory (in vitro fertilisation embryo transfer, sperm banks, etc.) —Chairperson of professional associations — journalist for the magazine Athena (new technologies).

Institut de morphologie pathologique
41, allée des Templiers
6280 Gerpinnes (Belgium)
✆ (32) 71 - 43 79 01 Fax: (32) 71 - 47 15 20

■ AFFILIATIONS: Association internationale droit, éthique et science (groupe de Milazzo), Milazzo, Italy (*Member*).
■ WORKING ON: code of deontology [B60], couple [D26], biology [E5], scientific misconduct [E25], freezing [E29], biomedical technologies [E28], animal experimentation [F10], sperm banks [I11], donors [I14], beginning of life [L24], aid children [L34], embryos [M10], fetuses [M11], sperm [M14], procreation [M15], wish of children [M16], reproduction [M17], fertility [M18], infertility [M19], reproductive technologies [M26], ovum donors [M27], semen donors [M28], in vitro fertilization [M29], FIV centre [M30], artificial insemination [M31], AID [M32], AIH [M33], excess embryos [M34], host mothers [M35], embryo transfer [M36], GIFT (Gamete Intrafallopian Transfer) [M37], embryo donation [M38].
■ INTERESTED IN: traffic of organs [B15], Council of Europe [B25], code of deontology [B60], curriculum [B83], universities [B84], ethicists [B92], ethical review [B93], biology and human future [C43], speciesism [C48], tissue banking [I12], organ donation [I25], biotechnology — genetic engineering [Q20], sex determination [Q26], hybrids [Q27], genetically modified organisms [Q28], sex preselection [Q30], preimplantation genetic diagnosis [Q38].

B 13 **DELFOSSE** Marie-Luce

Teacher of Ethics, Faculty of Law — Director of Research at the CIDES — First assistant, Facultés Universitaires Notre-Dame-de-la-Paix (Namur).

Facultés universitaires Notre-Dame-de-la-Paix
61, rue de Bruxelles
5000 Namur (Belgium)
✆ (32) 81-72 41 11

■ AFFILIATIONS: Centre interfacultaire Droit, éthique et sciences de la santé — CIDES, Namur, Belgium (*Organiser and Direction of Research*).
■ TEACHING: Conferences on Bioethics Law (Facultés Notre-Dame de la Paix, Namur). — Interdisciplinary seminars on Bioethics Morals, organises and takes part in colloquia and seminars on bioethics (seconde candidature de droit, Facultés Notre-Dame-de-la-Paix, Namur).
■ PUBLICATIONS: yes.
■ WORKING ON: ethics [B48], codes of ethics [B49], moral policy [B55], deontology [B59], codes of biomedical ethics [B90], professional ethics [C51], medical ethics [C52], human experimentation [F11], research subjects [F12], healthy volunteers [F13].

B 14 **DIJON** Xavier

Professor of Law, Facultés Universitaires Notre-Dame-de-la-Paix, Namur (Belgium).

Faculté de droit
Rempart de la Vierge, n° 5

5000 Namur (Belgium)
✆ (32) 81 - 72 47 92 Fax: (32) 81 - 22 88 58

■ AFFILIATIONS: Centre interfacultaire Droit, éthique et sciences de la santé — CIDES, Namur, Belgium (*Member*).
■ TEACHING: Occasional conferences on Bioethics Law.
■ PUBLICATIONS: yes.
■ WORKING ON: human rights [B3], integrity [B8], human body commercialization [B13], ethics committees [B22], Declaration on Human and Citizen Rights [B38], patient advocacy [B42], morality [B61], ethical relativism [C10], Roman catholic ethics [C23], natural law [C24], scriptural interpretation [C40], nursing ethics [C53], handicapped [D43], legal personality [K2], potentiality of personhood [K5], medical law [K51], adoption [L32], aid children [L34], transsexualism [M24], AID [M32], withholding treatment [O10], active euthanasia [O16], suicide [O19].
■ INTERESTED IN: *idem*.

B 15 **DRUET** Pierre-Philippe

Professor of Philosophy and Psychology, Faculty of Namur, Catholic University of Louvain-la-Neuve.

Facultés universitaires de Namur
61, rue de Bruxelles
5000 Namur (Belgium)
✆ (32) 81 - 72 42 08 Fax: (32) 81 - 72 42 03

■ AFFILIATIONS: Association européenne pour les soins palliatifs (*Member*).
■ TEACHING: Introduction to philosophy, ethics — Occasional lectures on palliative treatments
■ PUBLICATIONS: yes.
■ WORKING ON: moral policy [B55], deontology [B59], morals [C12], social control [D3], death and resuscitation [O1], terminal care [O7], euthanasia [O15], suicide [O19], attitudes to death [O20], resuscitation [O21].
■ INTERESTED IN: *idem*.

B 16 **DUCHENE** Joseph

Professor of Philosophy, Faculty of Notre-Dame-de-la-Paix à Namur, Department Head (Sciences-Philosophie-Societés). — Director of CIDES.

Centre interfacultaire droit, éthique et sciences de la santé (CIDES)
61, rue de Bruxelles
5000 Namur (Belgium)
✆ (32) 81 - 72 41 30 or 72 47 89 or 72 41 17 Fax: (32) 81 - 72 41 18

■ AFFILIATIONS: Société belge d'éthique et de morale médicale, Bruxelles, Belgium (*Member*). — Conseil général de l'apostolat des laïcs — CGAL, Bruxelles, Belgium (*Member*).
■ TEACHING: interdisciplinary seminars on bioethics for students in medicine, sciences, law economy and philosophy . (30 hours, different subjects every year: genetics and therapeutics, experiments and human rights, mental disorder and society, etc.) —Moral bioethics.
■ PUBLICATIONS: yes.
■ WORKING ON: human rights [B3], human person [B12], World Medical Assembly [B20], Helsinki Declaration [B37], Universal Declaration of Human Rights [B39], due process [B40], public advocacy [B41], ethics [B48], solidarity [B65], interdisciplinary communication [B70], public debates [B75], education [B76], biomedical ethics education [B78], students [B82], curriculum [B83], universities [B84], CCNE advice (France) [B91], Bioethics bill, 1992 (France) [B98], philosophical ethics [C2], ethical relativism [C10], morals [C12], religion and religious ethics [C19], medical ethics [C52], social control [D3], handicapped [D43], cultural pluralism [D51], social discrimination [D53], democracy [D82], history of science [E3], biology [E5], genetics [E6], research policy [E18], computerized file [E35], data protection [E52], social control of science [E64], nontherapeutic research [F5], cognitive research [F7], human experimentation [F11], research subjects [F12], healthy volunteers [F13], control groups [F17], random selection [F18], selection of subjects [F19], group of vulnerable subjects [F20], medical informatics [F21], WHO — World Health Organization [G36], risks and benefices [H11], prenatal diagnosis [I3], placebos [I36], patient access [J5], disclosure [J7], informed consent [J9], patient participation [J16], investigator subject relationship [J29], personhood [K3], legally incompetent person [K4].
■ INTERESTED IN: bibliography [A4], book review [A6], documentation [A8], publications [A9].

B 17 **EGGERMONT** Ephrem

Head of the Paediatrics and Neonatology Department, University Hospital of Gasthuisberg, Leuven. — Professor of Paediatrics and Medical Deontology.

Université de Leuven
Grote Molenweg 90
3020 Herent (Belgium)
✆ (32) 16 - 34 38 42

■ AFFILIATIONS: Conseil national de l'Ordre des médecins, Brussels, Belgium (*Member*).
■ TEACHING: Medical Deontology.
■ PUBLICATIONS: yes.
■ WORKING ON: professional deontology organs [B26], deontology [B59], bioethics [B89], sex offenses [D92], biomedical research [F2], maternal health [G3], physicians [G40], pediatrics [G68], childhood difficulties, diseases, protection [L21], terminal care [O7], genetic defects and hereditary diseases [Q2].
■ INTERESTED IN: World Medical Assembly [B20], professional deontology organs [B26], patient advocacy [B42], deontology [B59], bioethics [B89], religious ethics [C21], Christian ethics [C22], medical ethics [C52], nursing ethics [C53], sex offenses [D92], rape [D93], biomedical research [F2], maternal health [G3], family planning [G8], physicians [G40], pediatrics [G68], health care professionals, researchers and patient relationship [J1], childhood difficulties, diseases, protection [L21], pregnancy and childbirth [M39], fetal development [M54], fetal therapy [M57], physically [N8], terminal care [O7], attitudes to death [O20], genetic defects and hereditary diseases [Q2].

B 18 **ENGLERT** Yvon
Head of the Unité de Fertilité. — Deputy Chairperson of the Local Ethics Committee. — Member of the Board of the Université Libre de Bruxelles (ULB). — Health Counsellor to the Secretary of State.

Unité de Fertilité
Hôpital Erasme
Route de Lennik 808
1070 Bruxelles (Belgium)
✆ (32) 2 - 555 36 89 Fax: (32) 2 - 555 38 21

■ AFFILIATIONS: Centre de recherche interdisciplinaire en bioéthique — CRIB, Bruxelles, Belgium (*Administrator*).
■ TEACHING: bioethics (with Professor Hottois), Faculté de Philosophie (ULB).
■ PUBLICATIONS: yes.
■ WORKING ON: fundamental rights of the individuals [B2], institutions involved in the protection of rights [B17], ethical rules and principles [B47], medical ethics [C52], social issues [D2], military personnel [D66], life sciences [E4], investigators [E15], biomedical technologies [E28], clinical trials [F3], experimentation [F9], health policy [G13], WHO — World Health Organization [G36], obstetrics and gynecology [G66], sperm banks [I11], physicians and researchers accountability [J23], interpersonal relationship [J27], beginning of life [L24], sexuality and procreation [M2], reproductive technologies [M26], embryo donation [M38], abortion [M47].
■ INTERESTED IN: ethical rules and principles [B47], bioethics [B89], professional ethics [C51], law, legislation and jurisprudence [K1].

B 19 **ETIENNE** Jacques
Professor Emeritus, Université Catholique de Louvain. — Academic Secretary of the Revue Philosophique de Louvain (Louvain-la-Neuve) and Répertoire Bibliographique de la Philosophie.

Institut Supérieur de Philosophie
Collège Thomas-More
Chemin d'Aristote 1
1348 Louvain-la-Neuve (Belgium)
✆ (32) 10 - 47 48 27

■ AFFILIATIONS: Centre d'études bioéthiques, Bruxelles, Belgium (*Member*).
■ WORKING ON: human rights [B3], dignity [B5], normative ethics [C5], teleological ethics [C7], conscience [C13], morals [C12].
■ INTERESTED IN: ethics committees [B22].

B 20 **FELTZ** Bernard
Lecturer

Institut supérieur de philosophie
Chemin d'Aristote, 1
1348 Louvain-La-Neuve (Belgium)
✆ (32) 10 - 47 46 11

■ TEACHING: Philosophy (30 hours). — Ethics in natural sciences (30 hours) for undergraduates and agronomics. — Seminar on philosophy of natural science (30 hours) for post graduates.

■ PUBLICATIONS: yes.
■ WORKING ON: philosophy of biology [C42], philosophy [C61], determinism [C62], normality [C78], intelligence [C85], science [E2], history of science [E3], life sciences [E4], molecular biology [E8], ecology [E43], decision making [E54], evaluation [E61], biomedical research [F2], experimentation [F9], methodology of research and experimentation [F15], human characteristics [L19], mind [L20].
■ INTERESTED IN: *idem, plus:* philosophical ethics [C2], science, technology, methods [E1], containment [E38], methods [E53], biomedical research and experimentation [F1], biomedical research [F2], experimentation [F9], methodology of research and experimentation [F15]

B 21 **FOUREZ** Gérard
University Teacher.

Facultés Universitaires
Département Sciences, Philosophies, Sociétés
Boulevard Cauchy, 24
5000 Namur (Belgium)
✆ (32) 81 - 72 41 04 Fax: (32) 81 - 72 41 18

■ TEACHING: course in ethics for Science students.
■ PUBLICATIONS: yes.
■ WORKING ON: fundamental rights of the individuals [B2], justice [B9], ethics [B48], education [B76], philosophical ethics [C2], ethical analysis [C3], Christian ethics [C22], Roman Catholic ethics [C23], religious sciences [C39], social impact [D49], history of science [E3], research policy [E18], computerized file [E35], methods [E53], decision making [E54], decision analysis [E55], social control of science [E64], technology assessment [E65].
■ INTERESTED IN: fundamental rights of the individuals [B2], justice [B9], ethics [B48], education [B76], philosophical ethics [C2], ethical analysis [C3], Christian ethics [C22], Roman Catholic ethics [C23], religious sciences [C39], social impact [D49], history of science [E3], research policy [E18], computerized file [E35], methods [E53], decision making [E54], decision analysis [E55], social control of science [E64], technology assessment [E65], allowing to die [O11], quality adjusted life years [O14].

B 22 **GEETS** Claude
In charge of the ethics department at the Fédération des Institutions Hospitalières de Wallonie (FIHW).

Fédération des Institutions Hospitalières de Wallonie
Rue Notre-Dame n°9
5000 Namur (Belgium)
✆ (32) 81 - 22 18 22

■ AFFILIATIONS: Fédération des Institutions hospitalières de Wallonie — FIHW, Namur, Belgium (*in charge of Ethics Department*).
■ TEACHING: Ethics for health care professionals at the FIHW.
■ WORKING ON: fundamental rights of the individuals [B2], ethics and humanities [C54], patients' rights [J2], physicians and researchers accountability [J23], aged — problems related with aging [L45], terminal care [O7], euthanasia [O15], suicide [O19], resuscitation [O21], behavioral and mental disorders [P2], outpatient commitment [P18].
■ INTERESTED IN: palliative care [O8], life-sustaining treatment [O9], withholding treatment [O10], allowing to die [O11], terminally ill [O12], prolongation of life [O13], quality adjusted life years [O14].

B 23 **GILLOT-DE VRIES** Francin
Professor, Deputy Director of the "psychologie du développement", faculté des sciences psychologiques et pédagogique department", Université libre de Bruxelles.

Université Libre de Bruxelles
Avenue Franklin Roosevelt 50
CP 122
1050 Bruxelles (Belgium)
✆ (32) 2 - 650 32 90 Fax: (32) 2 - 650 31 36

■ AFFILIATIONS: Centre de recherche interdisciplinaire en bioéthique — CRIB, Bruxelles, Belgium (*Member*).
■ TEACHING: Bioethics (with Professeur G. Hottois), Faculté de Philosophie et Lettres.
■ PUBLICATIONS: yes.
■ WORKING ON: psychology [C68], personality [C82], beginning of life [L24], children's rights [L25], childhood defense [L26], children [L27], infants [L28], newborns [L29], prematurity [L30], aid children [L34], child diseases [L43], procreation [M15], wish of children [M16], reproduction [M17],

fertility [M18], infertility [M19], homosexuality [M22], in vitro fertilization [M29], artificial insemination [M31], AIH [M33], pregnancy and childbirth [M39], childbirth [M40], caesarean section [M41], pregnant women [M42], mother fetus relationship [M43], pregnancy [M44], pregnancy in adolescence [M45], multiple pregnancy [M46], abortion [M47], therapeutic abortion [M50], abortion on demand [M52], fetal therapy [M57].
■ INTERESTED IN: *idem*.

B 24 **GODFRAIND** Théophile
Professor of Pharmacology, Director of the Laboratory of Pharmacology, Faculté de Médecine de l'Université catholique de Louvain (UCL). — Executive Secretary of the International Union of Pharmacology (IUPHAR).

Centre d'études bioéthiques
bte UCL 73-50
Avenue Emmanuel Mounier 73
1200 Bruxelles (Belgium)
✆ (32) 2 - 764 73 50 Fax: (32) 2 - 764 73 08

■ AFFILIATIONS: Centre d'Etude bioéthique affilié à l'Université catholique de Louvain, Bruxelles, Belgium (*Chairperson*).
■ TEACHING: Seminars in medical ethics.
■ PUBLICATIONS: yes.
■ WORKING ON: biomedical research [F2], clinical trials [F3], therapeutic research [F6], cognitive research [F7], experimentation [F9], human experimentation [F11], animal testing alternatives [F14].
■ INTERESTED IN: philosophical ethics [C2], religion and religious ethics [C19], philosophy of biology [C42], medical ethics [C52], neurosciences — psychiatry [P1].

B 25 **GUIGUI** Albert
Chief Rabbi of Brussels and of the Consistoire Central Israélite de Belgique.

Consistoire central israélite de Belgique
rue Joseph-Dupont 2
1000 Bruxelles (Belgium)
✆ (32) 2 - 512 21 90

■ PUBLICATIONS: yes.
■ WORKING ON: religion and religious ethics [C19], Jewish ethics [C29], Judaism [C34], scriptural interpretation [C40].
■ INTERESTED IN: respect of human dignity and human rights [B1], stages of life — problems proper to childhood and elderly [L1], sexuality and procreation — unborn child [M1], health care professionals, researchers and patient relationship [J1].

B 26 **HAYEZ** Jean-Yves
Head of the Department of Pédopsychiatrie à l'Université Catholique de Louvain.

Département de pédopsychiatrie
Université catholique de Louvain
Clos Chapelle-aux-Champs 30
1200 Bruxelles (Belgium)
✆ (32) 2 - 764 31 42 or 764 31 20 or 764 19 60 Fax: (32) 2 - 764 39 55

■ AFFILIATIONS: Centre de bioéthique at the Université catholique de Louvain, Belgium (*External Consultant*).
■ TEACHING: Seminars and conferences on: Medical care for ill-treated and abused teenagers, children as subjects or objects of treatment, attitudes towards teenagers in clinics. — Organisation of miscellaneous activities (teaching, debates).
■ PUBLICATIONS: yes.

B 27 **HENNAU-HUBLET** Christiane
Professor, Université catholique de Louvain (Law Faculty).

Département de droit pénal — Université catholique de Louvain
Collège Thomas-More
Place Montesquieu 2
1348 Louvain-La-Neuve (Belgium)
✆ (32) 10 - 47 46 71 Fax: (32) 10 - 88 01 30

■ AFFILIATIONS: Centre d'études bioéthiques, Bruxelles, Belgium (*Administration*).
■ TEACHING: "Law and Ethics in Care", to undergraduates (3rd year), Health course, Université catholique de Louvain.
■ PUBLICATIONS: yes.
■ WORKING ON: respect of human dignity and human rights [B1], deontology [B59], code of deontology [B60], values [C15], biomedical research and experimentation [F1], health personnel [G37], health care, medical acts [I1], organization of health care; facilities, manpower and services; health occupations [G1], potentiality of personhood [K5], jurisprudence — accountability [K8], medical law [K51], criminal law [K53], organ transplantation [I24], sexuality and procreation — unborn child [M1], death and resuscitation [O1], psychoactive drugs [P33], AID [M32].
■ INTERESTED IN: medical education [B80], war [D87], torture [D94], pollution [E47], health hazards [E49], sexuality and procreation — unborn child [M1].

B 28 **HENNAUX** Jean -Marie

Priest. — Professor of Moral Theology at the Institut d'études théologiques de Bruxelles.

Institut d'études théologiques
Rue du Collège Saint-Michel 60
1150 Bruxelles (Belgium)
✆ (32) 2-739 34 51

■ AFFILIATIONS: Société française de réflexion bioéthique — SFRB, Paris (*Member*).
■ TEACHING: Director of the Seminar on Bioethics, Institut d'études théologiques (IET), Bruxelles.
■ PUBLICATIONS: yes.
■ WORKING ON: documentation [A8], fundamental rights of the individuals [B2], due process [B40], moral policy [B55], biomedical ethics education [B78], universities [B84], bioethical issues [B96], morals [C12], Roman catholic ethics [C23], double effect [C46], medical ethics [C52], family members [D24], family planning [G8], prenatal diagnosis [I3], conscience clause [K63], embryos [M10], in vitro fertilization [M29], excess embryos [M34], abortion [M47], determination of death [O2], euthanasia [O15].
■ INTERESTED IN: *idem*.

B 29 **HOTTOIS** Gilbert

Professor, Université libre de Bruxelles. — Deputy Director of the CRIB.

Centre de Recherches Interdisciplinaires en Bioéthique (CRIB)
Université libre de Bruxelles
145, avenue Adolphe-Buyl
1050 Bruxelles (Belgium)
✆ (32) 2 - 650 26 28 or 650 23 43 Fax: (32) 2 - 650 36 47

■ AFFILIATIONS: Centre de recherches interdisciplinaires en bioéthique — CRIB, Bruxelles, Belgium (*Co-Director*).
■ TEACHING: University courses of collective participation.
■ PUBLICATIONS: yes.
■ WORKING ON: fundamental rights of the individuals [B2], ethics [B48], bioethics [B89], philosophical ethics [C2], evolution [C47], humanism [C66].
■ INTERESTED IN: respect of human dignity and human rights [B1], specific approaches to ethics [C1].

B 30 **JACOB** Marie-Jeanne

Medical Director of a centre of palliative care (treatment of cancer symptoms, care to terminally ill patients).

Foyer Saint-François
Maison de soins palliatifs
39A, rue Loiseau
5000 Namur (Belgium)
✆ (32) 81 - 74 13 00

■ WORKING ON: quality of life [C16], value of life [C18], hospices [G34], health care delivery [G48], patient compliance [G49], medical records [G50], medical evaluation [G51], selection for treatment [G52], home care [G55], physician's role [J26], physician nurse relationship [J31], physician patient relationship [J32], professional patient relationship [J33], disease [N1], suffering [N11], pain [N12], follow-up studies [N13], cancer [N17], death and resuscitation [O1], terminal care [O7], palliative care [O8], life-sustaining treatment [O9], allowing to die [O11], terminally ill [O12], quality adjusted life years [O14], attitudes to death [O20].

■ INTERESTED IN: biomedical ethics education [B78], medical education [B80], situational ethics [C9], quality of life [C16], value of life [C18], religious ethics [C21], Roman catholic ethics [C23], natural law [C24], medical ethics [C52], dehumanization [D86], alternative therapies [I33], euthanasia [O15].

B 31 **JACQUEMIN** Dominique

Research fellow, bioethics unit of the FIMD, UCL. — Episcopal Delegate to the Pastorale Familiale Church. — Counsellor of the ACN (Catholic Association of Nursing).

Centre d'études bioéthiques de l'Université catholique de Louvain
bte UCL 43/4534
51 Promenade de l'Alma
1200 Bruxelles (Belgium)
✆ (32) 2-762 02 97

■ AFFILIATIONS: Centre d'études bioéthiques — CEB, Bruxelles, Belgium (*Member*).
■ TEACHING: Religion, bioethics.
■ PUBLICATIONS: yes.
■ WORKING ON: integrity [B8], ethics committees [B22], humanity heritage [B46], ethics [B48], solidarity [B65], biomedical ethics education [B78], nursing education [B79], ethicists [B92], bioethics movement [B95], bioethical issues [B96], morals [C12], quality of life [C16], clergy [C20], religious ethics [C21], Christian ethics [C22], Roman Catholic ethics [C23], theology [C41], medical ethics [C52], nursing ethics [C53], social groups [D22], biomedical technologies [E28], quality of environment [E41], ecology [E43], pollution [E47], health hazards [E49], decision making [E54], human experimentation [F11], research subjects [F12], mental health [G4], diagnosis [I2], required request [I10], donors [I14], transplantation [I22], referral and consultation [J24], nurse patient relationship [J30], physician nurse relationship [J31], physician patient relationship [J32], professional patient relationship [J33], reproductive technologies [M26], abortion [M47]palliative care [O8], life-sustaining treatment [O9], withholding treatment [O10], allowing to die [O11], quality adjusted life years [O14], behavioral and mental disorders [P2].
■ INTERESTED IN: ethics committees [B22], humanity heritage [B46], ethics [B48], biomedical ethics education [B78], nursing education [B79], codes of biomedical ethics [B90], ethicists [B92], moral policy [B55], altruism [B56], beneficence [B57], authoritarianism [B58], morals [C12], medical ethics [C52], nursing ethics [C53], social groups [D22], biomedical technologies [E28], quality of environment [E41], ecology [E43], pollution [E47], radiation [E48], decision making [E54], human experimentation [F11], research subjects [F12], developing countries [H21], population control [H22], elderly [H23], diagnosis [I2], donors [I14], transplantation [I22], referral and consultation [J24], sexuality and procreation [M2], reproductive technologies [M26], abortion [M47], brain death [O6], palliative care [O8], life-sustaining treatment [O9], withholding treatment [O10], allowing to die [O11], euthanasia [O15], behavioral and mental disorders [P2].

B 32 **KOULISCHER** Lucien

Professor, Head of the Department of the CHU de l'université de Liège (Belgique).

Université de Liège
CHU - Tour de Pathologie (B23) SART TILMAN
4000 Liège (Belgium)
✆ (32) 41 - 5625 60 or 41 - 56 25 62 Fax: (32) 41 - 56 29 74

■ AFFILIATIONS: Centre de réflexion bioéthique, Liège, Belgium (*Member*).
■ PUBLICATIONS: yes.
■ WORKING ON: chromosomal disorders [Q4], down's syndrome [Q6], genetic counseling [Q36], genetic screening [Q37], carriers [Q46].

B 33 **LAHAYE-BEKAERT** Nicole

Executive Secretary of the Centre national de criminologie.

Université Libre de Bruxelles (ULB)
Centre de Criminologie
44, avenue Jeanne
1050 Bruxelles (Belgium)
✆ (32) 2- 650 34 08

■ AFFILIATIONS: Centre de Recherches Interdisciplinaires en Bioéthique. — CRIB, Bruxelles, Belgium (*Member of the Scientific Committee*).
■ TEACHING: Collective course on bioethics, Faculté de Philosophie de l'Université libre de Bruxelles.
■ PUBLICATIONS: yes.

■ WORKING ON: information centers [A5], publications [A9], freedom [B10], human body commercialization [B13], ethics committees [B22], CCNE — Comité Consultatif National d'Ethique (France) [B23], Council of Europe [B25], codes of ethics [B49], moral policy [B55], interdisciplinary communication [B70], universities [B84], quality of life [C16], Roman catholic ethics [C23], social control [D3], couple [D26], sex offenses [D92], rape [D93], research institutes [E17], self regulation [E62], peer review [E63], social control of science [E64], prenatal diagnosis [I3], organ transplantation [I24], organ donation [I25], informed consent [J9], spousal consent [J12], medical secrecy [J22], physician patient relationship [J32], potentiality of personhood [K5], legitimacy [K21], child's interest [K24], criminal law [K53], criminal code [K54], bill [K60], children's rights [L25], procreation [M15], infertility [M19], in vitro fertilization [M29], artificial insemination [M31], excess embryos [M34], host mothers [M35], embryo transfer [M36], multiple pregnancy [M46], abortion on demand [M52], persistent vegetative state [N42], brain death [O6], quality adjusted life years [O14], attitudes to death [O20], biotechnology [Q21], eugenics [Q39].

■ INTERESTED IN: documentation [A8], women's rights [B6], privacy [B11], traffic of organs [B15], ethics committees [B22], CCNE — Comité Consultatif National d'Ethique (France) [B23], humanity heritage [B46], code of deontology [B60], mass media [B72], health and biology mediatisation [B85], codes of biomedical ethics [B90], CCNE advice (France) [B91], European Convention of Bioethics [B97], Bioethics bill, 1992 (France) [B98], quality of life [C16], medical ethics [C52], formal social control [D10], model legislation [D14], democracy [D82], genetics [E6], research institutes [E17], biomedical technologies [E28], data protection [E52], human genome project [F8], human experimentation [F11], mass screening [I6], organ donors [I15], organ procurement [I26], patient information [J4], consent to treatment [J8], living wills [J19], wrongful life [K33], criminal law [K53], trade secrets [K61], conscience clause [K63], beginning of life [L24], aging [L50], sexuality [M20], in vitro fertilization [M29], abortion [M47], fetal therapy [M57], venereal diseases [N31], persistent vegetative state [N42], determination of death [O2], withholding treatment [O10], euthanasia [O15], suicide [O19], alcohol abuse [P4], drug abuse [P6], genetic defects [Q3], genome mapping [Q23], genetic screening [Q37], human genome [Q47].

B 34 **LEROY,** O.G. Fernand

Clinic Director (artificial procreation), Professor of gynaecology obstetrics (Université libre de Bruxelles).

Université libre de Bruxelles
Hôpital Saint-Pierre
Clinique de procréation assistée
1000 Bruxelles (Belgium)
✆ (32) 2-535 34 05 Fax: (32) 2-537 59 26

■ AFFILIATIONS: Centre de Recherches Interdisciplinaires en Bioéthique — CRIB, Belgium (Founding *Member*).

■ TEACHING: interdisciplinary course on bioethics for philosophy students (ULB).

■ PUBLICATIONS: yes.

■ WORKING ON: life sciences [E4], research [E14], biomedical research [F2], experimentation [F9], methodology of research and experimentation [F15], sexuality and procreation [M2], reproductive technologies [M26], fetal development [M54], twinning [M55], genetic defects and hereditary diseases [Q2].

■ INTERESTED IN: biology [E5], genetics [E6], molecular biology [E8], epidemiology [E9], biomedical research [F2], clinical trials [F3], experimentation [F9], animal experimentation [F10], human experimentation [F11], sexuality and procreation [M2], reproductive technologies [M26], fetal development [M54], twinning [M55], genetic defects and hereditary diseases [Q2], sex preselection [Q30], preimplantation genetic diagnosis [Q38], gene therapy [Q42].

B 35 **LEVAUX** Paul

Executive Secretary of the Fonds de la recherche scientifique médicale (FNRS/NFWO).

Fonds de la recherche scientifique médicale (FRSM)
5, rue d'Egmont
1050 Bruxelles (Belgium)
✆ (32) 2 - 5049211 Fax: (32) 2 - 5049292

■ AFFILIATIONS: Commission d'éthique médicale du Fonds de la recherche scientifique médicale — FRSM, Bruxelles, Belgium (*Secretary General*).

B 36 **MAERTENS** Guido

Professor, Institut supérieur de philosophie, Leuven. — Honorary Director of the Kortrijk Campus. — Chairperson of the Overlegcentrum Christelijke Ethiek, Kath. Univ. Leuven.

Overlegcentrum Christelijke Ethiek
Kathol. Univ. Leuven
Rijselstraat 38/11
8500 Kortrijk (Belgium)
✆ (32) 56 - 21 49 69

■ AFFILIATIONS: Centre Bioéthique, Leuven (*Member*).
■ TEACHING: Ethics and Bioethics, 3rd year medical students, Leuven-Kortrijk.
■ PUBLICATIONS: yes.
■ WORKING ON: human person [B12], ethical rules and principles [B47], deontological ethics [C6], teleological ethics [C7], utilitarianism [C8], situational ethics [C9], religious ethics [C21], Christian ethics [C22], Roman catholic ethics [C23].
■ INTERESTED IN: fundamental rights of the individuals [B2], human rights [B3], self determination [B4], Helsinki Declaration [B37], Universal Declaration of Human Rights [B39], common good [B45], biomedical ethics education [B78], history of biomedical ethics [B94], European Convention of Bioethics [B97], biological specimens procurement, blood transfusion, organ transplantation [I8], in vitro fertilization [M29].

B 37 **MISSA** Jean-Noël
Research Fellow, Fonds national belge de la recherche scientifique (FNRS).

Centre de recherche interdisciplinaire en bioéthique (CRIB)
Université libre de Bruxelles CP 188
Avenue Adolphe-Buyl, 145
1050 Bruxelles (Belgium)
✆ (32) 2 - 650 26 28

■ AFFILIATIONS: Centre de recherche interdisciplinaire en bioéthique — CRIB, Bruxelles, Belgium (*Secretary and Member of the Board*).
■ TEACHING: Participation in a bioethics course called "éthique et technique", faculté de Philosophie et Lettres de Bruxelles.
■ PUBLICATIONS: yes.
■ WORKING ON: cognitive research [F7], human experimentation [F11], neurosciences — psychiatry [P1].
■ INTERESTED IN: genetics [E6], cognitive research [F7], human experimentation [F11], neurosciences — psychiatry [P1].

B 38 **MOULIN** Didier
Paediatrician, Director of the Department of Intensive Care for children. — Lecturer at the Université catholique de Louvain. — Teacher, Institut supérieur of nursing (Bruxelles).

Cliniques universitaires Saint-Luc
Avenue Hippocrate 10
1200 Bruxelles (Belgium)
✆ (32) 2 - 764 27 01 or 2 - 764 11 11 Fax: (32) 2 - 764 37 03

■ TEACHING: Seminars at the Centre de bioéthique de l'Université catholique de Louvain (UCL).
■ PUBLICATIONS: yes.
■ WORKING ON: self determination [B4], integrity [B8], emergency care [G54], pediatrics [G68], donors [I14], organ donors [I15], transplantation [I22], transplant recipients [I23], organ transplantation [I24], organ donation [I25], organ procurement [I26], parental notification [J6], disclosure [J7], nurse patient relationship [J30], physician nurse relationship [J31], physician patient relationship [J32], professional patient relationship [J33], parent child relationship [J34], beginning of life [L24], newborns [L29], prematurity [L30], handicapped children [L38], abortion [M47], selective abortion [M49], therapeutic abortion [M50], abortion on demand [M52], fetal development [M54], viability [M56], iatrogenic disease [N3], suffering [N11], pain [N12], prognosis [N14], transmission [N15], heart diseases [N20], nervous system diseases [N40], coma [N41], persistent vegetative state [N42], determination of death [O2], death [O5], brain death [O6], terminal care [O7], euthanasia [O15], resuscitation [O21].
■ INTERESTED IN: nursing education [B79], medical education [B80], codes of biomedical ethics [B90], CCNE advice (France) [B91], European Convention of Bioethics [B97], Bioethics bill, 1992 (France) [B98], medical devices [G23], artificial organs [G24], biological specimens procurement, blood transfusion, organ transplantation [I8], health care professionals, researchers and patient relationship [J1], childhood difficulties, diseases, protection [L21], abortion [M47], fetal development [M54], iatrogenic disease [N3], suffering [N11], pain [N12], prognosis [N14], nervous system diseases [N40], coma [N41], death and resuscitation [O1].

B 39 **MOULIN** Madeleine
Deputy Director of the Centre de sociologie de la santé. — Professor, Université libre de Bruxelles (Lecturer).

Centre de sociologie de la santé
Institut de sociologie, Université libre de Bruxelles
44, avenue Jeanne
1050 Bruxelles (Belgium)
✆ (32) 2 - 650 34 78 or 650 34 51 Fax: (32) 2 - 650 33 35

■ TEACHING: Occasional lectures, (faculté de Philosophie et Lettres de l'Université libre de Bruxelles; Université de Montréal, Université de Laval).
■ PUBLICATIONS: yes.
■ WORKING ON: fundamental rights of the individuals [B2], human rights [B3], dignity [B5], privacy [B11], ethics committees [B22], International Charter of Human Rights [B34], Nuremberg Code [B35], European Convention on Human Rights [B36], Helsinki Declaration [B37], Declaration on Human and Citizen Rights [B38], Universal Declaration of Human Rights [B39], patient advocacy [B42], ethics [B48], codes of ethics [B49], obligations of society [B53], moral policy [B55], deontology [B59], code of deontology [B60], solidarity [B65], interdisciplinary communication [B70], public debates [B75], health education [B77], health and biology mediatisation [B85], codes of biomedical ethics [B90], bioethical issues [B96], ethical relativism [C10], values [C15], quality of life [C16], social worth [C17], medical ethics [C52], nursing ethics [C53], attitudes [C74], intention [C75], motivation [C76], normality [C78], emotions [C79], love [C80], trust [C81], social sciences [C89], social control [D3], informal social control [D4], exclusion [D9], formal social control [D10], handicapped [D43], social impact [D49], social interaction [D50], cultural pluralism [D51], social discrimination [D53], sociology of medicine [D55], democracy [D82], dehumanization [D86], quality of environment [E41], social control of science [E64], maternal health [G3], maternal welfare [G10], health services [G25], residential facilities [G26], nursing home [G27], hospital [G28], hospices [G34], social workers [G44], medical records [G50], home care [G55], occupational medicine [G65], elderly [H23], mass screening [I6], mandatory screening [I7], alternative therapies [I33], patients' rights [J2], patient association [J3], patient information [J4], informed consent [J9], confidentiality [J20], referral and consultation [J24], investigator subject relationship [J29], physician nurse relationship [J31], physician patient relationship [J32], professional patient relationship [J33], legally incompetent person [K4], age [L13], aged — problems related with aging [L45], disease and aged [L46], life extension [L47], health care services and aged [L48], aged [L49], aging [L50], old person abuse [L51], single gene defects [Q7], pregnant women [M42], pregnancy [M44], death [O5], palliative care [O8], withholding treatment [O10], euthanasia [O15], active euthanasia [O16], involuntary euthanasia [O17], voluntary euthanasia [O18], attitudes to death [O20], institutionalized persons [P23], genome mapping [Q23].
■ INTERESTED IN: *idem, plus:* integrity [B8], traffic of organs [B15], institutions involved in the protection of rights [B17], CNIL — Commission Nationale Informatique et Libertés (France) [B19], World Medical Assembly [B20], Council of Europe [B25], College of Physicians [B27], office of science and technology assessment [B28], United Nations [B29], manifests and declarations concerning human rights [B33], common good [B45], humanity heritage [B46], obligations to society [B54], beneficence [B57], biomedical ethics education [B78], universities [B84], CCNE advice (France) [B91], history of biomedical ethics [B94], European Convention of Bioethics [B97], Bioethics bill, 1992 (France) [B98], normative ethics [C5], deontological ethics [C6], situational ethics [C9], value of life [C18], natural law [C24], Protestant ethics [C27], Protestantism [C35], philosophy of biology [C42], evolution [C47], sociobiology [C63], humanism [C66], self concept [C83], dehumanization [D86], history of science [E3], epidemiology [E9], research policy [E18], research design [E19], health hazards [E49], self regulation [E62], peer review [E63], health services research [F4], human experimentation [F11], WHO — World Health Organization [G36], nurse midwives [G43], resource allocation [H4], health care costs [H9], costs and benefits [H10], risks and benefices [H11], placebos [I36], presumed consent [J10], right to treatment [J17], living wills [J19], privileged communication [J21], medical secrecy [J22], professional deontology [J25], nurse patient relationship [J30], potentiality of personhood [K5], moratory [K25], therapeutic risk [K32], wrongful life [K33], fertility [M18], infertility [M19], iatrogenic disease [N3], physically [N8], venereal diseases [N31], syphilis [N32], acquired immunodeficiency syndrome [N35], brain death [O6], allowing to die [O11], terminally ill [O12], quality adjusted life years [O14], suicide [O19], deinstitutionalized persons [P22], genetic counseling [Q36], eugenics [Q39]

B 40 **NYS** Herman
Professor of Medical Law, Faculté de médecine et droit, Katholieke Universiteit Leuven.

Centrum voor Bio-Medische Ethiek en Recht
Kapucijnenvoer 35
3000 Leuven (Belgium)
✆ (32) 16 33 69 51 Fax: (32) 16 33 69 52

■ AFFILIATIONS: Centre de bioéthique et droit, Leuven, Belgium (*Administrator*).
■ TEACHING: postgraduate.
■ WORKING ON: respect of human dignity and human rights [B1], privacy [B11], protection of rights —

involved institutions [B16], ethics committees [B22], Helsinki Declaration [B37], biomedical research and experimentation [F1], health policy [G13], health legislation [G14], health occupations [G61], health care, medical acts [I1], biological specimens procurement, blood transfusion, organ transplantation [I8], biological substances contamination [I9], health care professionals, researchers and patient relationship [J1], patients' rights [J2], confidentiality [J20], physicians and researchers accountability [J23], law, legislation and jurisprudence [K1], legal personality [K2], jurisprudence — accountability [K8], legislation and law [K34], reproductive technologies [M26], abortion [M47], death and resuscitation [O1], determination of death [O2], euthanasia [O15], suicide [O19], outpatient commitment [P18], genetics and applications — biotechnology [Q1], medical genetics [Q35].
■ INTERESTED IN: *idem.*

B 41 **ROLIES** J.

Burgemeesterstraat 14
3000 Leuven (Belgium)
✆ (32) 16 -22 58 77 Fax: (32) 2 - 384 58 84

■ PUBLICATIONS: yes.
■ WORKING ON: reproductive technologies [M26], euthanasia [O15], outpatient commitment [P18], handicapped children [L38].
■ INTERESTED IN: reproductive technologies [M26], euthanasia [O15], outpatient commitment [P18], handicapped children [L38].

B 42 **SCHOTSMANS** Paul
Professor of Medical Ethics, faculté de médecine (KUL, Leuven). — Director of the Centrum voor Bio-Medische Ethiek en Recht (KUL, Leuven). — Member of the board, Séminaire Jean-XXIII (diocèse de Malines-Bruxelles).

Centrum voor Bio-Medische Ethiek en Recht
Kapucijnenvoer 35
3000 Leuven (Belgium)
✆ (32) 16 - 33 69 51 Fax: (32) 16 - 33 69 52

■ AFFILIATIONS: European Association of Centres of Medical Ethics — Association européenne des centres d'éthique médicale — EACME - AECEM, Leuven, France (*Treasurer, Secretary General, ad interium*).
■ TEACHING: Teaching Medical Ethics at the University of Leuven: third and seventh year of Medical School, third and fourth year of the Institute for Familial and Sexological Sciences, third and fourth year of Theology and Philosophy, third year of Hospital Management.
■ PUBLICATIONS: yes.
■ WORKING ON: human person [B12], bioethics [B89], religion and religious ethics [C19], ethics and humanities [C54], biological specimens procurement, blood transfusion, organ transplantation [I8], patients' rights [J2], inter-personal relationship [J27], sexuality and procreation [M2], reproductive technologies [M26], abortion [M47], fetal development [M54], fetal therapy [M57], terminal care [O7], euthanasia [O15], suicide [O19], attitudes to death [O20], resuscitation [O21], genetics and applications — biotechnology [Q1].
■ INTERESTED IN: *idem.*

B 43 **SOKAL** Gérard
Head of the Emeritus Department of Hematology. — Professor of Medical Deontology.

Université catholique de Louvain
Clos Chapelle-aux-Champs 30
BP 3052
1200 Bruxelles (Belgium)
✆ (32) 2 - 764 33 32 Fax: (32) 2 - 762 58 55

■ AFFILIATIONS: Conseil national de l'Ordre des médecins, Bruxeles, Belgium (*Member*).
■ TEACHING: Medical Deontology.
■ PUBLICATIONS: yes.
■ WORKING ON: human rights [B3], Helsinki Declaration [B37], Universal Declaration of Human Rights [B39], codes of ethics [B49], bioethics [B89], codes of biomedical ethics [B90], philosophy of biology [C42].
■ INTERESTED IN: *idem.*

B 44 **SUSANNE** Charles
Professor — Director of the Centre de bioéthique (Centrum Bioethiek- VUB).

Laboratoire de Génétique Humaine
Université libre de Bruxelles
50, avenue Franklin-Roosvelt (CP 192)
1050 Bruxelles (Belgium)
✆ (32) 2 - 650 37 79

■ AFFILIATIONS: Centre de Recherches interdisciplinaires en bioéthique — CRIB, Bruxelles, Belgium (*Chairperson*). — CRIB
■ TEACHING: bioethics.
■ PUBLICATIONS: yes.
■ WORKING ON: education [B76], universities [B84], bioethics [B89], philosophy of biology [C42], biology and human future [C43], evolution [C47], speciesism [C48], biological life [C49], primates [C50], biology [E5], genetics [E6], molecular biology [E8], epidemiology [E9], research team [E16], quality of environment [E41], ecology [E43], nuclear energy [E44], pollution [E47], health hazards [E49], prenatal diagnosis [I3], nutrition [L16], sexuality and procreation [M2], reproductive technologies [M26], abortion [M47], twinning [M55], congenital defects [N23], genetic defects and hereditary diseases [Q2], biotechnology — genetic engineering [Q20], population genetics [Q43].
■ INTERESTED IN: education [B76], universities [B84], bioethics [B89], philosophy of biology [C42], biology and human future [C43], evolution [C47], speciesism [C48], biological life [C49], primates [C50], biology [E5], molecular biology [E8], epidemiology [E9], research team [E16], quality of environment [E41], ecology [E43], nuclear energy [E44], pollution [E47], health hazards [E49], prenatal diagnosis [I3], nutrition [L16], sexuality and procreation [M2], reproductive technologies [M26], abortion [M47], twinning [M55], congenital defects [N23], genetic defects and hereditary diseases [Q2], biotechnology — genetic engineering [Q20], population genetics [Q43], genetics [E6].

B 45 **VAMOS** Ester

Clinic at the Laboratory of Cytogenetics, University Hospitals Brugmann and Erasme (Bruxelles).

Université libre de Bruxelles
Hôpital universitaire Brugmann
4, place Van-Gehuchten
1020 Bruxelles (Belgium)
✆ (32) 2 - 477 2217 Fax: (32) 2 - 479 8197

■ PUBLICATIONS: yes.
■ WORKING ON: genetic defects and hereditary diseases [Q2], genetic defects [Q3], chromosomal disorders [Q4], prenatal diagnosis [I3], amniocentesis [I4], chorionic villus sampling [I5], medical genetics [Q35], genetic counseling [Q36], genetic screening [Q37].
■ INTERESTED IN: human body commercialization [B13], European Convention of Bioethics [B97], Bioethics bill, 1992 (France) [B98], genetic screening [Q37], preimplantation genetic diagnosis [Q38], eugenics [Q39], preconception injuries [K27], wrongful life [K33], selective abortion [M49].

B 46 **VAN BORTEL** Paulus

Teacher at the Rijksuniversiteit Limburg (Maastricht NL). — Research (doctoral thesis) in Philosophy on medical ethics for the Higher Institute for Philosophy of the Catholic University of Leuven (Belgium). — Ethicist for the ethics committee of the Catholic Hospital in Duffel (Antwerp, Belgium).

Centrum voor Bio-Medische Ethiek
Kapucijnenvoer 35
3000 Leuven (Belgium)
✆ (32) 16 - 21 69 51 Fax: (32) 16 - 21 69 52

■ AFFILIATIONS: Vakgroop Gezondheidsethiek en Wijsbegeerte (RL- Maastricht), Leuven, Netherlands (*Researcher/teacher*). — Centrum voor Biomedische ethiek en recht (K.U. Leuven), Belgium (*Researcher/teacher*).
■ TEACHING: Case reviews with medical students.
■ PUBLICATIONS: yes.
■ WORKING ON: dignity [B5], human person [B12], CCNE — Comité Consultatif National d'Ethique (France) [B23], Council of Europe [B25], manifests and declarations concerning human rights [B33], Nuremberg Code [B35], European Convention on Human Rights [B36], Helsinki Declaration [B37], common good [B45], ethics [B48], codes of ethics [B49], moral policy [B55], codes of biomedical ethics [B90], history of biomedical ethics [B94], bioethics movement [B95], ethical analysis [C3], metaethics [C11], value of life [C18], medical ethics [C52], humanities [C55], philosophy [C61], judgement [C71], public opinion [D7], regulation [D16], government regulation [D17], cultural pluralism [D51], EC — European Communities [D62], advisory committees [D69], politics [D76], public policy [D77], democracy [D82], decision making [E54], standards [E60], therapeutic research [F6], public hospitals [G30], private hospitals [G31], medicine [G63], resource allocation [H4], physician patient relationship [J32], personhood [K3], conflict of interest [K9], generalization of expertise [K17], expert testimony [K18], embryos [M10], fetuses [M11].

■ INTERESTED IN: European Convention of Bioethics [B97], therapeutic research [F6], human genome project [F8], patient care [G53], home care [G55], resource allocation [H4], amniocentesis [I4], placebos [I36], European law [K49], medical law [K51], newborns [L29], disease and aged [L46], genetic defects and hereditary diseases [Q2], population genetics [Q43], genetic identity [Q44].

B 47 **VAN CUTSEM** Benoît

Research and teaching activities in nursing, Centre d'études bioéthiques de l'Université catholique de Leuven.

Centre d'études bioéthiques
bte UCL 43 / 4534
51, Promenade de l'Alma
1200 Bruxelles (Belgium)
✆ (32) 2 - 764 43 30

■ AFFILIATIONS: Centre d'études bioéthiques a. s. b. l — CEB, Bruxelles, Belgium (*Member*).
■ TEACHING: teacher training.
■ PUBLICATIONS: yes.
■ WORKING ON: ethics committees [B22], professional deontology organs [B26], ethics [B48], codes of ethics [B49], deontology [B59], interdisciplinary communication [B70], biomedical ethics education [B78], nursing education [B79], bioethical issues [B96], situational ethics [C9], Christian ethics [C22], theology [C41], nursing ethics [C53], decision making [E54], teaching methods [E59], nurses [G39], patient information [J4], disclosure [J7], consent to treatment [J8], nurse patient relationship [J30], physician nurse relationship [J31], abortion on demand [M52].
■ INTERESTED IN: women's rights [B6], human equality [B7], ethics committees [B22], professional deontology organs [B26], ethics [B48], codes of ethics [B49], deontology [B59], interdisciplinary communication [B70], biomedical ethics education [B78], nursing education [B79], bioethical issues [B96], ethical analysis [C3], deontological ethics [C6], situational ethics [C9], Christian ethics [C22], theology [C41], nursing ethics [C53], informal social control [D4], coercion [D5], social dominance [D6], decision making [E54], teaching methods [E59], nurses [G39], patient information [J4], disclosure [J7], consent to treatment [J8], nurse patient relationship [J30], physician nurse relationship [J31], abortion on demand [M52].

B 48 **VAN DEN BERGHE** Herman

M. D.

Centre for Human Genetics
Campus Gasthuisberg O en N
3000 Leuven (Belgium)
✆ (32) 16 - 21 58 78 Fax: (32) 16 - 21 59 92

■ INTERESTED IN: religion [C36], prenatal diagnosis [I3], gene therapy [Q42].

B 49 **VERELLEN** Gaston

Head of the Department of Neonatology, Cliniques universitaires Saint-Luc.

Cliniques universitaires Saint-Luc
Université catholique de Louvain (UCL)
Avenue Hippocrate, 10
1200 Bruxelles (Belgium)
✆ (32) 2 - 764 11 11 Fax: (32) 2 - 764 37 03

■ TEACHING: Clinical case studies in neonatology, Centre d'études bioéthiques de la faculté de médecine de l'UCL.
■ WORKING ON: professional competence [B52], nursing education [B79], medical education [B80], decision making [E54], human experimentation [F11], animal testing alternatives [F14], intensive care units [G33], patient care team [G38], patient care [G53], emergency care [G54], pediatrics [G68], prenatal diagnosis [I3], mass screening [I6], extraordinary treatment [I39], parental notification [J6], disclosure [J7], patient information [J4], parent child relationship [J34], newborns [L29], prematurity [L30], handicapped children [L38], procreation [M15], reproduction [M17], fertility [M18], infertility [M19], in vitro fertilization [M29], excess embryos [M34], fertility [M18], infertility [M19], in vitro fertilization [M29], excess embryos [M34], multiple pregnancy [M46], therapeutic abortion [M50], viability [M56], fetal therapy [M57], terminal care [O7], resuscitation [O21].
■ INTERESTED IN: ethics committees [B22], medical ethics [C52], clinical trials [F3], health services research [F4], control groups [F17], random selection [F18], maternal welfare [G10], parental consent [J13], treatment refusal [J15], physician nurse relationship [J31], wrongful life [K33], adoption [L32], abandoned (child) [L33], unwanted children [L35], host mothers [M35], embryo donation [M38],

pregnancy in adolescence [M45], abortion on demand [M52], acquired immunodeficiency syndrome [N35], persistent vegetative state [N42], prenatal injuries [N50], autopsies [O3], euthanasia [O15], gene therapy [Q42].

B 50 **VERMYLEN** J.

Professor of Internal Medicine, Katholieke Universiteit Leuven, Chairperson of the Ethics Commission.

Wildenhoge 9
3020 Winksele (Belgium)
✆ (32) 16 23 31 17

■ INTERESTED IN: self determination [B4], bioethics [B89], medical ethics [C52], clinical trials [F3], biological specimens procurement, blood transfusion, organ transplantation [I8], patients' rights [J2], life-sustaining treatment [O9], withholding treatment [O10], sex linked defects [Q19].

B 51 **WATTIAUX** Henri

Professor, Faculté de théologie de l'Université de Louvain-la-Neuve.

Allée de la Relevée 21
1400 Nivelles (Belgium)
✆ (32) 67 -21 49 70

■ AFFILIATIONS: Faculté de médecine, Unité de philosophie des sciences biomédicales, Bruxelles, Belgium (*Member*).
■ TEACHING: Professor, Faculté de théologie.
■ PUBLICATIONS: yes.
■ WORKING ON: codes of biomedical ethics [B90], CCNE advice (France) [B91], European Convention of Bioethics [B97], Bioethics bill, 1992 (France) [B98], conscience [C13], quality of life [C16], Christian ethics [C22], human genome project [F8], persistent vegetative state [N42], dominant genetic conditions [Q8], Huntington's chorea [Q9], recessive genetic conditions [Q10], sickle cell anemia [Q11], Duchenne muscular dystrophy [Q12], cystic fibrosis [Q15].
■ INTERESTED IN: codes of biomedical ethics [B90], CCNE advice (France) [B91], European Convention of Bioethics [B97], Bioethics bill, 1992 (France) [B98], Roman catholic ethics [C23], medical ethics [C52], regulation [D16], cohabitation [D27], biomedical technologies [E28], social control of science [E64], human genome project [F8], prenatal diagnosis [I3], sperm banks [I11], organ transplantation [I24], intracerebral fetal tissue transplantation [I28], informed consent [J9], infanticide [L42], handicapped children [L38], contraception [M3], embryos [M10], procreation [M15], artificial insemination [M31], embryo donation [M38], mother fetus relationship [M43], selective abortion [M49], mifepristone [M53], neural tube defects [N24], anencephaly [N25], spina bifida [N26], acquired immunodeficiency syndrome [N35], HIV seropositivity [N36], coma [N41], persistent vegetative state [N42], brain death [O6], palliative care [O8], life-sustaining treatment [O9], withholding treatment [O10], allowing to die [O11], terminally ill [O12], terminally ill [O12], euthanasia [O15], suicide [O19], resuscitation [O21], genetic defects [Q3], chromosomal disorders [Q4], down's syndrome [Q6], Huntington's chorea [Q9], sickle cell anemia [Q11], Duchenne muscular dystrophy [Q12], cystic fibrosis [Q15], phenylketonuria [Q16], thalassemia [Q17], hereditary diseases [Q18], sex linked defects [Q19], biotechnology [Q21], genetic intervention [Q22], DNA linkage [Q33], genetic counseling [Q36], preimplantation genetic diagnosis [Q38], gene therapy [Q42], human genome [Q47].

B 52 **ZORRILLA** Sergio

Research Fellow, Centre d'études bioéthiques, Medical School of the Université catholique de Louvain.

Centre d'études bioéthiques
bte UCL 43 /4534
Promenade de l'Alma 51
1200 Bruxelles (Belgium)
✆ (32) 2 - 764 43 30

■ TEACHING: Intensive seminars on ethics and medical philosophy.
■ PUBLICATIONS: yes.
■ WORKING ON: ethics committees [B22], CCNE — Comité Consultatif National d'Ethique (France) [B23], ethics [B48], codes of ethics [B49], biomedical ethics education [B78], nursing education [B79], bioethics [B89], bioethics movement [B95], specific approaches to ethics [C1], philosophy [C61], social issues [D2], socioeconomic factors [D18], hospital [G28], resource allocation [H4], health care costs [H9].
■ INTERESTED IN: ethics committees [B22], CCNE — Comité Consultatif National d'Ethique (France) [B23], Council of Europe [B25], ethics [B48], biomedical ethics education [B78], medical education [B80], bioethics [B89], bioethics movement [B95], specific approaches to ethics [C1], philosophy [C61], social issues [D2], socioeconomic factors [D18], hospital [G28], resource allocation [H4], health care costs [H9], neurosciences — psychiatry [P1], biotechnology — genetic engineering [Q20].

Denmark

Area: 43 069 km²
Population: 5.1 m
Gross domestic product: DKr 884bn; US$ 135 bn
Gross domestic product per head: 26 150 $
Gross domestic product growth: 1991: 1.2%
1992: 2%

Organisations

O DK 1 **CENTRAL SCIENTIFIC-ETHICAL COMMITTEE OF DENMARK**
40, blvd H.C. Andersens
1553 Copenhagen V (Denmark)
✆ (45) 33 11 43 00

O DK 2 **COMITÉ D'ETHIQUE DE L'ASSOCIATION DANOISE DES INFIRMIERES**
Vimmelskaftet 38
P.O. Box 1084
1008 Copenhagen K (Denmark)
✆ (45) 33 15 15 55 Fax: (45) 33 15 24 55

O DK 3 **CONSEIL NATIONAL D'ETHIQUE — DET ETISKE RAD**
Ravnsborggade 2-4
2200 Copenhagen N (Denmark)
✆ (45) 35 37 58 33 Fax: (45) 35-37 57 55

O DK 4 **COUNCIL OF THE DANISH BISHOPS**
Nørregade 11
1165 Copenhagen K. (Denmark)
✆ (45) 33 13 35 08

O DK 5 **DANISH MEDICAL ASSOCIATION BOARD OF ETHICS**
Trondhjemsgade 9
2100 Copenhagen Ø (Denmark)
✆ (45) 31 38 55 00

O DK 6 **DANISH SOCIETY OF MEDICAL RESOURCE ALLOCATION**
Amtsgården
Dammaven 26
7100 Vejle (Denmark)
✆ (45) 75 83 53 33

O DK 7 **INSTITUTE OF HUMAN GENETICS — CLINICAL GENETICS UNIT**
University of Aarhus - Bartholinbygningen
Wilhem Meyers Alle
8000 Aarhus C (Denmark)
✆ (45) 35 37 58 33 Fax: (45) 35 37 57 55

O DK 8 **REGIONAL SCIENTIFIC-ETHICAL COMMITTEE OF THE COUNTIES OF RIVE RINGKØLING AND ØNDERJNNAND**
Østergade 80
6700 Esbjerg (Denmark)
✆ (45) 75 18 12 00

O DK 9 **REHABILITERINGS-OG FORSKINGSCENTRET FOR TORTUROFRE (REHABILITATION AND RESEARCH CENTRE FOR TORTURE VICTIMS)** (RCT)
P.O. Box 2672
2100 Copenhagen Ø (Denmark)
✆ (45) 31 39 46 94 Fax: (45) 31 39 50 20

■ OTHER FIELDS OF ACTIVITIES: rehabilitation of persons who have been subjected to torture, and rehabilitation of their families. — Instruction of Danish health professionals in the examination and treatment of persons who have been subjected to torture. — Teaching in a wider forum to contribute to the spread of knowledge about torture, torture methods, and the possibilities for rehabilitating persons who have been subjected to torture. — Research on torture and the nature and extent of the consequences of torture.
■ DATE OF FOUNDATION: 1982.
■ EXECUTIVE BOARD: Dr. Inge Genefke (*Medical Director*), Dr. Søren Bøjholm (*Chief psychiatrist*), Jens Andersen (*Director of Finance and Planning*).

■ TEACHING: RCT regularly (twice a year) conducts training seminars in Denmark and several times abroad, on rehabilitation of torture survivors and the ethical aspects involved in this work.
■ PUBLICATIONS: Quarterly journal on rehabilitation of torture victims and prevention of torture, ISBN: 0904-1982.
■ DOCUMENTATION CENTRE: open to the public.
■ COLLOQUIUMS, SYMPOSIUMS: doctors, ethics and torture, yearly, medical and legal profession.
■ DECISIONS, ADVICE: WMA: Declaration of Tokyo (1975). — Standing Committee of Doctors of the EC : Statement of Madrid (1989). — UN: Principles of Medical Ethics (1982). — UN: Convention Against Torture (1985).

O DK 10 **THE REGIONAL ETHICAL COMMITTEE OF THE COUNTY OF AARHUS**
Lyseng Allee 1
8240 Højbjerg (Denmark)
✆ (45) 86 27 30 40

O DK 11 **UNIT OF MEDICAL PHILOSOPHY AND CLINICAL THEORY**
Pancem Institute
Blegdamsvej 3
2200 N Copenhagen (Denmark)
✆ (45) 31 35 75 00

Individuals

DK 1 **ANDERSEN** Daniel
Head of department of surgical gastroenterology (Odense University Hospital). — Professor of surgery.

Odense University Hospital
Surgical Department K
5000 Odense C (Denmark)
✆ (45) 66 11 33 33 Fax: (45) 66 13 28 54

■ AFFILIATIONS: Danish Society of Medical Resource Allocation, Vejle, Denmark (*Vice Chairperson*).
■ TEACHING: Leader of a postgraduate one-week in Medical. Research Ethics for Ph.D. students.
■ PUBLICATIONS: yes.
■ WORKING ON: justice [B9], Helsinki Declaration [B37], beneficence [B57], deontology [B59], history of biomedical ethics [B94], ethical analysis [C3], religion and religious ethics [C19], humanities [C55], selection for treatment [G52], donors [I14], professional deontology [J25], deception [K14].
■ INTERESTED IN: *idem.*

DK 2 **ANDERSEN** Svend
Professor.

Institute of Ethics and Philosophy of Religion
Aarhus Universiteit
Building 410
8000 Aarhus C (Denmark)
✆ (45) 86 13 67 11 Fax: (45) 86 13 04 90

■ AFFILIATIONS: centre for Bioethics, Aarhus C, Denmark (*Chairperson*).
■ TEACHING: as part of teaching theology students.
■ PUBLICATIONS: yes.
■ WORKING ON: dignity [B5], reification [B14], ethics committees [B22], ethics [B48], philosophical ethics [C2], Christian ethics [C22], Protestant ethics [C27], history of science [E3], semen donors [M28], biotechnology [Q21].
■ INTERESTED IN: data bases [A3], bibliography [A4], European Convention of Bioethics [B97], philosophy of biology [C42], scarcity [D21], North South relationships [D60], resource allocation [H4], developing countries [H21], euthanasia [O15], human genome [Q47].

DK 3 **BOLUND** Lars
Professor of Clinical Genetics. — Research leader in the Danish Centre for Human Genome Research.

University of Åarhus — Institute of Human Genetics
Bartholinbygningen
Wilhem Meyers Alle
8000 Aarhus C. (Denmark)
✆ (45) 86 12 31 73 Fax: (45) 86 19 12 77

■ AFFILIATIONS: E.C. Ethical Committee on Human Embryos and Research, Brussels, Belgium (*Chairperson*).
■ PUBLICATIONS: yes.
■ WORKING ON: self determination [B4], ethics committees [B22], codes of ethics [B49], medical education [B80], universities [B84], biology and human future [C43], future generations [C44], social control [D3], informal social control [D4], EC — European Communities [D62], ECC — European Community Commission [D63], genetics [E6], research design [E19], biological containment [E39], data protection [E52], nontherapeutic research [F5], therapeutic research [F6], human genome project [F8], research subjects [F12], healthy volunteers [F13], medical records [G50], mass screening [I6], tissue banking [I12], fetal tissue transplantation [I27], patient information [J4], informed consent [J9], parental consent [J13], medical secrecy [J22], investigator subject relationship [J29], physician patient relationship [J32], property rights [K46], patents [K47], body parts and fluids [L4], human development [L12], embryos [M10], fetuses [M11], ovum [M12], sperm [M14], ovum donors [M27], semen donors [M28], in vitro fertilization [M29], artificial insemination [M31], excess embryos [M34], selective abortion [M49], therapeutic abortion [M50], aborted fetuses [M51], cancer [N17], cardiovascular diseases [N19], congenital defects [N23], genetic defects [Q3], chromosomal disorders [Q4], hereditary diseases [Q18], genetic intervention [Q22], genome mapping [Q23], recombinant DNA research [Q31], DNA fingerprinting [Q32], DNA linkage [Q33], genetic counseling [Q36], genetic screening [Q37], gene therapy [Q42], genetic identity [Q44], gene pool [Q45], carriers [Q46], human genome [Q47].
■ INTERESTED IN: self determination [B4], ethics committees [B22], humanity heritage [B46], codes of ethics [B49], biology and human future [C43], future generations [C44], social control [D3], informal social control [D4], research design [E19], human genome project [F8], human development [L12], preimplantation genetic diagnosis [Q38], negative eugenics [Q40], positive eugenics [Q41], gene therapy [Q42].

DK 4 **BONNELYCKE** Per Kleis
SAS DATA, EDP Department Head, Copenhagen Airport.

Glamsbjergvej 49
2770 Kastrup (Denmark)
✆ (45) 31 51 04 81

■ WORKING ON: respect of human dignity and human rights [B1], fundamental rights of the individuals [B2], human rights [B3], self determination [B4], human equality [B7], justice [B9], Helsinki Declaration [B37], ethics [B48], codes of ethics [B49], prisoners [D12], legislation [D13], data protection [E52], patients' rights [J2], legislation and law [K34], criminal law [K53], criminal code [K54], sexual behavior [M21], homosexuality [M22], homosexuals [M23], transsexualism [M24], prostitution [M25], HIV seropositivity [N36], life-sustaining treatment [O9], withholding treatment [O10], allowing to die [O11].
■ INTERESTED IN: minority groups [D42], data protection [E52], sexual behavior [M21], homosexuality [M22], homosexuals [M23], HIV seropositivity [N36].

DK 5 **BOYSEN** Gudrun
Chief Consultant Neurologist. — Associate Professor.

Rigshospitalet
Dept. of Neurology
Blegdamsvej 9
2100 Copenhagen Ø (Denmark)
✆ (45) 35 45 20 87

■ AFFILIATIONS: Rehabilitation Centre for Torture Victims — RCT, Copenhagen, Denmark (*Vice-Chairperson*).
■ PUBLICATIONS: yes.
■ WORKING ON: fundamental rights of the individuals [B2].
■ INTERESTED IN: *idem, plus:* women's rights [B6]

DK 6 **DRAMINSKY PETERSEN** Hans
M.D., specialist in medicine and gastroenterology.

Nordborggade 10, 2. TH
2100 Copenhagen Ø (Denmark)

■ AFFILIATIONS: Physicians for Human Rights, Aarhus C, Denmark (*Board member*).
■ TEACHING: Seminars arranged by Physicians for Human Rights.
■ PUBLICATIONS: yes.
■ WORKING ON: Torture [D94].
■ INTERESTED IN: *idem, plus:* injuries [N49]

DK 7 **EHMER** Elisabeth

Strandvejen 96,
7120 Vejle Øst (Denmark)
✆ (45) 75 81 58 65 Fax: (45) 75 81 58 65

■ AFFILIATIONS: The Regional Ethical Committee for the Counties Vejle and Fyner and the Central Ethical Committee, Denmark (*Member*).
■ TEACHING: yes.
■ PUBLICATIONS: yes.

DK 8 **ENGELBRECHT** Nils
General Practitioner.

Tårbæk Strandvej 93
2930 Klampenborg (Denmark)

■ AFFILIATIONS: The Danish Society of Medical Philosophy, Ethics and Methodology, Denmark (*Member of the Board*). — Member of The Scientific-Ethical Committee of the County of Copenhagen, Denmark (*Member*).
■ PUBLICATIONS: yes.

DK 9 **FASTING** Ulla
Senior Health Researcher.

Nerredige 20,
6950 Ringkøbing (Denmark)
✆ (45) 97 32 17 22 Fax: (45) 97 32 49 31

■ AFFILIATIONS: The Danish Medical Council, Copenhagen N, Denmark (*Member*).
■ TEACHING: Teaching nurses, physicians. — Teaching the public.
■ PUBLICATIONS: yes.

DK 10 **FATUM** Lone
Teacher, author, minister (Danish Church).

Faculty of Theology - University of Copenhagen
Købmagergade 44-46
1150 Copenhagen K (Denmark)
✆ (45) 45 33 15 28 11

■ TEACHING: Advanced teaching of nurses. — Training of Commission Minister of the Danish Church in matter of bioethical critique.
■ PUBLICATIONS: yes.

DK 11 **FJALLAND** Bjarne
Head of Department of Biology, Royal Danish School of Pharmacy.

Danmarks Farmaceutiske Højskole
Universitetsparken 2
2100 Copenhagen Ø (Denmark)
✆ (45) 35 37 44 57

■ WORKING ON: life sciences [E4], toxicity [E50], safety [E51], nontherapeutic research [F5], animal experimentation [F10], animal testing alternatives [F14], drugs [I34].
■ INTERESTED IN: animal experimentation [F10], animal testing alternatives [F14], biological substances contamination [I9], drugs [I34].

DK 12 **FOLTMANN** Bent
Professor Emeritus.

University of Copenhagen
Institute of Biochem. Genetics

Øster Farimagsgade 2A
1353 Copenhagen K (Denmark)
✆ (45) 33 15 88 42 Fax: (45) 33 93 52 20

■ TEACHING: Adult Education. — Hoved stadens Oplysnings forbund (Metropolitan Association of Education).
■ PUBLICATIONS: yes.
■ WORKING ON: moral development [C14], evolution [C47], philosophy [C61], molecular biology [E8], brain [L6].
■ INTERESTED IN: moral development [C14], evolution [C47], determinism [C62], sociobiology [C63], the brain [L6], genetic intervention [Q22], genome mapping [Q23].

DK 13 **GARDE** Karin
Head of Psychiatric Department U, Sct. Hans Hospital (mental hospital for Copenhagen community).

Head of Psychiatric Department for young psychotic patients
Sct. Hans hospital
4000 Roskilde (Denmark)
✆ (45) 46 33 46 33

■ AFFILIATIONS: Group of medical women's research, Copenhagen N, Denmark (*Member*).
■ TEACHING: Guest Lecturer at Copenhagen University, at the lectures for women's medical research.
■ PUBLICATIONS: yes.
■ WORKING ON: patient advocacy [B42], public debates [B75], health and biology mediatisation [B85], uncontrolled information [B88], medical ethics [C52], females [D35], biomedical technologies [E28], psychiatry [G69], mass screening [I6].
■ INTERESTED IN: patient advocacy [B42], health and biology mediatisation [B85], advertising [B87], uncontrolled information [B88], females [D35], mass screening [I6].

DK 14 **GENEFKE** Inge Kemp
Medical Director of the International Rehabilitation and Research Centre for Torture Victims (RCT).

International Rehabilitation and Research Centre for Torture Victims (RCT)
34, Juliane Maries Vej
2100 Copenhagen Ø (Denmark)
✆ (45) 31 39 46 94 Fax: (45) 31 39 50 20

■ TEACHING: Teaching within all aspects of torture: methods, aims, sequels, treatment, prevention, challenge to the health professions, threat to democracy — nationally and internationally.
■ PUBLICATIONS: yes.
■ WORKING ON: fundamental rights of the individuals [B2], integrity [B8], justice [B9], freedom [B10], human person [B12], institutions involved in the protection of rights [B17], professional deontology organs [B26], International Court [B30], International Charter of Human Rights [B34], European Convention on Human Rights [B36], Declaration on Human and Citizen Rights [B38], Universal Declaration of Human Rights [B39], codes of ethics [B49], code of deontology [B60], codes of biomedical ethics [B90], bioethical issues [B96], professional ethics [C51], medical ethics [C52], nursing ethics [C53], coercion [D5], dissent [D58], democracy [D82], dehumanization [D86], torture [D94], suffering [N11], pain [N12], psychological stress [P17].
■ INTERESTED IN: *idem.*

DK 15 **GYLDENHOLM** A. G.
Head of Department of Zoology, Institute of Biological Sciences (Aarhus).

Institute of Biological Sciences
Department of Zoology
Building 135, Universitetsparken
8000 Aarhus C (Denmark)
✆ (45) 86 20 27 11 Fax: (45) 86 12 51 75

■ AFFILIATIONS: Danish Council of Ethics, Copenhagen N, Denmark (*Member*).
■ TEACHING: Lectures (schools, public, examinations for university students) and articles.
■ PUBLICATIONS: yes.

DK 16 **HANSEN** Jørgen
General Practitioner.

Søndervang 12
2670 Greve (Denmark)

■ AFFILIATIONS: Regional Scientific-Ethical Committee of Denmark (*Member*).
■ WORKING ON: terminal care [O7], patient information [J4], informed consent [J9], patient participation [J16], confidentiality [J20], physician's role [J26], physician patient relationship [J32].
■ INTERESTED IN: *idem*.

DK 17 **HOLLNAGEL** Hanne
General Practitioner. — Clinical Lecturer. — Ph.D.

Danstrupvej 8,
3480 Fredensborg (Denmark)
✆ (45) 42 28 17 10 Fax: (45) 42 28 18 55

■ PUBLICATIONS: yes.
■ WORKING ON: mass media [B72], public debates [B75], medical ethics [C52], socioeconomic factors [D18], social groups [D22], preventive medicine [G58], family practice [G60], professional patient relationship [J33], follow-up studies [N13], heart diseases [N20].
■ INTERESTED IN: preventive medicine [G58], family practice [G60].

DK 18 **HOLME HANSEN** Ebba
Professor of Social Pharmacy (practice combining theory and research methods from Natural, Human and Social Sciences), Royal Danish School of Pharmacy.

The Royal Danish School of Pharmacy
Universitetsparken 2
2100 Copenhagen (Denmark)
✆ (45) 35 37 08 50 Fax: (45) 35 37 57 44

■ AFFILIATIONS: Health Care Technology Assessment, the Danish Research Councils, Copenhagen, Denmark (*Member*).
■ TEACHING: Teach a.o. the subjects clinical trials and information to patients in obligatory and optional courses at the royal Danish School of Pharmacy. — Post-graduate courses for health professionals in Denmark and Scandinavia.
■ PUBLICATIONS: yes.
■ WORKING ON: human rights [B3], self determination [B4], dignity [B5], women's rights [B6], human equality [B7], integrity [B8], justice [B9], freedom [B10], privacy [B11], human person [B12], office of science and technology assessment [B28], Nuremberg Code [B35], Helsinki Declaration [B37], competence [B51], authoritarianism [B58], deontology [B59], respect [B63], virtues [B64], communication [B69], information dissemination [B71], mass media [B72], advertising [B87], medical ethics [C52], informal social control [D4], coercion [D5], females [D35], associations [D47], sociology of medicine [D55], international organizations [D61], research design [E19], biomedical technologies [E28], drug industry [E37], review committees [E42], survey [E58], teaching methods [E59], self regulation [E62], peer review [E63], social control of science [E64], technology assessment [E65], clinical trials [F3], health services research [F4], therapeutic research [F6], cognitive research [F7], human experimentation [F11], research subjects [F12], control groups [F17], random selection [F18], selection of subjects [F19], medical informatics [F21], physicians [G40], medical staff [G41], pharmacists [G42], patient compliance [G49], remuneration [H14], medical fees [H15], developing countries [H21], alternative therapies [I33], drugs [I34], placebos [I36], patient association [J3], patient information [J4], patient access [J5], consent to treatment [J8], informed consent [J9], presumed consent [J10], patient participation [J16], investigator subject relationship [J29], professional patient relationship [J33], malpractice [K19], therapeutic risk [K32], mind [L20], contraception [M3], central nervous system diseases [N43], substance dependence [P3], drug abuse [P6], institutionalized persons [P23], psychoactive drugs [P33].
■ INTERESTED IN: information sources, data bases [A1], human rights [B3], self determination [B4], dignity [B5], women's rights [B6], human equality [B7], integrity [B8], justice [B9], freedom [B10], privacy [B11], human person [B12], ethics committees [B22], office of science and technology assessment [B28], Nuremberg Code [B35], Helsinki Declaration [B37], Declaration on Human and Citizen Rights [B38], Universal Declaration of Human Rights [B39], public advocacy [B41], patient advocacy [B42], codes of ethics [B49], communication [B69], information dissemination [B71], ethical review [B93], history of biomedical ethics [B94], professional ethics [C51], medical ethics [C52], sociobiology [C63], mental processes [C69], social control [D3], informal social control [D4], coercion [D5], regulation [D16], socioeconomic factors [D18], social groups [D22], females [D35], sociology of medicine [D55], international organizations [D61], research design [E19], biomedical technologies [E28], drug industry [E37], review committees [E42], teaching methods [E59], self regulation [E62], peer review [E63], social control of science [E64], technology assessment [E65], clinical trials [F3], health services research [F4], therapeutic research [F6], cognitive research [F7], human experimentation [F11], research subjects [F12], debriefing [F16], control groups [F17], random selection [F18], selection of subjects [F19], group of vulnerable subjects [F20], medical informatics

[F21], maternal health [G3], mental health [G4], physicians [G40], medical staff [G41], pharmacists [G42], patient compliance [G49], medical records [G50], remuneration [H14], medical fees [H15], developing countries [H21], alternative therapies [I33], drugs [I34], placebos [I36], patient association [J3], patient information [J4], patient access [J5], consent to treatment [J8], informed consent [J9], presumed consent [J10], consent forms [J14], treatment refusal [J15], patient participation [J16], right to treatment [J17], confidentiality [J20], investigator subject relationship [J29], professional patient relationship [J33], fraud [K13], malpractice [K19], misconduct [K23], therapeutic risk [K32], mind [L20], cancer [N17], HIV seropositivity [N36], central nervous system diseases [N43], substance dependence [P3], drug abuse [P6], outpatient commitment [P18], psychiatric technics [P24], psychoactive drugs [P33].

DK 19 **ILSØE** Grethe

Maribovej 6,
2500 Valby (Denmark)

■ AFFILIATIONS: Det Elisher Råd, Kubenhavn N, Denmark (*Member*).
■ WORKING ON: respect of human dignity and human rights [B1], fundamental rights of the individuals [B2].

DK 20 **KROGH JENSEN** Kirsten

County Councillor.

County Council
Rosenlunden 15
9000 Aalborg (Denmark)
✆ (45) 98 12 02 70

■ AFFILIATIONS: Regional Ethical Committee, Aalborg, Denmark (*Chairperson*). — Central Ethical Committee (Copenhagen), Denmark (*Member*).

DK 21 **JØRGENSEN** Niels Åge

General Practitioner.

Dalgasvej 5,
6623 Vorbasse (Denmark)
✆ (45) 75 33 34 44 Fax: (45) 75 33 38 58

■ AFFILIATIONS: Regional Scientific-Ethical Committee of the counties of Ribe, Rinkøbing and South Jutland, Esbjerg, Denmark (*Member*).
■ WORKING ON: ethics committees [B22], Helsinki Declaration [B37].

DK 22 **KARDEL** Troels

General Practitioner.

Østerbrogade 54 A
2100 Copenhagen Ø. (Denmark)
✆ (45) 31 42 65 16 Fax: (45) 32 96 01 12

■ AFFILIATIONS: Rehabilitation Centre for Torture Victims — RCT, Copenhagen, Denmark (*Member*).
■ TEACHING: External examiner in medical philosophy and bioethics, Medical Faculty (University of Copenhagen).
■ WORKING ON: History before the 20th century [C58].
■ INTERESTED IN: *idem*.

DK 23 **KEMP** Peter

Professor of Philosophy.

University of Copenhagen
Philosophy Institute
Njalsgade 80,
2300 Copenhagen S (Denmark)
✆ (45) 31 54 22 11 Fax: (45) 32 54 04 11

■ PUBLICATIONS: yes.
■ WORKING ON: publications [A9], respect of human dignity and human rights [B1], protection of rights — involved institutions [B16], ethical rules and principles [B47], information — communication —

media [B68], bioethics [B89], specific approaches to ethics [C1], social and political issues [D1], political issues [D56], violence [D85], science, technology, methods [E1], biomedical research and experimentation [F1], organization of health care; facilities, manpower and services; health occupations [G1], health economics, population characteristics [H1], health care professionals, researchers and patient relationship [J1], law, legislation and jurisprudence [K1], legislation and law [K34], sexuality and procreation — unborn child [M1], terminal care [O7], euthanasia [O15], mentally handicapped [P7], schizophrenia [P12], biotechnology — genetic engineering [Q20], medical genetics [Q35].

DK 24 **KIRKEBÆK** Birgit
Associate Professor of Special Education.

The Royal Danish University of Educational Studies
Emdrupvej 101,
2400 Copenhagen NV (Denmark)
✆ (45) 31 69 66 33 Fax: (45) 39 66 00 81

■ PUBLICATIONS: yes.
■ WORKING ON: history of biomedical ethics [B94], historical aspects [C57], history before the 20th century [C58], history of the 20th century [C60], sociobiology [C63], normality [C78], handicapped [D43], government and political systems [D64], professional-patient relationship [J33], parent child relationship [J34], prematurity [L30], child and family [L31], child diseases and problems [L37], handicapped children [L38], mentally retarded [P8], eugenics [Q39], negative eugenics [Q40].
■ INTERESTED IN: historical aspects [C57], history before the 20th century [C58], history of the 20th century [C60], mentally retarded [P8], negative eugenics [Q40].

DK 25 **KJÆRULFF** Erling
General Practitioner. — Senior Lecturer, University of Aarhus, Denmark.

Sosvinget 22,
8250 Egå (Denmark) Fax: (45) 31 38 55 00

■ AFFILIATIONS: Danish Medical Association, Board of Ethics , Copenhagen Ø, Denmark (*Member*).
■ TEACHING: Senior Lecturer, Department of General Practice, University of Aarhus.
■ WORKING ON: fundamental rights of the individuals [B2], health education [B77], bioethics [B89].
■ INTERESTED IN: fundamental rights of the individuals [B2], health education [B77], bioethics [B89], professional ethics [C51], research [E14], biomedical research [F2], organization of health care [G35].

DK 26 **KNUDSEN** Knud Olav
Veterinary Surgeon.

Solbakken 20
6100 Haderslev (Denmark)
✆ (45) 74 53 15 65 Fax: (45) 74 52 53 86

■ AFFILIATIONS: Scientific-ethical committee of the counties of Ribe, Ringkøbing and South Jutland, Esbjerg, Denmark (*Member*).
■ PUBLICATIONS: yes.

DK 27 **KOCH** Lene
Research Fellow at the Institute of Social Medicine, University of Copenhagen.

Institute of Social Medicine
University of Copenhagen
Blegdamsvej 3
2200 Copenhagen N. (Denmark)
✆ (45) 31 35 79 00

■ PUBLICATIONS: yes.
■ WORKING ON: history of science [E3], genetics and applications — biotechnology [Q1], genetic counseling [Q36], genetic screening [Q37], eugenics [Q39], negative eugenics [Q40], positive eugenics [Q41], preimplantation genetic diagnosis [Q38], human genome [Q47], reproductive technologies [M26], in vitro fertilization [M29].
■ INTERESTED IN: eugenics [Q39], history of science [E3], reproductive technologies [M26].

DK 28 **LARSEN** Niels A.
Head of Nephrology Service.

Medicinsk Afdelinf
7500 Holstebro (Denmark)
✆ (45) 97 41 42 00

■ AFFILIATIONS: Den Almindelige Danske Lægeforening (Danish Medical Society)., Denmark (*Member*). — Danish Society of Nephrology, Denmark (*Vice Chairperson*). — European Renal Association/European Dialys and Transplant Association (*Member*).
■ WORKING ON: fundamental rights of the individuals [B2], human rights [B3], integrity [B8], human person [B12], traffic of organs [B15], College of Physicians [B27], Helsinki Declaration [B37], biomedical research [F2], methodology of research and experimentation [F15], artificial organs [G24], health services [G25], hospital [G28], public hospitals [G30], patient care team [G38], physicians [G40], medical staff [G41], health care [G45], medicine [G63], organ donors [I15], donor cards [I16], directed donation [I18], transplant recipients [I23], organ transplantation [I24], organ donation [I25], organ procurement [I26], renal dialysis [I32], kidney diseases [N39].

DK 29 **LIND** Vincent
Bishop (Lutheran Church).

Bispegården
5000 Odense C (Denmark)
✆ (45) 66 14 52 96 Fax: (45) 66 12 35 24

■ AFFILIATIONS: Council of the Danish Bishops, Copenhagen K., Denmark (*Informing the rest of the members*).
■ TEACHING: Seminars,congresses, negotiations arranged by Ethics Council, Danish Medical Society, Public Health sector.
■ PUBLICATIONS: yes.
■ WORKING ON: situational ethics [C9], decision making [E54].
■ INTERESTED IN: physician patient relationship [J32], research [E14], confidentiality [J20].

DK 30 **LINDHARDT** Nina

Østergade 80,
6700 Esbjerg (Denmark)

■ AFFILIATIONS: Regional Scientific-Ethical Committee of the counties of Rive, Ringkøling and Sønderjnnand, Esljerg, Denmark (*Secretary*).

DK 31 **LINNEMAN** Dorte
Scientist, M.D., DMSC.

University of Copenhagen
The Protein Laboratory
3C, Blegdamsvej
2200 Copenhagen (Denmark)
✆ (45) 31 35 79 00 Fax: (45) 35 36 01 16

■ AFFILIATIONS: Danish Council of Ethics, Copenhagen N, Denmark (*Member*).
■ WORKING ON: fundamental rights of the individuals [B2], ethical rules and principles [B47], information — communication — media [B68], codes of biomedical ethics [B90], ethics committees [B22], Helsinki Declaration [B37], philosophical ethics [C2], Christian ethics [C22], Roman catholic ethics [C23], philosophy of biology [C42], historical aspects [C57], life sciences [E4], research [E14], technology [E27], evaluation [E61], biomedical research and experimentation [F1], diagnosis [I2], biological specimens procurement, blood transfusion, organ transplantation [I8], renal dialysis [I32], confidentiality [J20], reproductive technologies [M26], abortion [M47], fetal development [M54], fetal therapy [M57], terminal care [O7], euthanasia [O15], biotechnology — genetic engineering [Q20], medical genetics [Q35], population genetics [Q43].

DK 32 **MARCKMANN** Anton
Chief Surgeon.

Kirurgisk Afdeling
Vesterbakken 45 Høruphav
6470 Sydals (Denmark)
✆ (45) 74 43 03 11

■ AFFILIATIONS: The Central Scientific-Ethical Committee of Denmark, Denmark (*Chairperson*). — The Regional Scientific-Ethical Committee of the Southern and Western part of Jutland, Denmark (*Member*).

DK 33 **MIKKELSEN** Margareta
Chairperson of Department of Medical Genetics, John F. Kennedy Institute.

Department of Medical Genetics
J.F. Kennedy Institute
7GL.Landevej
2600 Glostrup (Denmark)
✆ (45) 42 45 22 28 Fax: (45) 43 43 11 30

■ AFFILIATIONS: Danish Council of Ethics, Copenhagen N, Denmark (*Member*)
■ TEACHING: Information to public.
■ PUBLICATIONS: yes.
■ WORKING ON: genetics and applications — biotechnology [Q1].
■ INTERESTED IN: *idem.*

DK 34 **MÜNSTER** Ole
Executive Director of the Danish Council of Ethics.

Danish Council of Ethics
2-4, Ravnsborggade
2200 Copenhagen N (Denmark)
✆ (45) 35 37 58 33 Fax: (45) 35 37 57 55

■ PUBLICATIONS: yes.
■ INTERESTED IN: Council of Europe [B25], codes of ethics [B49], interdisciplinary communication [B70], press conference [B73], public debates [B75], bioethical issues [B96], European Convention of Bioethics [B97], ethical analysis [C3], utilitarianism [C8], situational ethics [C9], consequences [C45], medical ethics [C52], research policy [E18], data protection [E52], prenatal diagnosis [I3], patents [K47], in vitro fertilization [M29], genome mapping [Q23], genetic screening [Q37], human genome [Q47].

DK 35 **NIELSEN** Henning Merrild
Member of the Central Scientific Ethical Committee of Denmark.

Strandvej 17,
5700 Svendborg (Denmark)
✆ (45) 62 21 45 04

DK 36 **NORBY** Søren
Full Professor of Human Genetics.

University of Genetics
Frederik V's Vej 11
2100 Copenhagen Ø (Denmark)
✆ (45) 35 37 32 22 Fax: (45) 35 37 16 51

■ AFFILIATIONS: Det Etiske Råd (National ethics committee) — DER, Copenhague N, Denmark (*Member*).
■ TEACHING: Conferences for laboratory techicicians working in hospitals. — Public conferences.
■ WORKING ON: philosophical ethics [C2], ethical analysis [C3], normative ethics [C5], situational ethics [C9], philosophy of biology [C42], biology and human future [C43], consequences [C45], evolution [C47], biological life [C49], genetic defects and hereditary diseases [Q2], genetic defects [Q3], single gene defects [Q7], biotechnology — genetic engineering [Q20], genetic intervention [Q22], genome mapping [Q23], sex determination [Q26], genetic counseling [Q36], genetic screening [Q37], eugenics [Q39], gene therapy [Q42], medical genetics [Q35], human genome project [F8], abortion [M47], selective abortion [M49], euthanasia [O15], eye diseases [N38], nervous system diseases [N40], hereditary diseases [Q18], recombinant DNA research [Q31], behavioral genetics [Q34], population genetics [Q43].
■ INTERESTED IN: fundamental rights of the individuals [B2], ethical rules and principles [B47], bioethics [B89], attitudes to death [O20].

DK 37 **PETERSEN** Ulrich Horst

Ministry of Health
Herluf Trollesgade 11
1052 Copenhagen K. (Denmark)
✆ (45) 33 92 33 60

■ PUBLICATIONS: yes.

DK 38 **PETERSSON** Birgit

Associate Professor. — Member of the Ministry of Justice Appeal Commission for Abortion and Sterilisation.

Institute of Social Medicine and Psychiatry
Panum Institutet
Blegdamsvej 3
2200 Copenhagen N. (Denmark)
✆ (45) 31 35 79 00 Fax: (45) 31 35 11 81

■ TEACHING: Educating medical students.
■ PUBLICATIONS: yes.
■ WORKING ON: ethics and humanities [C54], reproductive technologies [M26], pregnancy and childbirth [M39], medical ethics [C52], bioethical issues [B96], philosophical ethics [C2], attitudes to death [O20], medical genetics [Q35], population genetics [Q43].
■ INTERESTED IN: *idem.*

DK 39 **RASMUSSEN** Kirsten

Medical Doctor, Clinical Geneticist. — Member of the Danish Council of Ethics.

Clinical Genetics Unit, Institute of Human Genetics
University of Aarhus
8000 Aarhus (Denmark)
✆ (45) 86 13 97 11 Fax: (45) 86 12 31 73

■ AFFILIATIONS: Danish Council of Ethics, Copenhagen N, Denmark (*Member*).
■ TEACHING: To medical students, nurses, general public.

DK 40 **RASMUSSEN** Ole Vedel

Medical Consultant, Rehabilitation Centre for Torture Victims (RCT) Copenhagen. — Editor, "Torture", quarterly journal. — Member of the IRCI board (International Rehabilitation Council for Torture Victims).

Rehabilitation Centre for Torture Victims (RCT)
34, Juliane Maries Vej
2100 Copenhagen Ø. (Denmark)
✆ (45) 31 39 46 94 et 31 39 50 20

■ TEACHING: Medical aspects of torture.
■ PUBLICATIONS: yes.
■ WORKING ON: publications [A9], human equality [B7], institutions involved in the protection of rights [B17], medical ethics [C52], torture [D94], investigators [E15], cognitive research [F7].
■ INTERESTED IN: traffic of organs [B15], torture [D94].

DK 41 **REHOF** Lars Adam

Associate Professor of Public International Law, Department of Public International Law, University of Copenhagen.

Department of Public International Law and E.C. Law
University of Copenhagen 6, Studiestraede
1455 Copenhagen (Denmark)
✆ (45) 33 91 21 66 Fax: (45) 33 91 05 52

■ AFFILIATIONS: The Danish Council of Ethics, Copenhagen N., Denmark (*Member*).
■ TEACHING: The Legal Position of the Patient and the Physician.
■ PUBLICATIONS: yes.
■ WORKING ON: constitutional amendments [K36], legal rights [K45], European law [K49], international law [K50], legal obligation [K59], conscience clause [K63].
■ INTERESTED IN: expert evaluation [K15], malpractice [K19], legal rights [K45], European law [K49], international law [K50], legal obligation [K59], conscience clause [K63].

DK 42 **RIIS** Povl

Head of Department of Medical Gastroenterology, Herlev University Hospital. — Full Professor of Medicine, University of Copenhagen.

Herlev University Hospital
Københavns Amts Sygehus i Herlev Herlev University Hospital
2730 Herlev (Denmark)
✆ (45) 44 53 53 00 Fax: (45) 44 53 53 32

■ AFFILIATIONS: Central Scientific-Ethical Comittee of Denmark, Copenhagen V, Denmark (*Chairperson*).
■ TEACHING: A number of national and international courses in bioethics, as an organizer and teacher.
■ PUBLICATIONS: yes.
■ WORKING ON: fraud [K13], ethics committees [B22], Council for International Organization of Medical Sciences [B24], Council of Europe [B25], Helsinki Declaration [B37], Red Cross [B67], biomedical ethics education [B78], ethical review [B93], medical ethics [C52], epidemiology [E9], scientific misconduct [E25], review committees [E42], clinical trials [F3], human experimentation [F11], medicine [G63], informed consent [J9], physician patient relationship [J32].

DK 43 **SCOCOZZA** Lone
Associate Professor of Sociology of Medicine. — Political appointed member of the Danish Ethical Committee.

Institute of Social Pharmacy
Unversitets Parken 2
2100 Copenhagen Ø (Denmark)
✆ (45) 31 37 08 50 Fax: (45) 35 37 57 44

■ AFFILIATIONS: The Medical Ethical Committee of Copenhagen (Den Videnskabsetiske Komité for Københshvnj og Frederiksberg Kommoner), Copenhagen, Denmark (*Member*).
■ TEACHING: Danish Royal School of Pharmacy. — University of Lund, Department of Sociology, Sweden. — University of Copenhagen, Institute of Political Science. — Workshops for Health Personnel. — Nurses Schools in Copenhagen, Aarhus, Hillerød, etc.
■ PUBLICATIONS: yes.
■ WORKING ON: review [A2], book review [A6], documentation [A8], publications [A9], fundamental rights of the individuals [B2], human rights [B3], self determination [B4], dignity [B5], human equality [B7], integrity [B8], justice [B9], freedom [B10], human person [B12], human body commercialization [B13], ethics committees [B22], Nuremberg Code [B35], Helsinki Declaration [B37], codes of ethics [B49], obligations of society [B53], obligations to society [B54], moral policy [B55], altruism [B56], deontology [B59], paternalism [B62], respect [B63], communication [B69], public debates [B75], health education [B77], biomedical ethics education [B78], medical education [B80], codes of biomedical ethics [B90], history of biomedical ethics [B94], normative ethics [C5], deontological ethics [C6], teleological ethics [C7], utilitarianism [C8], ethical relativism [C10], metaethics [C11], values [C15], value of life [C18], medical ethics [C52], coercion [D5], social dominance [D6], public opinion [D7], legislation [D13], model legislation [D14], socioeconomic factors [D18], social groups [D22], sociology of medicine [D55], history of science [E3], research policy [E18], research design [E19], scientific misconduct [E25], biomedical technologies [E28], drug industry [E37], review committees [E42], biomedical research [F2], clinical trials [F3], human experimentation [F11], research subjects [F12], control groups [F17], random selection [F18], patients' rights [J2], patient information [J4], disclosure [J7], informed consent [J9], presumed consent [J10], consent forms [J14], physician's role [J26], investigator subject relationship [J29], physician patient relationship [J32], professional patient relationship [J33], conflict of interest [K9], fraud [K13], deception [K14], expert evaluation [K15], technical expertise [K16], malpractice [K19], cancer [N17].
■ INTERESTED IN: self determination [B4], patient advocacy [B42], equal protection [B43], interdisciplinary communication [B70], information dissemination [B71], mass media [B72], education [B76], ethical review [B93], Bioethics bill, 1992 (France) [B98], social control [D3], progress [E26], health services research [F4], therapeutic research [F6], cognitive research [F7], prematurity [L30].

DK 44 **SIMONSEN** Jørn
Chairperson of the Danish Medico-Legal Council. — Chairperson of Institute of Forensic Pathology, University of Copenhagen.

Institute of Forensic Pathology
Frederik V-svej 11
2100 Copenhagen Ø (Denmark)
✆ (45) 35 37 32 22 Fax: (45) 35 37 16 51

■ AFFILIATIONS: Danish Rehabilitation Centre for Torture Victims — RCT, Copenhagen, Denmark (*Board member*).
■ TEACHING: Postgraduate teaching of medical students.
■ PUBLICATIONS: yes.
■ INTERESTED IN: respect for human dignity and human rights [B1].

DK 45 **SKOU** Ellen Margrethe

Rislundsvej 9,
8240 Risskov (Denmark)
✆ (45) 86 17 79 18

■ AFFILIATIONS: The Danish Central Scientific Ethical Review Committee, Copenhagen V, Denmark (*Vice-Chairman*).
■ TEACHING: Medical doctors, nurses.
■ PUBLICATIONS: yes.
■ WORKING ON: human rights [B3], self determination [B4], dignity [B5], women's rights [B6], human equality [B7], integrity [B8], justice [B9], freedom [B10], ethics committees [B22], Council for International Organization of Medical Sciences [B24], Nuremberg Code [B35], Helsinki Declaration [B37], patient advocacy [B42], ethical rules and principles [B47], health education [B77], biomedical ethics education [B78], nursing education [B79], medical education [B80], ethical review [B93], history of biomedical ethics [B94], ethical analysis [C3], quality of life [C16], professional ethics [C51], medical ethics [C52], nursing ethics [C53], philosophy [C61], social control [D3], political activity [D57], public policy [D77], community services [D78], democracy [D82], national socialism [D83], methodology of research and experimentation [F15], medical informatics [F21], maternal health [G3], mental health [G4], nursing home [G27], hospital [G28], hospices [G34], WHO — World Health Organization [G36], nurses [G39], psychiatry [G69], health economics, population characteristics [H1], diagnosis [I2], biological specimens procurement, blood transfusion, organ transplantation [I8], organ donors [I15], donor cards [I16], anonymous donation [I17], directed donation [I18], blood transfusions [I19], blood donation [I20], blood substitutes [I21], transplantation [I22], transplant recipients [I23], organ transplantation [I24], organ donation [I25], organ procurement [I26], fetal tissue transplantation [I27], intracerebral fetal tissue transplantation [I28], bone marrow transplantation [I29], renal dialysis [I32], health care professionals, researchers and patient relationship [J1], stages of life — problems proper to childhood and elderly [L1].
■ INTERESTED IN: human rights [B3], self determination [B4], dignity [B5], women's rights [B6], human equality [B7], integrity [B8], justice [B9], freedom [B10], ethics committees [B22], Council for International Organization of Medical Sciences [B24], Nuremberg Code [B35], Helsinki Declaration [B37], patient advocacy [B42], ethical rules and principles [B47], health education [B77], biomedical ethics education [B78], nursing education [B79], medical education [B80], ethical review [B93], history of biomedical ethics [B94], ethical analysis [C3], professional ethics [C51], nursing ethics [C53], philosophy [C61], social control [D3], political activity [D57], democracy [D82], national socialism [D83], genetics [E6], microbiology [E7], molecular biology [E8], epidemiology [E9], health services research [F4], human genome project [F8], human experimentation [F11], methodology of research and experimentation [F15], organization of health care; facilities, manpower and services; health occupations [G1], maternal health [G3], mental health [G4], institutional policies [G15], institutional obligations [G16], artificial organs [G24], nursing home [G27], hospital [G28], hospices [G34], WHO — World Health Organization [G36], nurses [G39], emergency care [G54], home care [G55], patient transfer [G56], psychiatry [G69], health economics, population characteristics [H1], economics [H2], administrators [H3], resource allocation [H4], insurance [H5], life insurance [H8], health care costs [H9], costs and benefits [H10], risks and benefices [H11], compensation [H12], public participation [H13], remuneration [H14], medical fees [H15], price [H16], profit [H17], financial support [H18], elderly [H23], diagnosis [I2], biological substances contamination [I9], renal dialysis [I32], health care professionals, researchers and patient relationship [J1], stages of life — problems proper to childhood and elderly [L1], sexuality and procreation — unborn child [M1], death and resuscitation [O1], neurosciences — psychiatry [P1], genetics and applications — biotechnology [Q1].

DK 46 **SØRENSEN** Bent

Rehabilitation and Research Centre for Torture Victims (RCT)
34, Juliane Maries Vej
2100 Copenhague Ø (Denmark)
✆ (45) 31 39 46 94 Fax: (45) 31 39 50 20

■ AFFILIATIONS: United Nations Committee against Torture — CAT, Geneva 10, Switzerland (*Member*).
■ TEACHING: Teaching about torture inside the above mentioned committees. — Teaching medical ethics and the problem of doctors as torturers, at international seminars. — Teaching prison medicine at the international seminars.
■ PUBLICATIONS: yes.
■ WORKING ON: fundamental rights of the individuals [B2], freedom [B10], protection of rights — involved institutions [B16], Council of Europe [B25], United Nations [B29], Helsinki Declaration [B37], codes of ethics [B49], professional competence [B52], deontology [B59], medical ethics [C52],

torture [D94], physical containment [E40], right to treatment [J17], medical secrecy [J22], medical law [K51].
■ INTERESTED IN: medical ethics [C52], torture [D94].

DK 47 **THOMSEN** Karen
Deputy Director in School of Nursing (Aalborg).

Sygeplejeskolen i Aalborg
Søndre Skovvej
9000 Aalborg (Denmark)
✆ (45) 99 32 34 67

■ AFFILIATIONS: The Danish Nurses Association, Council of Nursing Ethics (Dansk Sygeplejerå, Det sygeplejeetiske Råd), Copenhagen K, Denmark (*Member*).
■ TEACHING: student and professional nurses. — Public lectures.
■ PUBLICATIONS: yes.
■ WORKING ON: self determination [B4], dignity [B5], traffic of organs [B15], ethics committees [B22], Helsinki Declaration [B37], ethics [B48], codes of ethics [B49], nursing education [B79], students [B82], curriculum [B83], nursing ethics [C53], love [C80], trust [C81], biological containment [E39], nurses [G39], nurse patient relationship [J30], brain death [O6], biotechnology [Q21].
■ INTERESTED IN: self determination [B4], dignity [B5], ethics committees [B22], Helsinki Declaration [B37], ethics [B48], nursing education [B79], students [B82], curriculum [B83], nursing ethics [C53], love [C80], trust [C81], biological containment [E39], nurses [G39], nurse patient relationship [J30], brain death [O6], biotechnology [Q21], traffic of organs [B15], codes of ethics [B49].

DK 48 **TIEDEMANN** Erling
County Mayor.

Danish Society for Medical Priority Setting
Amtsgarden
7100 Vejle (Denmark)
✆ (45) 75 83 53 33

■ WORKING ON: state government [D67], state responsibility [D75], community services [D78].

DK 49 **VILIEN** Mogens
Head of Medical Department, Gastroenterology section.

Medical Department
Central Hospital Holbaek
4300 Holbaek (Denmark)
✆ (45) 42 96 43 33

■ AFFILIATIONS: Danish Society of Gastroenterology, Holbaek, Denmark (*General secretary*).

DK 50 **WOLF** Hans
Professor of Urology.

Skejby Sygehus
Department of Urology
8200 Aarhus N (Denmark)
✆ (45) 45 78 45 11

■ AFFILIATIONS: The Regional Ethical Committee of the County of Aarhus, Højbjerg, Denmark (*Chairperson*).

DK 51 **WULFF** Henrik R.
Professor of Clinical Theory and Clinical Ethics at the Medical Faculty of the University of Copenhagen.

Herlev Hospital
Herlev Ringvej 75
2730 Herlev (Denmark)
✆ (45) 44 53 53 00 Fax: (45) 44 53 53 32

■ AFFILIATIONS: Unit of Medical Philosophy and Clinical Theory, Copenhagen, Denmark (*Professor*).
■ TEACHING: undergraduate and postgraduate teaching.
■ PUBLICATIONS: yes.
■ WORKING ON: health services [G25].

France

Area: 547 062 km²
Population: 56.8m
Gross domestic product: FFr 7.16trn; US$ 1.3trn
Gross domestic product per head: $ 22,900
Gross domestic product growth: 1991: 1.6%
1992: 2.5%

Organisations

OF1 **ACADÉMIE NATIONALE DE MÉDECINE**
16, rue Bonaparte
75006 Paris (France)

OF2 **APRSP ALTAIR**
30, rue Salvador-Allende
92000 Nanterre (France)
✆ (33) 1 - 47 21 46 46 Fax: (33) 1 - 47 29 06 62

■ OTHER FIELDS OF ACTIVITIES: Preventive actions for, and rehabilitation of, adult prostitutes, social action on AIDS, actions in the field of sexual identity (transsexuals, transvestites).
■ OFFICIALLY REGULATED ORGANISATION: yes.
■ DATE OF FOUNDATION: 1984.
■ EXECUTIVE BOARD: Pr. Victor Girard (*Chairperson*), Marc Chapiro (*Vice-Chairperson*), Marie Wargnier (*Treasurer*), Viviane Dubol (*Executive Secretary*), Félix Marchand (*Director*).

OF3 **ASSOCIATION DESCARTES**
1, rue Descartes
75231 Paris Cedex 05 (France)
✆ (33) 1 -46 34 32 96 Fax: (33) 1 - 46 34 33 40

OF4 **ASSOCIATION DU GRIEPS**
74, rue du Grand-Roule
69110 Sainte-Foy-lès-Lyon (France)
✆ (33) 78 50 86 86 Fax: (33) 78 51 11 23

OF5 **ASSOCIATION FRANÇAISE DE DÉPISTAGE ET PRÉVENTION DU SIDA**
Hôtel-Dieu
58, quai Jules-Courmont
69002 Lyon (France)

OF6 **ASSOCIATION FRANÇAISE DE DROIT DE LA SANTÉ**
CERAF faculté de droit
Avenue Léon-Duguit
33604 Pessac Cedex (France)
✆ (33) 56 84 84 31 Fax: (33) 56 84 84 32

OF7 **ASSOCIATION GÉNÉTIQUE ET LIBERTÉ**
45, rue d'Ulm
75005 Paris (France)

■ OFFICIALLY REGULATED ORGANISATION: Legislation on associations (1901).
■ DATE OF FOUNDATION: June 1989.
■ EXECUTIVE BOARD: Patricia Baldaci (*Member of the board*), Franck Bourrat (*Member of the board*), André Klarsfeld (*Member of the board*).
■ PUBLICATIONS: Newsletter (January 1991 to April 1992).
■ DOCUMENTATION CENTRE: open to the public.
■ COLLOQUIUMS, SYMPOSIUMS: Monthly Conferences at the ENS (45, rue d'Ulm Paris).
■ WORKING ON: human equality [B7], freedom [B10], privacy [B11], CNIL — Commission Nationale Informatique et Libertés (France) [B19], information dissemination [B71], health and biology mediatisation [B85], Bioethics bill, 1992 (France) [B98], human genome project [F8], preventive medicine [G58], sexuality and procreation — unborn child [M1], genetics and applications — biotechnology [Q1].

OF8 **ASSOCIATION INTERNATIONALE DROIT, ÉTHIQUE ET SCIENCE (GROUPE DE MILAZZO)**
6, boulevard Gambetta

67000 Strasbourg (France)
✆ (33) 88 41 28 20 Fax: (33) 88 41 27 99

■ OTHER FIELDS OF ACTIVITIES: Science Ethics.
■ DATE OF FOUNDATION: July 1989.
■ EXECUTIVE BOARD: Christian Byk (*Executive Secretary*), Georges Biname (*Treasurer*), Pr. Jiri Haderka, Paula Martenko da Silva, Pr. Marco Milani-Compantti, Pr. Amos Shapiro, Dr. Antonio Piga, Laurence O'Connell, Pr. Marcel Mélanson .
■ TEACHING: Seminars (Eastern Europe).
■ PUBLICATIONS: Bilingual quarterly (F/E): Journal International de Bioéthique. published since 1992 — Worldwide national ethics committees' advisory notices. — Colloquia proceedings. — 1989 book on Assisted Medical Procreation.
■ DOCUMENTATION CENTRE: internal.
■ COLLOQUIUMS, SYMPOSIUMS: International statutory conferences (every two years): 1989 (Assisted Medical Procreation); 1991 (Genetics); 1993 (Bioethics and power). — Others: 1992 (Bioethics and power); National colloquiums organized by the various departments: France (1991, 1993) "L'expérience biomédicale"; Belgium (1992) "Les biotechnologies"; Canada (1993) "Bioéthique et femmes".
■ DECISIONS, ADVICE: Declaration of Milazzo on ethical and legal aspects of the human genome analysis programme (although the Association does not usually produce official texts).
■ WORKING ON: humanity heritage [B46], moral obligations [B50], obligations of society [B53], obligations to society [B54], information — communication — media [B68], information dissemination [B71], mass media [B72], biomedical ethics education [B78], health and biology mediatisation [B85], codes of biomedical ethics [B90], CCNE advice (France) [B91], religion and religious ethics [C19], religion [C36], biology and human future [C43], historical aspects [C57], social and political issues [D1], social control [D3], legislation [D13], regulation [D16], government regulation [D17], family members [D24], females [D35], cultural pluralism [D51], sociology of medicine [D55], advisory committees [D69], public policy [D77], science, technology, methods [E1], life sciences [E4], biology [E5], genetics [E6], research [E14], research policy [E18], industrial research [E22], international research [E23], biomedical technologies [E28], freezing [E29], drug industry [E37], quality of environment [E41], data protection [E52], methods [E53], decision making [E54], standards [E60], social control of science [E64], biomedical research and experimentation [F1], biomedical research [F2], clinical trials [F3], human genome project [F8], experimentation [F9], human experimentation [F11], public health [G5], health policy [G13], health legislation [G14], WHO — World Health Organization [G36], patients' rights [J2], medical secrecy [J22], physician's role [J26], physician patient relationship [J32], potentiality of personhood [K5], legislation and law [K34], developing countries [H21], health care, medical acts [I1], diagnosis [I2], prenatal diagnosis [I3], amniocentesis [I4], mass screening [I6], biological specimens procurement, blood transfusion, organ transplantation [I8], blood transfusions [I19], blood donation [I20], transplantation [I22], organ transplantation [I24], fetal tissue transplantation [I27], international law [K50], European law [K49], stages of life — problems proper to childhood and elderly [L1], anatomy, physiology, development [L2], human body [L3], aid children [L34], sports [L53], sexuality and procreation [M2], embryos [M10], fetuses [M11], procreation [M15], reproductive technologies [M26], in vitro fertilization [M29], artificial insemination [M31], neurosciences — psychiatry [P1], behavioral and mental disorders [P2], genetics and applications — biotechnology [Q1], genetic defects and hereditary diseases [Q2], cystic fibrosis [Q15], biotechnology — genetic engineering [Q20], biotechnology [Q21], medical genetics [Q35], genetic screening [Q37], population genetics [Q43], genetic identity [Q44], gene pool [Q45], human genome [Q47].
■ INTERESTED IN: codes of biomedical ethics [B90], CCNE advice (France) [B91], religion and religious ethics [C19], professional ethics [C51], ethics and humanities [C54], social issues [D2], legislation [D13], model legislation [D14], regulation [D16], government regulation [D17], socioeconomic factors [D18], social groups [D22], social impact [D49], social interaction [D50], stigmatization [D54], political issues [D56], international aspects [D59], North-South relationships [D60], EC — European Communities [D62], ECC — European Community Commission [D63], advisory committees [D69], health legislation [G14], self regulation [E62], social control of science [E64], biomedical research and experimentation [F1], public health [G5], health policy [G13], artificial organs [G24], WHO — World Health Organization [G36], sports medicine [G64], economics [H2], health insurance [H6], health care costs [H9], remuneration [H14], biological substances contamination [I9], blood transfusions [I19], drugs [I34], jurisprudence — accountability [K8], malpractice [K19], child's interest [K24], therapeutic risk [K32], court decision [K40], patents [K47], criminal law [K53], bill [K60], conscience clause [K63], animal organs [L18], prematurity [L30], sports [L53], contraception [M3], transsexualism [M24], abortion [M47], fetal therapy [M57], nervous system diseases [N40], brain diseases [N44], neurosciences — psychiatry [P1], behavioral and mental disorders [P2], biotechnology [Q21], genome mapping [Q23], sex determination [Q26], transgenic animals [Q29], eugenics [Q39].

OF9 **ASSOCIATION POUR LE DÉVELOPPEMENT DE LA PHARMACOLOGIE CLINIQUE**

J.-P. Demarez
2, place de Lassue
78260 Achères (France)

■ OTHER FIELDS OF ACTIVITIES: Methodology of Human Experimentation. — Legal aspects. — Medical deontology and clinical trials.
■ DATE OF FOUNDATION: 1985.
■ EXECUTIVE BOARD: Pr. Patrice Jaillon (*Chairperson*), Dr. Jean Thébault (*Vice-Chairperson*), Pr. Michel Ollagnier (*Treasurer*), Dr. Jean-Paul Demarez (*Executive Secretary*).
■ TEACHING: Annual one-day meeting. — Postgraduate DEA grant (FF 60,000).
■ PUBLICATIONS: "La lettre du pharmacologue".
■ DOCUMENTATION CENTRE: internal.
■ COLLOQUIUMS, SYMPOSIUMS: Annual colloquium.

OF10 **ASSOCIATION POUR LE DROIT DE MOURIR DANS LA DIGNITÉ** (ADMD)

103, rue La Fayette
75010 Paris (France)
✆ (33) 1 42 85 12 22 Fax: (33) 1 45 96 00 50

■ OTHER FIELDS OF ACTIVITIES: The aim of the Association is to bring individuals and the goverment to change their attitudes and the legislation in order to allow terminally ill patients to die in dignity. Our field of action includes: relieving-pain methods, fighting against life-sustaining treatments, palliative care, and the right to active euthanasia. (The law should provide this possibility for patients having, themselves, repeatedly expressed the wish to die).
■ OFFICIALLY REGULATED ORGANISATION: Legislation on associations (1901).
■ DATE OF FOUNDATION: 18/04/1980.
■ EXECUTIVE BOARD: Jacques Pohier (*Chairperson*), Henri Caillavet (*Honorary Chairperson*), Paul Chauvert (*Honorary Chairperson*), Rolande Tourteau-Fontenelle (*Executive Secretary*), Anne-Marie Dourlen-Rollier (*Vice-Chairperson*), Madeleine Gavelle (*Vice-Chairperson*), Aimée Candau (*Deputy Secretary*), Hubert Moreau (*Treasurer*), Régine Grassano (*Deputy Treasurer*).
■ PUBLICATIONS: Quarterly Newsletter (25,000 copies) with articles concerning the theory and practice of bioethics regarding the right to die in dignity. 1992: "Les droits des vivants sur la fin de leur vie. Pourquoi la loi doit les protéger" (The right of individuals as regards the end of their lives. Why should the law protect them ?).
■ COLLOQUIUMS, SYMPOSIUMS: 16-17 /11/1991 (Paris): "Mourir dans la dignité. Vivre sa mort: un choix, un droit" (Dying in dignity. Living one's death: a choice and a right).
■ DECISIONS, ADVICE: Development of laws, regulations, and deontology in various countries, particularly in the 18 countries represented in the "World federation of right-to-die societies".
■ WORKING ON: self determination [B4], dignity [B5], ethics committees [B22], Council of Europe [B25], patient advocacy [B42], obligations of society [B53], deontology [B59], biomedical ethics education [B78], nursing education [B79], Bioethics bill, 1992 (France) [B98], quality of life [C16], medical ethics [C52], nursing ethics [C53], socioeconomic factors [D18], residential facilities [G26], nursing home [G27], intensive care units [G33], hospices [G34], patient care team [G38], physicians [G40], patient admission [G47], medical records [G50], selection for treatment [G52], patient care [G53], emergency care [G54], home care [G55], health care costs [H9], elderly [H23], patients' rights [J2], physician patient relationship [J32], personhood [K3], legally incompetent person [K4], economic value of life [K28], wrongful life [K33], aged — problems related with aging [L45], disease and aged [L46], life extension [L47], health care services and aged [L48], aged [L49], aging [L50], old person abuse [L51], death and resuscitation [O1].

OF12 **CECOS**

124, rue Camille-Desmoulins
80000 Amiens (France)
✆ (33) 22 53 37 77 Fax: (33) 22 53 37 90

OF13 **CENTRE D'ÉTHIQUE MÉDICALE**

60, boulevard Vauban
BP 109
59016 Lille Cedex (France)
✆ (33) 20 42 09 12 et 20 30 88 27 Fax: (33) 20 78 26 45

■ AFFILIATION: Institut catholique de Lille (Catholic institution composed of 35 institutes and 5 faculties).
■ DATE OF FOUNDATION: 1984.
■ EXECUTIVE BOARD: Pr. Jacques Liefooghe (*Chairperson*), Sr. Marie-Louise Lamau (*Director*), Dr. Bruno

Cadore (*Deputy-Director*).
■ TEACHING: yes.

0 F 14 **CENTRE DE BIOÉTHIQUE**
Université catholique de Lyon
25, rue du Plat
69288 Lyon Cedex 02 (France)
✆ (33) 72 32 50 22 Fax: (33) 72 32 50 19

0 F 15 **CENTRE DE DOCUMENTATION ET D'INFORMATION EN ÉTHIQUE DE L'INSTITUT NATIONAL DE LA SANTÉ ET DE LA RECHERCHE MÉDICALE** (CDIE-INSERM)
101, rue de Tolbiac
75654 Paris Cedex 13 (France)
✆ (33) 1 - 44 23 60 92 Fax: (33) 1 - 45 85 68 56

■ AFFILIATION: INSERM.
■ OFFICIALLY REGULATED ORGANISATION: the CDIE was created by the INSERM in order to provide technical and administrative support to the CCNE (Comité consultatif national d'éthique pour les sciences de la vie et de la santé).
■ DATE OF FOUNDATION: 1983.
■ EXECUTIVE BOARD: Dr. Paulette Dostatni (*Director*).
■ TEACHING: technical support in ethics teaching (documentary files).
■ PUBLICATIONS: The CDIE coordinates the editing of the CCNE Newsletter: "Lettre du Comité consultatif national d'éthique pour les sciences de la vie et de la santé" published by the Documentation française. — Monthly review of the 80 magazines recieved by the CDIE library. — Quarterly bibliography on current event subjects.
■ DOCUMENTATION CENTRE: open to the public.
■ COLLOQUIUMS, SYMPOSIUMS: Co-organisers of annual ethics meetings of the CCNE.
■ WORKING ON: information sources, data bases [A1], review [A2], data bases [A3], bibliography [A4], information centers [A5], book review [A6], literature [A7], documentation [A8], publications [A9], editorial policies [A10], information — communication — media [B68], information dissemination [B71], students [B82], universities [B84], CCNE advice (France) [B91], history of biomedical ethics [B94].
■ INTERESTED IN: bioethics [B89], codes of biomedical ethics [B90], legislation and law [K34], guidelines [K55], law [K56].

0 F 16 **CENTRE DE DROIT ET ÉTHIQUE DE LA SANTÉ**
Hôpital Edouard-Herriot
Pavillon N4
69437 Lyon Cedex 03 (France)
✆ (33) 72 34 48 13 et78 77 70 00 Fax: (33) 78 77 71 60

0 F 17 **CENTRE DE RECHERCHE EN PHILOSOPHIE DES SCIENCES**
Université Paris-X
A 301
92001 Nanterre (France)
✆ (33) 1 - 40 97 73 53

■ OTHER FIELDS OF ACTIVITIES: Research in Philosophy and History of Science. Support to doctoral candidates.
■ AFFILIATION: Ecole doctorale de philosophie, psychologie et sciences de l'éducation.
■ DATE OF FOUNDATION: 1971. — Modified in 1987.
■ EXECUTIVE BOARD: Pr.. Anne Fagot-Largeault (*Coordinator*), Pr.. Jean Seidengart (*Coordinator*), Pr. Bernadette Bensaude-Vincent (*Coordinator*).
■ TEACHING: Postgraduate teaching (DEA, doctorat).
■ PUBLICATIONS: Postgraduate research dissertation.
■ DOCUMENTATION CENTRE: internal.
■ WORKING ON: informed consent [J9], prenatal diagnosis [I3], organ donors [I15], decision analysis [E55], quality of life [C16], ethics committees [B22], animal experimentation [F10], philosophical ethics [C2], embryos [M10], euthanasia [O15], research subjects [F12], group of vulnerable subjects [F20], pediatrics [G68], nurses [G39], ethical analysis [C3], morals [C12], Huntington's chorea [Q9], withholding treatment [O10], voluntary euthanasia [O18].
■ INTERESTED IN: information sources, data bases [A1], respect of human dignity and human rights [B1], specific approaches to ethics [C1], social and political issues [D1], science, technology, methods [E1], biomedical research and experimentation [F1], organization of health care; facilities, manpower and

services; health occupations [G1], health economics, population characteristics [H1], health care, medical acts [I1], law, legislation and jurisprudence [K1], stages of life — problems proper to childhood and elderly [L1], sexuality and procreation — unborn child [M1], disease [N1], death and resuscitation [O1], neurosciences — psychiatry [P1], genetics and applications — biotechnology [Q1].

O F 18 CENTRE DE RECHERCHE ET DE FORMATION SUR L'ACCOMPAGNEMENT EN FIN DE VIE (CREFAV)

Université Paris -XI
5, rue J.-B.-Clément
92296 Chatenay-Malabry Cedex (France)
✆ (33) 1 - 46 83 57 89

O F 19 CENTRE DE RECHERCHES ÉCONOMIQUES, SOCIOLOGIQUES ET DE GESTION

1, rue Norbert-Segard
B.P.109
59016 Lille Cedex (France)
✆ (33) 20 30 88 27 Fax: (33) 20 78 26 45

O F 20 CENTRE DE SOCIOLOGIE DE L'ÉTHIQUE

59, rue Pouchet
75849 Paris Cedex 17 (France)
✆ (33) 1-40 25 10 25 Fax: (33) 1-42 28 95 44

■ OTHER FIELDS OF ACTIVITIES: teaching and research in the sociology of ethics.
■ AFFILIATION: CNRS Research Unit.
■ OFFICIALLY REGULATED ORGANISATION: regulations on state institutions for science and technology.
■ DATE OF FOUNDATION: 01/01/1979.
■ EXECUTIVE BOARD: Dr. Paul Ladrière (*Director*).
■ TEACHING: P. Ladrière et S. Novaes: René-Descartes University (Paris); Necker School of Medicine (Paris); postgraduate DEA (1991-1992).
■ PUBLICATIONS: Cahiers du Centre de sociologie de l'éthique.
■ DOCUMENTATION CENTRE: internal.
■ WORKING ON: human rights [B3], women's rights [B6], human person [B12], ethics committees [B22], professional deontology organs [B26], Declaration on Human and Citizen Rights [B38], Universal Declaration of Human Rights [B39], patient advocacy [B42], ethical rules and principles [B47], public debates [B75], education [B76], biomedical ethics education [B78], health and biology mediatisation [B85], codes of biomedical ethics [B90], CCNE advice (France) [B91], history of biomedical ethics [B94], bioethics movement [B95], bioethical issues [B96], Bioethics bill, 1992 (France) [B98], specific approaches to ethics [C1], religion and religious ethics [C19], professional ethics [C51], sociobiology [C63], Marxism [C67], comprehension [C70], social sciences [C89], social issues [D2], political issues [D56], cryonic suspension [E31], diagnostic imaging [E32], computers [E33], computerized file [E35], decision analysis [E55], human experimentation [F11], prenatal diagnosis [I3], mass screening [I6], sperm banks [I11], donors [I14], blood donation [I20], patient association [J3], patient information [J4], disclosure [J7], informed consent [J9], presumed consent [J10], professional deontology [J25], physician's role [J26], inter-personal relationship [J27], legal personality [K2], personhood [K3], potentiality of personhood [K5], legislation and law [K34], beginning of life [L24], sexuality and procreation — unborn child [M1], sexuality and procreation [M2], reproductive technologies [M26], pregnancy and childbirth [M39], abortion [M47], terminal care [O7], euthanasia [O15], genetic defects and hereditary diseases [Q2], medical genetics [Q35].
■ INTERESTED IN: respect of human dignity and human rights [B1], protection of rights — involved institutions [B16], Declaration on Human and Citizen Rights [B38], Universal Declaration of Human Rights [B39], patient advocacy [B42], ethical rules and principles [B47], public debates [B75], education [B76], biomedical ethics education [B78], health and biology mediatisation [B85], codes of biomedical ethics [B90], CCNE advice (France) [B91], history of biomedical ethics [B94], bioethics movement [B95], bioethical issues [B96], European Convention of Bioethics [B97], specific approaches to ethics [C1], religion and religious ethics [C19], professional ethics [C51], social sciences [C89], social issues [D2], political issues [D56], cryonic suspension [E31], diagnostic imaging [E32], computers [E33], computerized file [E35], human experimentation [F11], prenatal diagnosis [I3], mass screening [I6], sperm banks [I11], donors [I14], blood donation [I20], patient association [J3], patient information [J4], disclosure [J7], informed consent [J9], presumed consent [J10], professional deontology [J25], physician's role [J26], inter-personal relationship [J27], legal personality [K2], personhood [K3], potentiality of personhood [K5], legislation and law [K34], beginning of life [L24], sexuality and procreation — unborn child [M1], sexuality and procreation [M2], reproductive technologies [M26], pregnancy and childbirth [M39], abortion [M47], terminal care [O7], euthanasia [O15], genetic defects and hereditary diseases [Q2], medical genetics [Q35].

O F 21 **CENTRE DES DROITS DE L'HOMME**
Faculté de droit - 47
38040 Grenoble Cedex 09 (France)
✆ (33) 76 82 54 00 Fax: (33) 76 82 56 54

O F 22 **CENTRE RACHI — CENTRE UNIVERSITAIRE D'ÉTUDES JUIVES** (CUEJ)
30, boulevard de Port-Royal
75005 Paris (France)
✆ (33) 1 - 43 31 75 47 Fax: (33) 1 - 43 37 58 49

■ OTHER FIELDS OF ACTIVITIES: Languages and literature.— History. — Sociology. — Philosophy.
■ AFFILIATION: Chair benjamine Edmond de Rothschild (biomedical ethics).
■ OFFICIALLY REGULATED ORGANISATION: legislation and associations (1901).
■ DATE OF FOUNDATION: 1973.
■ EXECUTIVE BOARD: Pr. Bernard Kanovitch (*Chairperson*).
■ TEACHING: annual congress.
■ PUBLICATIONS: proceedings of the annual congress.
■ COLLOQUIUMS, SYMPOSIUMS: annual congress.
■ WORKING ON: information sources, data bases [A1], respect of human dignity and human rights [B1], specific approaches to ethics [C1], stages of life — problems proper to childhood and elderly [L1], sexuality and procreation — unborn child [M1], death and resuscitation [O1], genetics and applications — biotechnology [Q1].

O F 23 **CENTRE THÉOLOGIQUE DE MEYLAN**
15, Chemin de la Carronnerie
38246 Meylan Cedex (France)
✆ (33) 76 90 41 51

■ OTHER FIELDS OF ACTIVITIES: Philosophy of Science. — Theology. — Study of the Bible. — Religion and spirituality. — Art and cultures.
■ OFFICIALLY REGULATED ORGANISATION: legislation on association (1901).
■ DATE OF FOUNDATION: 1971.
■ EXECUTIVE BOARD: Pr. Pierre Bréchon (*Chairperson*), Pr. Louis Boisset (*Director*).
■ TEACHING: lectures, seminars, conferences.
■ PUBLICATIONS: André Berrel Baron "Vivre, naître et mourir" Cerf 1988. — Cahiers de Meylan "Naître: techniques biologiques et enjeux humains" 1985/1, "La vie, quelle histoire" 1985/2, "La vie et après" 1985/3.
■ DOCUMENTATION CENTRE: open to the public.
■ COLLOQUIUMS, SYMPOSIUMS: "D'où vient la vie". — Participates in various European colloquiums.
■ WORKING ON: comprehension [C70], judgement [C71], behavior [C73], intention [C75], motivation [C76], love [C80], pastoral care [C88], social and political issues [D1], tissue donation [I13], directed donation [I18], health care professionals, researchers and patient relationship [J1], law, legislation and jurisprudence [K1], sexuality and procreation — unborn child [M1], terminal care [O7].
■ INTERESTED IN: protection of rights — involved institutions [B16], ethical rules and principles [B47], bioethics [B89], professional ethics [C51], love [C80], aliens [D23], family members [D24], single persons [D25], couple [D26], cohabitation [D27], married persons [D28], marital relationship [D29], sibling [D30], family relationship [D31].

O F 24 **CENTRE THOMAS-MORE**
La Tourette
69210 L'Arbresle (France)
✆ (33) 74 01 01 03 Fax: (33) 74 01 47 27

■ OTHER FIELDS OF ACTIVITIES: debates and colloquiums on humanities, philosophy, particularly with regard to religious matters.
■ DATE OF FOUNDATION: 1970.
■ EXECUTIVE BOARD: Hugues Puel (*Chairperson*), Antoine Lion (*Director*).
■ PUBLICATIONS: proceedings of colloquiums (irregular).
■ COLLOQUIUMS, SYMPOSIUMS: occasional colloquiums (examples of matters discussed: (1991): "Y a-t-il une morale catholique ? (Is there a Catholic morality ?); "La revendication de la dignité humaine" (Human dignity). — (1990): "La reconnaissance du père" (Paternal recognition); "Les chrétiens et le sida" (Christians and AIDS).
■ WORKING ON: respect of human dignity and human rights [B1], ethical rules and principles [B47], religion and religious ethics [C19], philosophical ethics [C2], ethics and humanities [C54], sexuality and procreation [M2], acquired immunodeficiency syndrome [N35], HIV seropositivity [N36].

O F 25 COMITÉ CONSULTATIF NATIONAL D'ÉTHIQUE POUR LES SCIENCES DE LA VIE ET DE LA SANTÉ

101, rue de Tolbiac
75654 Paris Cedex 13 (France)
✆ (33) 1 - 44 23 60 16 Fax: (33) 1 - 45 85 68 56

■ OFFICIALLY REGULATED ORGANISATION: decree N° 83-132, 23 February 1983.
■ DATE OF FOUNDATION: 23/02/1983.
■ MEMBRES: Pr. Jean Bernard *(Honorary Chairman)*, Pr. Jean-Pierre Changeux *(Chairman)*. *Personalities nominated by the French President and belonging to the main philosophical or spiritual communities* Mr. Mohammed Arkoun, Mr. Henri Atlan, Père Olivier de Dinechin, Mrs. France Quere, Mr. Lucien Sève. *Personnalities chosen for their qualifications and committment to ethical issues:* Madame Geneviève Barrier, Mr. Bernard Bioulac, Mr. Henri Caillavet, Mrs. Anne-Marie David, Mrs. Renée Dufourt, Mr. Pierre Dumayet, Mrs. Anne Fagot Largeault, Mrs. Dominique Frering, Mrs. Michelle Gobert, Mr. Pierre Laroque, Mr. Jean Michaud, Mr. Michel Miroudot, Mrs. Nicole Questiaux, Mr. Louis René, Mrs. Yvette Roudy, Mr. Pierre Royer, Mr. René Sautier, Mrs. Anne Vellay. *Personalities involved in research actions:* Mr. André Boué, Mrs. Béatrice Descamps-Latscha, Mrs. Catherine Driancourt, Mr. Georges Durry, Mrs. Nicole Echard, Mrs. Odile Fichot, Mr. François Gros, Mr. François Jacob, Mr. Axel Kahn, Mr. Yves Laporte, Mr. Claude Laroche, Mr. Joseph Lellouch, Mr. Philippe Lucas, Mrs. Simone Novaes, Mr. Jacques Seylaz.
■ TEACHING: secondary school and university students.
■ PUBLICATIONS: annual report of the board. — Research reports.

O F 26 COMITÉ D'ÉTHIQUE DE L'UNIVERSITÉ CLAUDE-BERNARD

Université Claude Bernard
Domaine Rockefeller
69008 Lyon (France)

O F 27 COMITÉ D'ÉTHIQUE DU CHU DE NANCY

Hôpital d'enfants - Hôpitaux du Brabois
rue du Morvan
54511 Vandoeuvre/Nancy (France)
✆ (33) 83 15 46 01 Fax: (33) 83 15 45 29

■ AFFILIATION: CHU (University hospital) of Nancy (Nancy I).
■ DATE OF FOUNDATION: 1991: replaces the ethics committee created in 1976.
■ TEACHING: inter-university diploma.
■ PUBLICATIONS: proceedings of two specialised meetings published in various magazines.
■ DOCUMENTATION CENTRE: internal.
■ COLLOQUIUMS, SYMPOSIUMS: annual meeting (Pont-à-Mousson).
■ DECISIONS, ADVICE: transplantation of children's organs . Rules for therapeutic trials (until 1988; Huriet law).
■ WORKING ON: fundamental rights of the individuals [B2], dignity [B5], ethics committees [B22], CCNE — Comité Consultatif National d'Ethique (France) [B23], professional deontology organs [B26], College of Physicians [B27], patient advocacy [B42], professional competence [B52], codes of ethics [B49], code of deontology [B60], education [B76], biomedical ethics education [B78], students [B82], CCNE advice (France) [B91], philosophical ethics [C2], metaethics [C11], morals [C12], religious ethics [C21], Christian ethics [C22], Roman catholic ethics [C23], Jewish ethics [C29], biology and human future [C43], medical ethics [C52], nursing ethics [C53], life sciences [E4], genetics [E6], research policy [E18], pilot projects [E20], drug industry [E37], biomedical research and experimentation [F1], biomedical research [F2], clinical trials [F3], health services research [F4], nontherapeutic research [F5], therapeutic research [F6], cognitive research [F7], research subjects [F12], health care, medical acts [I1], prenatal diagnosis [I3], mass screening [I6], biological specimens procurement, blood transfusion, organ transplantation [I8], patients' rights [J2], physicians and researchers accountability [J23], professional deontology [J25], physician nurse relationship [J31], physician patient relationship [J32], professional patient relationship [J33], parent child relationship [J34], stages of life — problems proper to childhood and elderly [L1], somatotropin [L11], human development [L12], age [L13], childhood difficulties, diseases, protection [L21], aged — problems related with aging [L45], sexuality and procreation — unborn child [M1], abortion [M47], death and resuscitation [O1], neurosciences — psychiatry [P1], behavioral and mental disorders [P2], outpatient commitment [P18], genetics and applications — biotechnology [Q1], genetic defects and hereditary diseases [Q2], medical genetics [Q35], genetic counseling [Q36].

O F 28 COMITÉ D'ÉTHIQUE MÉDICALE DU CHU DE POITIERS

CHU La Milétrie
Service de neurologie

86021 Poitiers Cedex (France)
✆ (33) 49 44 44 46

■ AFFILIATION: created by the CHU (University hospital) medical commission and Medical Faculty.
■ DATE OF FOUNDATION: 1984.
■ EXECUTIVE BOARD: Pr. Roger Gil (*Chairperson*).

O F 29 **COMITÉ D'ÉTHIQUE POUR LES TECHNIQUES DE PROCRÉATION ARTIFICIELLE**
4, rue Mably
33000 Bordeaux (France)
✆ (33) 56 81 35 90

O F 30 **COMITÉ DE RÉFLEXION D'ÉTHIQUE DE L'UNION NATIONALE DES ASSOCIATIONS DE PARENTS D'ENFANTS INADAPTÉS**
15, rue Coysevox
75018 Paris (France)
✆ (33) 1 - 42 63 84 33

O F 31 **COMITÉ MÉDICAL POUR LES EXILÉS**
Hôpital de Bicêtre
78, rue du Général-Leclerc
94272 Le Kremlin-Bicêtre Cedex (France)
✆ (33) 1 - 39 94 38 55

O F 32 **COMMISSION NATIONALE DE L'INFORMATIQUE ET DES LIBERTÉS — CNIL**
21, rue Saint-Guillaume
75340 Paris Cedex 07 (France)
✆ (33) 1 - 45 44 40 65 Fax: (33) 1 - 45 49 04 55

■ OTHER FIELDS OF ACTIVITIES: Data processing and Freedom — Respecting the privacy of individuals whose files have been processed. — Control of nominal records and the use of data processing.
■ OFFICIALLY REGULATED ORGANISATION: Law N° 78-17, 6/01/1978, concerning data processing freedom (Journal officiel, N° 1473).
■ DATE OF FOUNDATION: Implementation decree n° 78774 of 17 July 1978 (J.O. of 23/07/1978).
■ EXECUTIVE BOARD: Jacques Fauvet (*Chairperson*), Jacques Tyraud, Louise Cadoux (*State Councillor*), Hubert Bouchet (*Economic and Social Councillor*), Pierre Bracque (*Economic and Social Councillor*), Henri Caillavet (*Barrister-at-law, Court of Appeal of Paris, former secretary of state, honorary member of the parliament*), Michel Elbel, Jean Hernandez (*Referendary Advisor at the French Audit Office*), Gérard Jaquet (*Former Secretary of the State and Former Deputy Chairman of the European Parliament*), Jean-Pierre Michel (*MP for the Haute-Saône*)
■ PUBLICATIONS: annual report of the CNIL published by the Documentation française.
■ COLLOQUIUMS, SYMPOSIUMS: 10 th anniversary of the CNIL. — International conferences of data protection registrars.
■ DECISIONS, ADVICE: Advisory notices on data processing projects in medical research (cf. annual report 85-07). — Recommendation (19/02/1985) on the processing of individuals' medical data for research use.
■ WORKING ON: freedom [B10], privacy [B11], CNIL — Commission Nationale Informatique et Libertés (France) [B19], Bioethics bill, 1992 (France) [B98], social and political issues [D1], epidemiology [E9], computers [E33], computerized file [E35], safety [E51], medical informatics [F21], patient information [J4], patient access [J5], informed consent [J9], confidentiality [J20], medical secrecy [J22], legislation and law [K34], child development [L23], adoption [L32], growth disorders [L44].

O F 33 **CONSEIL D'ÉTHIQUE EN PSYCHIATRIE**
Hôpital Saint-Jean-de-Dieu
290, route de Vienne
69373 Lyon Cedex 08 (France)
✆ (33) 78 09 78 15 Fax: (33) 78 77 96 67

■ AFFILIATION: centre of activity (equivalent to a committee) of Saint-Jean-de-Dieu Hospital..
■ DATE OF FOUNDATION: 31/01/1991.
■ EXECUTIVE BOARD: Dr. Catherine Graber (*Coordinator*), Dr. Jean-Pierre Vignat (*Coordinator*).

O F 34 **CONSEIL DE L'EUROPE**
Département de santé publique
15, rue de l'Ecole-de-Médecine
75005 Paris (France)

O F 35 **CONSEIL NATIONAL DE L'ORDRE DES MÉDECINS**
60, boulevard Latour-Maubourg
75007 Paris (France)
✆ (33) 47 53 24 00

■ OTHER FIELDS OF ACTIVITIES: the French College of Physicians is in charge of deontology matters (Art. L 366 of the Code of Civil Health) and administrative matters as far as the registration of physicians is concerned. It also plays a disciplinary role via the Conseil Régional (professional court of first instance) and the disciplinary section of the Conseil National (court of appeals).
■ OFFICIALLY REGULATED ORGANISATION: Chapitre II du livre IV, titre 1er of the Code de la santé publique "organisation de la profession de médecin". — Sections I and II (art. L.381 to L.411): structures et missions de l'Ordre. — Section III (art. L. 412 to L. 416): inscription au tableau de l'Ordre. — Section IV: discipline.

O F 36 **CONSEIL NATIONAL DU SIDA**
7, rue d'Anjou
75008 Paris (France)
✆ (33) 1 - 40 07 01 06

O F 37 **DÉPARTEMENT D'ÉTHIQUE BIOMÉDICALE — CENTRE SEVRES**
12, rue d'Assas
75006 Paris (France)
✆ (33) 1 - 45 44 18 99

■ AFFILIATION: Centre Sèvres, 35, rue de Sèvres, 75006 Paris (School of Theology, Philosophy and Humanities).
■ DATE OF FOUNDATION: 1985.
■ EXECUTIVE BOARD: Patrick Verspieren (*Director*), Bernard Matray (*Member of the Board of Directors*), Olivier de Dinechin (*Member of the Board of Directors*).
■ TEACHING: 100 to 110 hours at the Centre Sèvres. — Teaching in universities and other institutions. — On request, training course for physicians and nurses.

O F 38 **ETHIQUE ET SOCIÉTÉ**
14, rue Eugène-Varlin
75010 Paris (France)
✆ (33) 1 - 42 34 30 24 Fax: (33) 1 - 42 34 34 72

■ DATE OF FOUNDATION: 1989.
■ EXECUTIVE BOARD: Franck Sérusclat (*Chairperson*), Vincent Moisselin (*Treasurer*), Elisabeth Aubeny (*Executive Secretary*).
■ PUBLICATIONS: "Ethica" review.
■ WORKING ON: ethical rules and principles [B47], bioethics [B89], political issues [D56], legislation and law [K34].

O F 39 **FÉDÉRATION DES ASSOCIATIONS JALMALV & ASSOCIÉES (PARIS-ILE-DE-FRANCE)**
36, rue de Prony
75017 Paris (France)

O F 40 **FÉDÉRATION DES ASSOCIATIONS JALMALV ET ASSOCIÉES**
Registered office: 4 bis, rue Hector Berlioz, 38000 Grenoble
Secretary: 21, rue Sainte-Anne, 45000 Orléans (France)

■ OTHER FIELDS OF ACTIVITIES: Public education, spiritual action. — Training. — Editing.
■ AFFILIATION: "Jusqu'à la mort accompagner la vie: Fédération des Associations Jalmalv et associées". — The Federation is composed of about 40 societies.
■ DATE OF FOUNDATION: Fédération Jalmalv: 1987. — Jalmalv Paris Ile-de-France: 1988.
■ EXECUTIVE BOARD: Dr. Jean-Michel Lassaunière (*Chairperson of the Jalmalv Paris Ile-de-France branch*), Pr. René Schaller (*Chairperson of the Jalmalv Fédération*), Geneviève Delachenal (*Vice-Chairperson*), Christiane Jomain (*Vice-Chairperson*), J.-F. Roche (*Executive Secretary*), Ginette Kruger (*Treasurer*), Dr. Véronique Blanchet (*Member*).
■ TEACHING: training of members and volunteers in medical and spiritual help for terminally ill patients (this training is available for anyone interested in it for professional or private reasons.).
■ PUBLICATIONS: quarterly newsletter Jalmalv since 1985 . The newsletters mainly deal with the various aspects of help to terminally ill patients with underlying reference to bioethics.
■ COLLOQUIUMS, SYMPOSIUMS: General Assembly of the Federation open to public audience.

■ WORKING ON: dignity [B5], obligations of society [B53], respect [B63], information dissemination [B71], nursing education [B79], medical education [B80], bioethical issues [B96], hospices [G34], bill [K60], suffering [N11], terminal care [O7], palliative care [O8], life-sustaining treatment [O9], terminally ill [O12], quality adjusted life years [O14], attitudes to death [O20].
■ INTERESTED IN: human person [B12], patient advocacy [B42], medical ethics [C52], nursing ethics [C53], attitudes [C74], counseling [C87], regulation [D16], patient care team [G38], nurses [G39], physicians [G40], patient care [G53], living wills [J19], professional patient relationship [J33].

O F 41 **FÉDÉRATION DES CENTRES D'ÉTUDE ET DE CONSERVATION DES ŒUFS ET SPERME HUMAINS** (CECOS)
CECOS Necker - hôpital Necker
149, rue de Sèvres
75015 Paris (France)
✆ (33) 1 - 42 73 88 56 Fax: (33) 1 - 40 56 05 98

■ OTHER FIELDS OF ACTIVITIES: artificial procreation, gift, gamete and embryo bank.
■ OFFICIALLY REGULATED ORGANISATION: registered centres (decree of 08/04/1988. — Organised according to the following legislation: law 91-406 of 31/12/1991, Art. 13 (DDPS) and decree 92-174 of 25/02/1992.
■ DATE OF FOUNDATION: 1973.
■ EXECUTIVE BOARD: Pr. Jacques Lansac (*Chairperson*), Pr. Pierre Jalbert (*Vice-Chairperson*), Dr. Marie-Odile Alnot (*Executive Secretary*), Pr. Pierre Jouannet (*Treasurer*).
■ TEACHING: the CEcOS Federation is used as a research laboratory for the biomedical section of the university of Paris V

O F 42 **FÉDÉRATION FRANÇAISE DES CECOS**
CECOS-Alpes
Faculté de médecine de Grenoble
38706 La Tronche Cedex (France)
✆ (33) 76 44 48 11 Fax: (33) 76 51 24 77

O F 43 **FÉDÉRATION INTERNATIONALE DES DROITS DE L'HOMME**
27, rue Jean-Dolent
75014 Paris (France)
✆ (33) 1-47 07 56 35 Fax: (33) 1-43 36 35 43

O F 44 **FONDATION MARCEL-MÉRIEUX**
Département médical
17, rue Bourgelat
69002 Lyon (France)
✆ (33) 72 73 79 68 Fax: (33) 72 73 79 93

■ OFFICIALLY REGULATED ORGANISATION: state approved Foundation.
■ DATE OF FOUNDATION: 1967.
■ EXECUTIVE BOARD: Dr. Charles Mérieux (*Chairperson*), Dr. Louis Valette (*Scientific Director*), Dr. Caroline Dupuy (*Medical Section*).
■ TEACHING: within the centre of bioethics of the Catholic University of Lyon; colloquiums.
■ PUBLICATIONS: Cahier de droit et d'éthique de la santé, 1990, issues N°1, 2, 3, 4. — Colloquium proceedings: "Le statut de l'Embryon humain", "Le diagnostic anténatal. Quels enjeux ?". "The Status of the Human Embryo", "The Prenatal Diagnosis. What is at stake ?"
■ COLLOQUIUMS, SYMPOSIUMS: the colloquiums organised by the Foundation usually also deal with ethics: 1987, the status of embryos, AIDS, epidemic and society; 1989, AIDS 2001; 1990, prenatal diagnosis; 1992, Fœtal and neonatal transplantation and retroviral gene therapy.
■ WORKING ON: science [E2], history of science [E3], life sciences [E4], biology [E5], genetics [E6], microbiology [E7], molecular biology [E8], epidemiology [E9], statistics [E10], incidence [E11], morbidity [E12], mortality [E13], pilot projects [E20], technology [E27], biomedical technologies [E28], freezing [E29], preservation [E30], cryonic suspension [E31], diagnostic imaging [E32], computerized file [E35], drug industry [E37], containment [E38], biological containment [E39], quality of environment [E41], pollution [E47], health hazards [E49], data protection [E52], decision making [E54], decision analysis [E55], [E56], goals [E57], survey [E58], teaching methods [E59], evaluation [E61], peer review [E63], biomedical research and experimentation [F1], maternal health [G3], accidents [G6], traffic accidents [G7], organization of health care [G35], home care [G55], preventive medicine [G58], immunization [G59], health economics, population characteristics [H1], resource allocation [H4], health care, medical acts [I1], genetics and applications — biotechnology [Q1], biotechnology — genetic engineering [Q20], gene therapy [Q42].

O F 45 **GROUPE D'ÉTUDE EN NÉONATOLOGIE ET URGENCES PÉDIATRIQUES DE LA RÉGION PARISIENNE**
Institut de puériculture
26,boulevard Brune
75014 Paris (France)
✆ (33) 1 - 40 44 39 39

O F 46 **GROUPE D'ÉTUDE ET DE RECHERCHE ETHIQUE ET MODERNITÉ** (GEREM)
Palais universitaire
9, place de l'Université
67000 Strasbourg (France)
✆ (33) 88 25 97 35

■ OTHER FIELDS OF ACTIVITIES: research on human rights and health care ethics (programmes: cancer and society, aging, the art of living and ways of dying, organ transplantation).
■ AFFILIATION: Faculty of Protestant Theology, Strasbourg University for Humanities.
■ DATE OF FOUNDATION: 1984.
■ EXECUTIVE BOARD: Pr. Jean-François Collange (*Director*), Pr. Bernard Kaempf, Dr. Isabelle Grellier, Pr. Gilbert Vincent, Pr. Freddy Raphael, Pr. Pottecher, Fritz Lienhard, Michel Cordier.

O F 47 **GROUPE D'ÉTUDES D'ÉTHIQUE BIOMÉDICALES DES PYRÉNÉES-ATLANTIQUES**
Complexe de la République
rue Carnot
64000 Pau (France)
✆ (33) 59 27 85 65 Fax: (33) 59 83 79 88

O F 48 **GROUPE DE RÉFLEXION SUR L'ÉTHIQUE BIOMÉDICALE DE BICETRE**
78, rue du Général-Leclerc
94270 Le Kremlin-Bicêtre (France)
✆ (33) 1 - 45 21 28 46 Fax: (33) 1 - 45 21 21 45

O F 49 **GROUPE DE TRAVAIL ÉTHIQUE**
Groupe hospitalier Saint-Vincent
26, rue du Faubourg-Narimal
67000 Strasbourg (France)
✆ (33) 88 64 70 70 Fax: (33) 88 64 71 70

O F 50 **GROUPE FRANCO-QUÉBÉCOIS DE RECHERCHE EN BIOÉTHIQUE**
8, rue du Mont-de-Veine
95350 St-Brice/Forêt (France)
✆ (33) 1 - 39 94 38 33

O F 51 **GROUPE FRANCO-QUÉBÉCOIS DE RECHERCHE EN BIOÉTHIQUE — ASSOCIATION DU B.O.F**
c/o Mme Myriam Le Sommer
29, rue de Canejan
33700 Mérignac (France)
✆ (33) 56 97 06 46

■ OFFICIALLY REGULATED ORGANISATION: legislation on associations (1901).
■ EXECUTIVE BOARD: Dr. Anne Fagot-Largeault (*Chairperson*), Laurence Clouet (*Treasurer*), Dr. Myriam Le Sommer (*Executive Secretary*), Geneviève Delaisi De Parseval (*Founding Member*), Dr. Nicole Lery (*Founding Member*), Hélène Parizeau (*Founding Member*), Jacqueline Fortin (*Member*), Anne Langlois (*Member*), André Jean (*Member*), Jean-Claude Péré (*Member*)
■ TEACHING: training sessions (e.g. Teaching ethics. Why ? University of Bordeaux II, February 1992.)
■ PUBLICATIONS: co-editing of two issues of medicine: "Le statut de l'enfant procréé artificiellement disparités internationale" (A. Fagot et G Delaisi).— "Les comités d'éthique: situation canadienne- débat français" (A Langlois). — Métaphysique et morale, one issue. — ARSI (magazine on nursing research): one issue in December 1989. — Special issue in September 1990 (N° 174) on human procreation and techno-scientific societies.
■ COLLOQUIUMS, SYMPOSIUMS: meeting of the Centre Thomas More (May 1987): "Enjeux psychologiques et éthiques des procréations médicalement assistées".
■ WORKING ON: respect of human dignity and human rights [B1], ethical rules and principles [B47], information — communication — media [B68], bioethics [B89], violence [D85], science, technology, methods [E1], biomedical research and experimentation [F1], organization of health care; facilities,

manpower and services; health occupations [G1], health care professionals, researchers and patient relationship [J1], stages of life — problems proper to childhood and elderly [L1], sexuality and procreation — unborn child [M1], death and resuscitation [O1].

O F 52 **INSTITUT DE RECHERCHES HOSPITALIERES**
253, cours Galliéni
33000 Bordeaux (France)
✆ (33) 56 24 49 39

■ OTHER FIELDS OF ACTIVITIES: counselling in hospitals. — Research on health care economics.
■ DATE OF FOUNDATION: 1988.
■ EXECUTIVE BOARD: Pr. Jean-Marie Clément (*Chairperson of the Board of Administration*), Maurice Rochaix (*Chairperson of the Scientific Board*).

O F 53 **INSTITUT DES DROITS DE L'HOMME**
Université catholique de Lyon
10-12, rue Alphonse-Fochier
69002 Lyon (France)
✆ (33) 72 32 50 50 Fax: (33) 72 32 50 19

O F 54 **INSTITUT DES SCIENCES DE LA FAMILLE**
Université catholique
30, rue Sainte-Hélène
69002 Lyon (France)
✆ (33) 78 92 91 24 Fax: (33) 72 32 50 19

■ OTHER FIELDS OF ACTIVITIES: interdisciplinary teaching and research on family.
■ AFFILIATION: Department of the Catholic University of Lyon.
■ DATE OF FOUNDATION: 1972.
■ EXECUTIVE BOARD: Xavier Lacroix (*Director*).
■ TEACHING: Human Fertilisation. — Assisted Medical Procreation. — Preventive action against AIDS.

O F 55 **INSTITUT FORMATION EN DROITS DE L'HOMME DU BARREAU DE PARIS**
Comité Droits de l'homme, Commission Française, Unisis
4, square La Bruyère
75009 Paris (France)
✆ (33) 1 - 42 80 49 54

O F 56 **LABORATOIRE D'ÉTUDE GÉNÉTIQUE DE LA PERSONNALITÉ ET DE PSYCHOPATHOLOGIE DE L'ENFANT ET DE L'ADOLESCENT**
Fondation Vallée
7, rue Benserade
94250 Gentilly (France)
✆ (33) 1 - 49 86 11 11 Fax: (33) 1 - 45 47 08 37

■ OTHER FIELDS OF ACTIVITIES: research on child psychiatry.
■ AFFILIATION: university laboratory and INSERM research network.
■ EXECUTIVE BOARD: Pr. Pierre Ferrari (*Co-Director*), Pr. Roger Mises (*Co-Director*), Pr. Roger Perron (*Co-Director*).
■ TEACHING: Teaching in, and research on, child psychiatry (mainly children's psychoses).

O F 57 **MOUVEMENT UNIVERSEL DE LA RESPONSABILITÉ SCIENTIFIQUE** (MURS)
127, boulevard Saint-Michel
75005 Paris (France)
✆ (33) 1-43 26 43 98 Fax: (33) 1-43 54 35 50

O F 58 **OFFICE PARLEMENTAIRE**
Palais du Luxembourg
15, rue de Vaugirard
75006 Paris (France)
✆ (33) 1 - 42 34 20 43 Fax: (33) 1 - 42 34 21 69

■ OTHER FIELDS OF ACTIVITIES: The Office Parlementaire informs the Parliament on scientific and technological matters.
■ AFFILIATION: Joint organ of the Assemblée Nationale and Sénat.
■ DATE OF FOUNDATION: 9/07/1983.

O F 59 **RÉSEAU EUROPÉEN DE COOPÉRATION SCIENTIFIQUE "MÉDECINE ET DROITS DE L'HOMME" DU CONSEIL DE L'EUROPE**
7, rue des Tilleuls
67800 Hoenheim (France)
✆ (33) 88 33 44 65 Fax: (33) 88 41 27 87 (c/o DR JP Massue)

O F 60 **"REVUE LAENNEC"**
12, rue d'Assas
75006 Paris (France)
✆ (33) 1 - 45 44 58 91

O F 61 **SANTÉ, ETHIQUE ET LIBERTÉS** (SEL)
95, boulevard Périel
69677 Bron Cedex (France)
✆ (33) 72 35 87 10 or 72 35 87 20 or 72 35 87 30 or 72 35 87 40

■ OFFICIALLY REGULATED ORGANISATION: legislation on associations (1901) (1984).
■ DATE OF FOUNDATION: 1976.
■ EXECUTIVE BOARD: Dr. Louis Lery (*Chairperson*), Dr. Nicole Lery (*Vice-Chairperson*), Dr. Paul Broussole (*Treasurer*), Jeannine Pelet (*Deputy Treasurer*).
■ TEACHING: to professionals. — Professional and public conferences.
■ PUBLICATIONS: special issue on Medicine and hygiene (every 2 years) and other special issues. — Articles published in various magazines.
■ DOCUMENTATION CENTRE: internal.
■ COLLOQUIUMS, SYMPOSIUMS: annual colloquium: journées d'éthique, meetings on ethics (last Thursday, Friday, Saturday of January), dixième journées in January 1993. — Participation aux colloques spécialisés (si publication à faire).
■ WORKING ON: professional ethics [C51], medical ethics [C52], nursing ethics [C53], aliens [D23], imprisonment [D11], international aspects [D59], sex offenses [D92], rape [D93], torture [D94], force feeding [D95], records [E34], computerized file [E35], data protection [E52], decision making [E54], biomedical research and experimentation [F1], organization of health care; facilities, manpower and services; health occupations [G1], health economics, population characteristics [H1], health care professionals, researchers and patient relationship [J1], law, legislation and jurisprudence [K1], stages of life — problems proper to childhood and elderly [L1], sexuality and procreation [M2], death and resuscitation [O1], neurosciences — psychiatry [P1].
■ INTERESTED IN: death and resuscitation [O1], behavioral and mental disorders [P2], outpatient commitment [P18].

O F 62 **SOCIÉTÉ FRANÇAISE D'ACCOMPAGNEMENT ET DE SOINS PALLIATIFS** (SFAP)
110, avenue Emile-Zola
75015 Paris (France)
✆ (33) 1 - 43 72 99 65

O F 64 **SOCIÉTÉ FRANÇAISE DE MÉDECINE LÉGALE ET DE CRIMINOLOGIE**
2, place Mazas
75006 Paris (France)
✆ (33) 1 - 43 43 42 54

O F 66 **SOCIÉTÉ FRANÇAISE DE RÉFLEXION BIOÉTHIQUE** (SFRB)
c/o M.-H. Congourdeau
30, rue d'Auteuil
75016 Paris (France)

■ DATE OF FOUNDATION: 1987.
■ EXECUTIVE BOARD: Marie-Hélène Congourdeau (*Chairperson*), Jean-Yves Lacoste (*Vice-Chairperson*), Dr. Alain Pinalie (*Vice-Chairperson*), Jean-Luc Meyer (*Vice-Chairperson*), Pr. Dominique Folscheid (*Executive Secretary*), Irène Fernandez (*Treasurer*).
■ PUBLICATIONS: Issue: "Ethique la vie en question".
■ WORKING ON: human person [B12], information dissemination [B71], history of biomedical ethics [B94], Christian ethics [C22], Roman catholic ethics [C23], eastern orthodox ethics [C26], history [C56], history before the 20th century [C58], ancient history [C59], history of science [E3], investigators [E15], newborns [L29], infanticide [L42], embryos [M10], fetuses [M11], procreation [M15], excess embryos [M34], embryo transfer [M36], embryo donation [M38], pregnant women [M42], mother fetus relationship [M43], illegal abortion [M48], selective abortion [M49], therapeutic abortion [M50], aborted fetuses [M51], abortion on demand [M52], fetal therapy [M57], sex determination [Q26], sex preselection [Q30], preimplantation genetic diagnosis [Q38], eugenics [Q39].

■ INTERESTED IN: Protestant ethics [C27], Islamic ethics [C28], Jewish ethics [C29], human experimentation [F11], prenatal diagnosis [I3], beginning of life [L24], newborns [L29], aid children [L34], unwanted children [L35], handicapped children [L38], infanticide [L42], determination of death [O2], autopsies [O3].

O F 67 UNION NATIONALE DES ASSOCIATIONS DE PARENTS D'ENFANTS INADAPTÉS — UNAPEI

46, rue de Rome
75008 Paris (France)
✆ (33) 1-42 94 93 01

O F 68 CONSULTATIVE COMMITTEES FOR PROTECTION OF PERSONS IN BIOMEDICAL RESEARCH (CCPPRB)

The amended law no. 88-1138 of the 20th of December 1988 (Journal Officiel of the 22nd of December 1988), entitled "Protection of persons taking part in biomedical research" specifies in article L 209-12: "Before carrying out research on human beings, any investigator is expected to submit the plans for such research for the opinion of a consultative committee for the protection of persons in biomedical research, whose headquarters is located in the region where he practises."
The decrees for application of the law (no. 90-872 of the 27th of September 1990 and no. 91-440 of the 14th of May 1991) define the constitution and functioning of these consultative committees for the protection of persons in biomedical research (CCPPRB). They comprise twelve full members: four persons, including at least three doctors, with qualifications and extensive experience in biomedical research; one general practitioner; two pharmacists, including at least one practising in a medical care establishment; a nurse as defined in articles L473 to L477 of the public health code; a person qualified by dint of his expertise on questions of ethics; a person qualified by dint of his activity in the social field; a person authorized to use the title of psychologist; and a person qualified by dint of his expertise on legal questions. The committees also comprise twelve deputy members meeting the same criteria. (Article R 2001: Constitution).

CCPPRB – Nice
Mr. Dumas
CHRU
06031 Nice Cedex

CCPPRB – Marseille 1
ProfessorGiraud
Assistance publique
9, rue Lafon
13006 Marseille

CCPPRB – Marseille 2
Professor Carcassonne
Assistance publique
9, rue Lafon
13006 Marseille

CCPPRB – Aix-en-Provence
Professor Airaudo
CHG
Avenue de Tamaris
13616 Aix-En-Provence
CCPPRB – Caen
Professor Lemenager
CHRU
14033 Caen Cedex

CCPPRB – DRASS de Corse
Mrs. Marchiset-Lecas
1, rue Colomba
20308 Ajaccio

CCPPRB – Dijon
Professor D. Lambert
Hôpital du Bocage
2, boulevard de Lattre-de-Tassigny
21034 Dijon Cedex
& 80 29 33 43

CCPPRB – Besançon
Professor Bechtel
CHR Jean-Minjoz
25030 Besançon Cedex

CCPPRB – Brest
Professor Mottier
CHR
29609 Brest

CCPPRB – de Nîmes
Mr. le professeur J. Fourcade
UFR de médecine Montpellier-Nîmes,
Faculté de médecine
Boulevard Kennedy
3000 Nîmes

CCPPRB – Toulouse-I
Doctor Marc-Vergnes
DRASS
71 bis, allées Jean-Jaurès
31050 Toulouse Cedex

CCPPRB – Toulouse-II
Professor Regnier
DRASS
71 bis, allées Jean-Jaurès
31050 Toulouse Cedex

CCPPRB – Bordeaux A
Professor B. Hoerni

DRASS
Rue Jules-Ferry – BP 100
33090 Bordeaux

CCPPRB – Bordeaux-B
Professor A. Taytard
DRASS
Rue Jules-Ferry – BP 100
33090 Bordeaux

CCPPRB – de Montpellier Saint-Eloi
Professor Mion
CHRU
2, avenue Bertin-Sans
34059 Montpellier Cedex

CCPPRB – Montpellier Lapeyronie
Professor Sany
CHR
555, route de Ganges
34059 Montpellier Cedex

CCPPRB – Rennes
Professor Gastard
CHRU
2, rue de l'Hôtel-Dieu
35033 Rennes Cedex
& 99 28 41 35

CCPPRB – Tours
Professor Breteau
CHRU
2, boulevard Tonnelé
37044 Tours

CCPPRB – Grenoble 1
Mr. Nicolas Drouet
CHRU
BP 217
38043 Grenoble Cedex 9
& 76 76 55 14

CCPPRB – Grenoble 2
Professor Barret
CHRU
BP 217
38043 Grenoble Cedex 9

CCPPRB – Rhône-Alpes Loire
Doctor Tempelhoff
CHRU
3, rue Claude-Lebois
42022 Saint-Etienne Cedex

CCPPRB – Pays-de-la-Loire n°2
Mr. Jean Guenel
CHRU
5, allée de l'Ile-Gloriette – BP 1005
44035 Nantes Cedex 01

CCPPRB – Orléans-La-Source
Doctor Archambault
CHR
BP 2439
45067 Orléans Cedex 2

CCPPRB – Pays-de-la-Loire n°1
Professor Renier
CHRU
49040 Angers Cedex

CCPPRB – Champagne-Ardennes
Doctor Pourny
CHRU
45, rue Cognacq-Jay
51092 Reims Cedex

CCPPRB – Nancy
Doctor Peton
29, avenue du Maréchal-de-Lattre-de-Tassigny
54035 Nancy Cedex

CCPPRB – Metz
Doctor Gille
28, rue du XXe Corps-Américain
57000 Metz
& 83 44 63 00

CCPPRB – Lille
Professor Hatron
Hôpital Huriez
Département de pharmacologie
2, avenue Oscar Lambret
59037 Lille Cedex

CCPPRB – Valenciennes
Doctor Dupont
CHG
59300 Valenciennes

CCPPRB – Lens
Doctor d'Hautefeuille
CH
62307 Lens Cedex

CCPPRB – Auvergne
Professor Lavarenne
CHRU (Administration centrale)
30, place Henri-Dunant
63003 Clermont-Ferrand Cedex

CCPPRB – Pays de l'Adour
Doctor M. Begorre
CHG
4, boulevard de Hauterive
64000 Pau

CCPPRB – Alsace-I Strasbourg
Professor Tempe
CHRU
1, place de l'Hôpital
67091 Strasbourg Cedex

CCPPRB – Alsace-II Colmar
Doctor Audhuy
CHG
39, avenue de la Liberté
68024 Colmar Cedex

CCPPRB – Lyon A
Doctor Clavel
Hôtel-Dieu
1, place de l'Hôpital
69288 Lyon Cedex 02

CCPPRB – Lyon B
Doctor Lachaux
Hôtel-Dieu
1, place de l'Hôpital
69288 Lyon Cedex 02
CCPPRB – Lyon C
Mr. Bourge
Hôtel-Dieu
1, place de l'Hôpital
69288 Lyon Cedex 02

CCPPRB – Centre régional de Lutte contre le cancer de Lyon
Mr. Sermanson
28, rue Laënnec
69373 Lyon Cedex 2
& 78 01 84 55

CCPPRB – Paris Hôtel-Dieu
Professor J-P Dadoune
Hôtel-Dieu
1, place du Parvis-Notre-Dame
75004 Paris

CCPPRB – Paris Saint-Antoine
Doctor Didier Kamioner
Hôpital Saint-Antoine
184, rue du Faubourg Saint-Antoine
75012 Paris

CCPPRB – Paris-Cochin
Professor Hazebroucq
27, rue du Faubourg-Saint-Jacques
75014 Paris

CCPPRB – Paris-Boucicaut
Doctor François-Noël Marie
Hôpital Boucicaut
78, rue de la Convention
75015 Paris

CCPPRB – Paris-Bichat-Claude Bernard
Mr. Alain Dauphin
Groupe hospitalier universitaire Bichat-Claude-Bernard
46, rue Henri-Huchard, service de pharmacie
75018 Paris

CCPPRB – Paris Saint-Louis
Professor Rouffy
Hôpital Saint-Louis
1, avenue Claude-Vellefaux
75475 Paris Cedex 10

CCPPRB – Paris Pitié-Salpêtrière
Doctor Martine Raphaël
Groupe hospitalier Pitié-Salpêtrière
47, boulevard de l'Hôpital
75651 Paris Cedex 13

CCPPRB – Paris Necker
Professor Hamon
Hôpital Necker-Enfants-Malades
149, rue de Sèvres
75743 Paris Cedex 15

CCPPRB – Haute-Normandie Rouen
Professor Leroy
Hôpital Charles-Nicolle
1, rue de Germont
76031 Rouen Cedex

CCPPRB – Saint-Germain-en-Laye
Doctor Beaufils
CHG
78100 Saint-Germain-en-Laye

CCPPRB – Versailles
Professor Brion
Centre hospitalier de Versailles
177, rue de Versailles
78157 Le Chesnay Cedex

CCPPRB – Picardie
Mr. Descoubes
DRASS
52, rue Daire
80037 Amiens Cedex

CCPPRB – Toulon
Doctor Balansard
CHI
1208, avenue Colonel-Picot
83056 Toulon Cedex

CCPPRB – Poitou-Charentes-Poitiers
Mr. Becq-Giraudon
DRASS
28, rue Gay-Lussac
86035 Poitiers
& 49 42 30 00

CCPPRB – Limousin
Doctor Blanc
CHR Dupuytren
2, avenue Alexis-Carrel
87042 Limoges Cedex

CCPPRB – Boulogne Ambroise-Paré
Professor Le Parc
Hôpital Ambroise-Paré
9, avenue Charles-de-Gaulle
92100 Boulogne

CCPPRB – Aulnay-Sous-Bois
Madame Fabreguettes
CHG Robert-Ballanger
93602 Aulnay-Sous-Bois

CCPPRB – Créteil Henri-Mondor
Professor Atlan
Hôpital Henri-Mondor
51, avenue du Maréchal-de-Lattre-de-Tassigny
94010 Créteil Cedex

CCPPRB – Paris Bicêtre
Professor Duroux
Centre hospitalier de Bicêtre – Cour de Sibérie
78, rue du Général Leclerc
94270 Kremlin-Bicêtre

Individuals

F 1 **ABIVEN** Maurice

Chairman of the Société française d'accompagnement et de soins palliatifs (French Society of Life-sustaining and palliative care).

119, boulevard de Grenelle
75015 Paris (France)
✆ (33) 1 - 47 34 44 13

■ AFFILIATIONS: Société française d'accompagnement et de soins palliatifs, Paris, France (*Chairperson*).
■ TEACHING: Academic and post-academic.
■ PUBLICATIONS: yes.
■ WORKING ON: terminal care [O7], palliative care [O8], life-sustaining treatment [O9], withholding treatment [O10], allowing to die [O11], terminally ill [O12], prolongation of life [O13], quality adjusted life years [O14], euthanasia [O15].
■ INTERESTED IN: CCNE advice (France) [B91], specific approaches to ethics [C1], religion and religious ethics [C19], terminal care [O7], palliative care [O8], life-sustaining treatment [O9], withholding treatment [O10], allowing to die [O11], terminally ill [O12], prolongation of life [O13], quality adjusted life years [O14], euthanasia [O15].

F 2 **AIRAUDO** Charles B.

Professor at the Faculty of Pharmacology of Marseille. — M.D. Pharmacist, Head of the Department of Pharmacology at the Hospital of Valvert, Marseille.

CCPPRB- Centre Hospitalier Général
Avenue des Tamaris
13616 Aix-en-Provence Cedex 1 (France)
✆ (33) 42 33 51 98 Fax: (33) 42 33 51 20

■ AFFILIATIONS: CCPPRB d'Aix-en-Provence, France (*Chairperson*).
■ PUBLICATIONS: yes.
■ WORKING ON: respect of human dignity and human rights [B1], protection of rights — involved institutions [B16], institutions involved in the protection of rights [B17], CCPPRB — French committees for the protection of human subjects of biomedical research [B32], due process [B40], patient advocacy [B42], ethical rules and principles [B47], codes of ethics [B49], obligations of society [B53], deontology [B59], communication [B69], interdisciplinary communication [B70], bioethics [B89], deontological ethics [C6], situational ethics [C9], quality of life [C16], professional ethics [C51], medical ethics [C52], nursing ethics [C53], biomedical research and experimentation [F1], biomedical research [F2], clinical trials [F3], organization of health care [G35], pharmacists [G42], patients' rights [J2], patient information [J4], parental notification [J6], informed consent [J9], third party consent [J11], parental consent [J13], consent forms [J14], professional deontology [J25], medical law [K51], conscience clause [K63].
■ INTERESTED IN: respect of human dignity and human rights [B1], protection of rights — involved institutions [B16], institutions involved in the protection of rights [B17], CCPPRB — French committees for the protection of human subjects of biomedical research [B32], due process [B40], patient advocacy [B42], ethical rules and principles [B47], codes of ethics [B49], obligations of society [B53], deontology [B59], communication [B69], public debates [B75], bioethics [B89], Bioethics bill, 1992 (France) [B98], deontological ethics [C6], situational ethics [C9], quality of life [C16], professional ethics [C51], medical ethics [C52], nursing ethics [C53], social and political issues [D1], legislation [D13], science, technology, methods [E1], scientific misconduct [E25], technology [E27], biomedical technologies [E28], industry [E36], drug industry [E37], organization of health care [G35], pharmacists [G42], patients' rights [J2], patient information [J4], parental notification [J6], informed consent [J9], third party consent [J11], parental consent [J13], consent forms [J14], professional deontology [J25], medical law [K51], conscience clause [K63].

F 3 **ALBY** J.M.

Head of the Psychiatry and Medical Psychology Service, Medical Faculty Saint-Antoine. — Director of Centre of Medical Ethopsychology. — President of Ethics Committee of Saint-Antoine University Hospital. — Vice-président of the Management Council of the A. et F. Minkowski Centre.

Centre d'éthopsychologie médicale, Département de santé publique
15, rue de l'Ecole-de-Médecine
75006 Paris (France)
✆ (33) 1 - 49 28 26 35 et 43 54 48 00 Fax: (33) 1 - 49 28 20 10

■ AFFILIATIONS: Council of Europe, Paris, France (*Co-Director*).
■ TEACHING: Training medicine, ethics and human rights, with J. Métrot.

F 4 **ALBY** Nicole

Clinical Psychologist. — Head of the Unit of Clinical Psychology, Saint-Louis Hospital, Department of Onco-Haematology.

Hôpital Saint-Louis
193, boulevard Saint-Germain
75007 Paris (France)
✆ (33) 1 - 45 48 62 54

■ TEACHING: Teaching in medical and nursing fields. — Conducts research in haematology and medullar transplants related to the psychological and ethical issues raised by pathologies and therapeutics.
■ PUBLICATIONS: yes.
■ INTERESTED IN: biomedical ethics education [B78], nursing education [B79], medical education [B80], health and biology mediatisation [B85], medical ethics [C52], behavioral research [E21], clinical trials [F3], therapeutic research [F6], research subjects [F12], methodology of research and experimentation [F15], biological specimens procurement, blood transfusion, organ transplantation [I8], transplant recipients [I23], organ transplantation [I24], organ donation [I25], fetal tissue transplantation [I27], bone marrow transplantation [I29], patient association [J3], patient information [J4], patient access [J5], parental notification [J6], consent to treatment [J8], informed consent [J9], consent forms [J14], physician patient relationship [J32], leukemia [N18].

F 5 **ALNOT** Marie-Odile

CECOS - Necker
Hôpital Necker
149, rue de Sèvres
75015 Paris (France)
✆ (33) 1 - 42 73 88 56 Fax: (33) 1 - 40 56 05 98

■ AFFILIATIONS: Commission d'éthique et déontologie de la Fédération des CECOS, Paris, France (*Member*).
■ TEACHING: Teaching in the laboratory of Bio-Medical Ethics at the University of Paris-V, member of the DEA Comittee
■ PUBLICATIONS: yes.
■ WORKING ON: review [A2], bibliography [A4], publications [A9], CCNE — Comité Consultatif National d'Ethique (France) [B23], biomedical ethics education [B78], CCNE advice (France) [B91], medical ethics [C52], freezing [E29], preservation [E30], cryonic suspension [E31], sperm banks [I11], anonymous donation [I17], directed donation [I18], parental consent [J13], potentiality of personhood [K5], legitimacy [K21], beginning of life [L24], aid children [L34], ovum [M12], volontary sterilization [M8], embryos [M10], sperm [M14], procreation [M15], ovum donors [M27], semen donors [M28], in vitro fertilization [M29], artificial insemination [M31], AID [M32], AIH [M33], excess embryos [M34], embryo transfer [M36], embryo donation [M38], acquired immunodeficiency syndrome [N35], HIV seropositivity [N36], genetic defects and hereditary diseases [Q2], medical genetics [Q35].
■ INTERESTED IN: review [A2], data bases [A3], bibliography [A4], publications [A9], integrity [B8], privacy [B11], human person [B12], human body commercialization [B13], reification [B14], CNIL — Commission Nationale Informatique et Libertés (France) [B19], CCNE — Comité Consultatif National d'Ethique (France) [B23], Council of Europe [B25], College of Physicians [B27], CCPPRB — French committees for the protection of human subjects of biomedical research [B32], patient advocacy [B42], humanity heritage [B46], obligations of society [B53], obligations to society [B54], biomedical ethics education [B78], CCNE advice (France) [B91], European Convention of Bioethics [B97], religious ethics [C21], medical ethics [C52], public opinion [D7], legislation [D13], regulation [D16], statistics [E10], freezing [E29], preservation [E30], cryonic suspension [E31], computerized file [E35], convenience medicine [G57], remuneration [H14], sperm banks [I11], anonymous donation [I17], directed donation [I18], patient access [J5], disclosure [J7], third party consent [J11], spousal consent [J12], parental consent [J13], potentiality of personhood [K5], conflict of interest [K9], legitimacy [K21], child's interest [K24], beginning of life [L24], aid children [L34], sexuality and procreation [M2], reproductive technologies [M26], selective abortion [M49], therapeutic abortion [M50], fetal

development [M54], mother-child transmission [N16], cancer [N17], hepatitis [N30], acquired immunodeficiency syndrome [N35], HIV seropositivity [N36], genetic defects and hereditary diseases [Q2], sex determination [Q26], sex preselection [Q30], DNA fingerprinting [Q32], medical genetics [Q35].

F 6 **ARTHUIS** Michel

Emeritus titular Professor of the Paediatric Clinical Chair at the René-Descartes University Paris V. — Honorary Head of the Department of Paediatric Neurology at the Saint-Vincent de Paul (AP). Hospital. At present Emeritus Professor, René-Descartes University, Paris V

46, rue de Rome
75008 Paris (France)
✆ (33) 16-1- 42 94 93 01

■ AFFILIATIONS: Comité de réflexion d'éthique à l'Union des associations de parents et amis des personnes handicapées mentales — UNAPEI, Paris Cedex 18, France (*Chairperson*). — Comité d'éthique des hôpitaux Cochin, Port-Royal, Saint-Vincent-de-Paul, Sainte-Anne à Paris (*Member*).
■ TEACHING: Teaching within the Associations for the Handicapped (epilepsy, physical disabilities, mental disabilities affecting children).
■ PUBLICATIONS: yes.
■ WORKING ON: children's rights [L25], handicapped children [L38], procreation [M15], sexuality [M20], behavioral and mental disorders [P2], diagnosis [I2], placebos [I36], surgery [I37].
■ INTERESTED IN: codes of biomedical ethics [B90], history of biomedical ethics [B94], bioethics movement [B95], bioethical issues [B96], European Convention of Bioethics [B97], Bioethics bill, 1992 (France) [B98], Christian ethics [C22], Roman catholic ethics [C23], eastern orthodox ethics [C26], Protestant ethics [C27], Islamic ethics [C28], Jewish ethics [C29].

F 7 **ASSENMACHER** Ivan

Professor of Psychology in Montpellier University. — Head of the investigation team of stress mechanisms, endocrinologic neurology laboratory (URA 1197, CNRS). — Member of Académie des Sciences, of l'Institut de France, Paris, in biology. Member of European Academy, Londres. — Member of Conseil supérieur de la Recherche et de la Technologie, ministère de la Recherche et de l'Espace, Paris.

Laboratoire de neurobiologie endocrinologique
Université de Montpellier
Place E. Bataillon
34095 Montpellier Cedex (France)
✆ (33) 67 14 36 17 Fax: (33) 67 14 42 51

■ AFFILIATIONS: CCNE, France (*ex-Member*). — Groupe de réflexion en bioéthique de Nîmes (Gard) (*Member*).
■ TEACHING: Montpellier University: BA in biochemistry and physiology: sexuality, contraception, assisted medical procreation: biotechnical aspects. Immaterial creation law: "Biotechnologies and bioethics".
■ PUBLICATIONS: yes.
■ WORKING ON: information sources, data bases [A1], review [A2], respect of human dignity and human rights [B1], fundamental rights of the individuals [B2], codes of ethics [B49], education [B76], universities [B84], bioethical issues [B96], religion and religious ethics [C19], Christian ethics [C22], Protestant ethics [C27], political issues [D56], advisory committees [D69], science, technology, methods [E1], biology [E5], research team [E16], biomedical technologies [E28], ecology [E43], teaching methods [E59], biomedical research and experimentation [F1], animal experimentation [F10], legislation and law [K34], rights [K42], law [K56], sexuality and procreation — unborn child [M1], procreation [M15], sexuality [M20], neurosciences — psychiatry [P1], psychological stress [P17], psychoactive drugs [P33], biotechnology [Q21].
■ INTERESTED IN: information sources, data bases [A1], review [A2], respect of human dignity and human rights [B1], fundamental rights of the individuals [B2], codes of ethics [B49], education [B76], universities [B84], bioethical issues [B96], religion and religious ethics [C19], Christian ethics [C22], Islamic ethics [C28], Jewish ethics [C29], ECC — European Community Commission [D63], public policy [D77], biology [E5], research policy [E18], military research [E24], biomedical technologies [E28], ecology [E43], animal experimentation [F10], authorization [K37], prohibition [K38], European law [K49], sexuality and procreation [M2], psychological stress [P17], psychoactive drugs [P33].

F 8 **ATLAN** Guy

Professor of Physiology. — M.D. Clinician. — Director of the Research Unit of Respiratory Physiology (U 296) of INSERM.

CCPPRB Hôpital Henri-Mondor
51, avenue du Maréchal-de-Lattre-de-Tassigny
94010 Créteil Cedex (France)
✆ (33) 49 81 21 51

■ AFFILIATIONS: CCPPRB Henri-Mondor — CCPPRB, Créteil, France (*Chairperson*).
■ TEACHING: Medical students (sensitising).
■ WORKING ON: experimentation [F9], animal experimentation [F10], human experimentation [F11], research subjects [F12], healthy volunteers [F13], methodology of research and experimentation [F15], control groups [F17], random selection [F18], selection of subjects [F19].
■ INTERESTED IN: health care professionals, researchers and patient relationship [J1], disclosure [J7], consent to treatment [J8], informed consent [J9], presumed consent [J10], third party consent [J11], spousal consent [J12], parental consent [J13], consent forms [J14], treatment refusal [J15], patient participation [J16], right to treatment [J17], advance directives [J18], living wills [J19], inter-personal relationship [J27], medical etiquette [J28], nurse patient relationship [J30], physician nurse relationship [J31], physician patient relationship [J32], professional patient relationship [J33], parent child relationship [J34].

F 9 **ATLAN** Henri

Professor of Biophysics at CHU Broussais-Hôtel Dieu of the Paris-VI University, Head of the Department of Biophysics and Nuclear Medicine of Hôtel-Dieu Hospital, Paris. — Professor of Bioethics, Head of Medical Biophysics Department, Scholar in Residence in Philosophy and Ethics of Biology, Hadassah- Hebrew University Hospital, Jerusalem.

Hôpital de l'Hôtel-Dieu
1, place du Parvis de Notre Dame
75004 Paris (France)
✆ (33) 1 - 42 34 82 45 et 41 Fax: (33) 1 - 43 35 53 09

■ AFFILIATIONS: Comité consultatif national d'éthique pour les sciences de la vie et de la santé — CCNE, Paris, France (*Member*).
■ PUBLICATIONS: yes.

F 10 **AUBENY** Elisabeth

Gynaecologist. — Head of Orthogenesis Centre, Broussais Hospital.

10, rue du docteur Lancereaux
75008 Paris (France)
✆ (33) 1 - 42 89 16 67 Fax: (33) 1 - 45 62 38 22

■ PUBLICATIONS: yes.
■ WORKING ON: women's rights [B6], utilitarianism [C8], contraception [M3].
■ INTERESTED IN: women's rights [B6], clinical trials [F3], family planning [G8], convenience medicine [G57], population control [H22], administrators [H3], insurance [H5], life insurance [H8], preimplantation genetic diagnosis [Q38].

F 11 **AUBY** Jean-Marie

Emeritus Professor, University of Bordeaux-I (France).

Faculté de droit
Avenue Léon-Duguit
33600 Pessac (France)
✆ (33) 56 84 84 45 Fax: (33) 56 84 84 32

■ AFFILIATIONS: Association française de droit de la santé, Pessac, France (*Chairperson*).
■ TEACHING: postgraduate DEA year in Health Care Law, University of Bordeaux-I.
■ PUBLICATIONS: yes.

F 12 **AUDHUY** Bruno

Head of the Department of Oncology-Haematology at Louis-Pasteur Hospital (Colmar). Chairperson at the CCPPRB Alsace-II Colmar.

CCPPRB Alsace-II Colmar
Centre hospitalier Louis-Pasteur
39, avenue de la Liberté
68024 Colmar Cedex (France)
✆ (33) 89 80 44 01 Fax: (33) 89 80 43 98

■ AFFILIATIONS: CCPPRB Alsace-II Colmar Cedex, France (*Chairperson*).
■ WORKING ON: human person [B12], CCPPRB — French committees for the protection of human subjects of biomedical research [B32], ethics [B48], bioethical issues [B96], medical ethics [C52], legislation [D13], research [E14], biomedical technologies [E28], drug industry [E37], health hazards [E49], toxicity [E50], safety [E51], clinical trials [F3], nontherapeutic research [F5], therapeutic research [F6], cognitive research [F7], human experimentation [F11], research subjects [F12], healthy volunteers [F13], methodology of research and experimentation [F15], selection for treatment [G52], health occupations [G61], medicine [G63], drugs [I34], placebos [I36], patient information [J4], informed consent [J9], consent forms [J14], patient participation [J16], therapeutic risk [K32], law [K56].
■ INTERESTED IN: tissue banking [I12], bone marrow transplantation [I29], alternative therapies [I33], cancer [N17], leukemia [N18], acquired immunodeficiency syndrome [N35], HIV seropositivity [N36], terminal care [O7], palliative care [O8], life-sustaining treatment [O9], withholding treatment [O10], allowing to die [O11], terminally ill [O12], euthanasia [O15], biotechnology — genetic engineering [Q20].

F 13 AUDRAS DE LA BASTIE Marguerite

Research group associated to the orthogenesis team of Dr. Quillet at Vaulx-en-Velin from 74 to 92. — Former Psychologist and Psychoanalyst at the Vinatier Hospital (Lyon-Bron), Pr. Daléry Department and at Vaulx-en-Velin (Rhône), Pr. Marie-Cardine.

API, SPP
3, quai Jules-Courmont
69002 Lyon (France)
✆ (33) 78 37 66 53 et 74 30 31 46

■ TEACHING: Working group for the Association Descartes. — Former lecturer at the University of Lyon I, Department of Human Biology (medical psychology). — Former Chairperson of the Société de psychosomatique, Rhône-Alpes. — Former lecturer for residents and non-residents of the Vinatier Bron Hospital.
■ PUBLICATIONS: yes.
■ WORKING ON: publications [A9], respect of human dignity and human rights [B1], dignity [B5], human equality [B7], protection of rights — involved institutions [B16], professional deontology organs [B26], patient advocacy [B42], ethical rules and principles [B47], professional competence [B52], respect [B63], information — communication — media [B68], public debates [B75], education [B76], nursing education [B79], medical education [B80], internship and residency [B81], students [B82], universities [B84], health and biology mediatisation [B85], bioethics [B89], bioethics movement [B95], European Convention of Bioethics [B97], Bioethics bill, 1992 (France) [B98], specific approaches to ethics [C1], situational ethics [C9], quality of life [C16], religion and religious ethics [C19], Christian ethics [C22], Islamic ethics [C28], religion and sects [C30], religious beliefs [C32], philosophy of biology [C42], professional ethics [C51], medical ethics [C52], nursing ethics [C53], ethics and humanities [C54], humanities [C55], humanism [C66], psychology [C68], mental processes [C69], comprehension [C70], judgement [C71], behavior [C73], attitudes [C74], motivation [C76], emotions [C79], love [C80], trust [C81], personality [C82], self concept [C83], intelligence [C85], social and political issues [D1], social dominance [D6], socioeconomic factors [D18], social groups [D22], aliens [D23], family members [D24], single persons [D25], couple [D26], cohabitation [D27], married persons [D28], marital relationship [D29], sibling [D30], family relationship [D31], parents [D32], mothers [D33], fathers [D34], females [D35], ethnic groups [D36], whites [D37], Blacks [D41], minority groups [D42], handicapped [D43], indigents [D48], social discrimination [D53], sociology of medicine [D55], political issues [D56], North-South relationships [D60], municipal government [D72], violence [D85], rape [D93], torture [D94].
■ INTERESTED IN: science, technology, methods [E1], biology [E5], genetics [E6], epidemiology [E9], statistics [E10], research [E14], investigators [E15], research team [E16], research institutes [E17], research policy [E18], research design [E19], pilot projects [E20], behavioral research [E21], progress [E26], technology [E27], biomedical technologies [E28], containment [E38], quality of environment [E41], health hazards [E49], evaluation [E61], biomedical research and experimentation [F1], biomedical research [F2], health services research [F4], therapeutic research [F6], human genome project [F8], experimentation [F9], methodology of research and experimentation [F15], debriefing [F16], medical informatics [F21], health [G2], health policy [G13], health legislation [G14], institutional policies [G15], institutional obligations [G16], health facilities [G22], organization of health care [G35], health care [G45], convenience medicine [G57], health occupations [G61], economics [H2], population [H19], stages of life — problems proper to childhood and elderly [L1], bone marrow [L8], kidneys [L9], hormones [L10], somatotropin [L11], human development [L12], age [L13], adolescents [L14], adults [L15], nutrition [L16], food [L17], animal organs [L18], human characteristics [L19], mind [L20], childhood difficulties, diseases, protection [L21], aged — problems related with aging [L45], human activities [L52], sexuality and procreation — unborn child [M1], disease [N1], acquired immunodeficiency syndrome [N35], nervous system diseases [N40], injuries,

occupational diseases, intoxications [N46], occupational diseases [N48], death and resuscitation [O1], brain death [O6], terminal care [O7], euthanasia [O15], suicide [O19], attitudes to death [O20], resuscitation [O21], behavioral and mental disorders [P2], outpatient commitment [P18], psychiatric technics [P24], behavior control [P25], psychotherapy [P30], hypnosis [P31], group therapy [P32], psychoactive drugs [P33], heroin [P34], genetics and applications — biotechnology [Q1], genetic defects and hereditary diseases [Q2], biotechnology — genetic engineering [Q20], behavioral genetics [Q34], population genetics [Q43], human genome [Q47].

F 14 **BAGROS** Ph.

Professor, School of Medicine. — Head of the Department of Nephrology at the CHU (Universtiy hospital).

Université François-Rabelais
Faculté de médecine de Tours
Service de Néphrologie
37000 Tours (France)
✆ (33) 47 47 37 46

■ TEACHING: methodology for the approach of ethical problems in hospital environments.
■ WORKING ON: biomedical ethics education [B78], medical education [B80].
■ INTERESTED IN: code of deontology [B60], biomedical ethics education [B78], medical education [B80], hospitalisation [G46], kidney diseases [N39], organ transplantation [I24], professional deontology [J25], physician patient relationship [J32], renal dialysis [I32], terminal care [O7], attitudes to death [O20].

F 15 **BALANSARD**

Head of the Department of Cardiology, Font-Pré Hospital.

CCPPRB Centre hospitalier intercommunal
Hôpital Font-Pré
1208, avenue Colonel-Picot
83056 Toulon Cedex (France)
✆ (33) 94 61 61 61

■ WORKING ON: health care professionals, researchers and patient relationship [J1], patients' rights [J2], patient association [J3], patient information [J4], patient access [J5], parental notification [J6], disclosure [J7], consent to treatment [J8], informed consent [J9], presumed consent [J10], third party consent [J11], spousal consent [J12], parental consent [J13], consent forms [J14], treatment refusal [J15], patient participation [J16], right to treatment [J17].

F 16 **BARRAL-BARON** André

Professor of Ethics at the Theological Centre of Meylan.

Centre théologique de Meylan
15, chemin de la Carronnerie
38246 Meylan Cedex (France)
✆ (33) 76 90 41 51

■ AFFILIATIONS: Theological Centre of Meylan — CIM, Meylan (*Professor of ethic*).
■ TEACHING: Theological Centre of Meylan. — School of Medicine,Grenoble. — CCPPRB. — Bioethical Sessions in France. — Health Pastoral. — Research fields: the couple, the family. — Group of Magistrates.
■ PUBLICATIONS: yes.
■ WORKING ON: review [A2], bibliography [A4], documentation [A8], publications [A9], respect of human dignity and human rights [B1], codes of ethics [B49], solidarity [B65], biomedical ethics education [B78], CCNE advice (France) [B91], Bioethics bill, 1992 (France) [B98], Christian ethics [C22], Protestant ethics [C27], theology [C41], professional ethics [C51], medical ethics [C52], nursing ethics [C53], psychology [C68], judgement [C71], intelligence [C85], pastoral care [C88], social sciences [C89], family members [D24], couple [D26], family relationship [D31], containment [E38], public hospitals [G30], developing countries [H21], prenatal diagnosis [I3], accountability [K29], conscience clause [K63], contraception [M3], infertility [M19], abortion [M47], euthanasia [O15].
■ INTERESTED IN: bibliography [A4], documentation [A8], human person [B12], traffic of organs [B15], CCNE — Comité Consultatif National d'Ethique (France) [B23], Council of Europe [B25], CCNE advice (France) [B91], Bioethics bill, 1992 (France) [B98], deontological ethics [C6], teleological ethics [C7], conscience [C13], Islam [C33], scriptural interpretation [C40], professional ethics [C51], motivation [C76], social sciences [C89], legislation [D13], married persons [D28], social

discrimination [D53], EC — European Communities [D62], biomedical technologies [E28], freezing [E29], biological containment [E39], data protection [E52], decision making [E54], decision analysis [E55], biomedical research [F2], experimentation [F9], hospital [G28], WHO — World Health Organization [G36], price [H16], profit [H17], financial support [H18], prenatal diagnosis [I3], mass screening [I6], biological specimens procurement, blood transfusion, organ transplantation [I8], organ transplantation [I24], fetal tissue transplantation [I27], informed consent [J9], presumed consent [J10], third party consent [J11], minors [K6], biotechnology — genetic engineering [Q20], gene pool [Q45].

F 17 **BARRET**
Professor of Forensic Medicine. — M.D. Clinician.

CCPPRB
CHRU
BP 217
38043 Grenoble Cedex 9 (France)
✆ (33) 76 76 57 83 Fax: (33) 76 76 51 77

■ TEACHING: teaching Forensic Medicine (University degree in Medical Law).
■ WORKING ON: CCPPRB — French committees for the protection of human subjects of biomedical research [B32], medical ethics [C52], biomedical research [F2], jurisprudence — accountability [K8].
■ INTERESTED IN: health care professionals, researchers and patient relationship [J1].

F 18 **BASTIAN** Bernard
M.D. Clinician, theologian, teacher of Biomedical Ethics.

12, rue des Dentelles
67000 Strasbourg (France)
✆ (33) 88 32 64 21

■ AFFILIATIONS: Groupe de travail éthique, groupe hospitalier Saint-Vincent (Strasbourg), Strasbourg, France (*Coordinator*).
■ TEACHING: School of Medicine of Strasbourg. — CECOS Alsace. — CES in biology. — The Saint-Vincent Hospital complex. — The private organism for the training of health care professionals. — Colloquium on "Enseignement de l'éthique et politique de la Cité" (The teaching of ethics and politics of the Authorities) at the Arche de la Défense (Paris, 1991).
■ PUBLICATIONS: yes.
■ WORKING ON: human person [B12], biomedical ethics education [B78], ethicists [B92], bioethical issues [B96], ethical analysis [C3], conscience [C13], Christian ethics [C22], Roman catholic ethics [C23], Protestant ethics [C27], Islamic ethics [C28], Jewish ethics [C29], religion and sects [C30], Jehovah's witnesses [C38], theology [C41], medical ethics [C52], nursing ethics [C53], decision making [E54], teaching methods [E59], religious hospitals [G32], sperm banks [I11], alternative therapies [I33], nurse patient relationship [J30], physician nurse relationship [J31], physician patient relationship [J32], professional patient relationship [J33], volontary sterilization [M8], embryos [M10], reproductive technologies [M26], ovum donors [M27], semen donors [M28], in vitro fertilization [M29], FIV center [M30], artificial insemination [M31], AID [M32], AIH [M33], excess embryos [M34], host mothers [M35], embryo transfer [M36], GIFT (Gamete Intrafallopian Transfer) [M37], embryo donation [M38], selective abortion [M49], therapeutic abortion [M50], aborted fetuses [M51], abortion on demand [M52], mifepristone [M53], suffering [N11], pain [N12], acquired immunodeficiency syndrome [N35], brain death [O6], withholding treatment [O10], attitudes to death [O20], preimplantation genetic diagnosis [Q38], eugenics [Q39].
■ INTERESTED IN: *idem*.

F 19 **BASTIN** Raymond
Honorary medical clinic Professor. — Assistant secretary of the Académie nationale de médecine, titular member of section 1 (medicine and medical specialisations).

Académie nationale de médecine
14, rue de Martignac
75007 Paris (France)
✆ (33) 1 - 43 26 69 80

■ PUBLICATIONS: yes.

F 20 **BEAUFILS**
Head of the Internal Medicine B Department, Saint-Germain-en-Laye Hospital..

CCPPRB Centre hospitalier général
20, rue Armagis
78100 Saint-Germain-en-Laye (France)
✆ (33) 1 - 39 73 92 01

■ WORKING ON: medical informatics [F21], state medicine [G9], health legislation [G14], hospital [G28], public hospitals [G30], physicians [G40], medical staff [G41], hospitalisation [G46], health care delivery [G48], patient compliance [G49], medical records [G50], medical evaluation [G51], selection for treatment [G52], patient care [G53], medicine [G63], administrators [H3], health care costs [H9], mass screening [I6], drugs [I34], insulin [I35], palliative care [O8], life-sustaining treatment [O9], allowing to die [O11], terminally ill [O12], quality adjusted life years [O14], substance dependence [P3], alcohol abuse [P4], drug abuse [P6], psychotherapy [P30].
■ INTERESTED IN: information sources, data bases [A1], data bases [A3], documentation [A8], law, legislation and jurisprudence [K1], terminal care [O7].

F 21 **BEAUFILS** Dominique

Head of the Department of Surgery of the Brisset Hospital.

32, rue de Neuve-Maison
02500 Hirson (France)
✆ (33) 23 98 68 99

■ TEACHING: lectures on medical ethics and bioethics for orthodox theological societies.
■ PUBLICATIONS: yes.
■ WORKING ON: human person [B12], Christian ethics [C22], eastern orthodox ethics [C26], theology [C41], medical ethics [C52], genetics [E6], freezing [E29], preservation [E30], diagnosis [I2], organ transplantation [I24], organ donation [I25], organ procurement [I26], beginning of life [L24], procreation [M15], reproductive technologies [M26], abortion on demand [M52], euthanasia [O15], biotechnology — genetic engineering [Q20], eugenics [Q39], negative eugenics [Q40], positive eugenics [Q41], gene therapy [Q42].
■ INTERESTED IN: human body commercialization [B13], traffic of organs [B15], Bioethics bill, 1992 (France) [B98], capital punishment [D15], genetics [E6], human genome project [F8], diagnosis [I2], organ transplantation [I24], organ donation [I25], organ procurement [I26], fetal tissue transplantation [I27], intracerebral fetal tissue transplantation [I28], child donation [K22], wrongful life [K33], aid children [L34], reproductive technologies [M26], genetic defects and hereditary diseases [Q2], biotechnology — genetic engineering [Q20], gene therapy [Q42], human genome [Q47].

F 22 **BECQ-GIRAUDON**

Head of Department of Internal Medicine/Infections and Tropical Diseases of the CHU (University Hospital) of Poitiers. — Professor of Tropical and Infectious Diseases at the School of Medicine of Poitiers.

CCPPRB- Poitou-Charentes
CHU la Milétrie
BP 577
86021 Poitiers Cedex (France)
✆ (33) 49 45 21 57 Fax: (33) 49 44 43 83

■ TEACHING: yes.

F 23 **BEGORRE** Michel

Full-time M.D. Clinician at the Oloron-Sainte-Marie Hospital, Department of Medicine.

CCPPRB des Pays de l'Adour
Centre hospitalier de Pau
64000 Pau (France)
✆ (33) 59 92 48 33

■ WORKING ON: integrity [B8], institutions involved in the protection of rights [B17], CCPPRB — French committees for the protection of human subjects of biomedical research [B32], therapeutic research [F6], human experimentation [F11], health care [G45], medicine [G63], aged — problems related with aging [L45], disease [N1].
■ INTERESTED IN: biomedical research and experimentation [F1], patients' rights [J2], disease and aged [L46], acquired immunodeficiency syndrome [N35].

F 24 **BEN HAMIDA** Faker
Honorary Research Director at the CNRS.

4, Allée des Tilleuls
92330 Sceaux (France)
✆ (33) 1 - 47 02 14 80 Fax: (33) 1 - 46 33 23 05

■ AFFILIATIONS: Réseau européen de Coopération scientifique médecin et droits de l'homme of the Council of Europe (*Member*).
■ TEACHING: Medical Faculty Paris-VI.
■ PUBLICATIONS: yes.

F 25 **BERNARD** Jean
Honorary Professor at University of Paris -III .

82, rue d'Assas
75006 Paris (France)
✆ (33) 1 - 43 26 41 27

■ AFFILIATIONS: Comité consultatif national d'éthique des sciences de la vie et de la santé — CCNE, Paris, France (*Honorary Chairperson*).
■ TEACHING: Séances du Comité consultatif national d'éthique auprès des universités, lycées et collèges.
■ PUBLICATIONS: yes.

F 26 **BESSIERE** Arlette
Responsible for Organization.

106, avenue du Général-Leclerc
75014 Paris (France)

■ AFFILIATIONS: B.O.F Groupe franco-québequois de bioéthique — B.O.F, Saint-Brice-Sous-Forêt, France (*Adhérente*).
■ PUBLICATIONS: yes.
■ WORKING ON: decision making [E54], decision analysis [E55], organizational policies [G17], public hospitals [G30].
■ INTERESTED IN: review [A2], book review [A6], ethics [B48], obligations of society [B53], obligations to society [B54], respect [B63], normative ethics [C5], situational ethics [C9], values [C15], nursing ethics [C53], philosophy [C61], informal social control [D4], socioeconomic factors [D18], social impact [D49], decision making [E54], [E56], technology assessment [E65], health care costs [H9], population control [H22], conflict of interest [K9].

F 27 **BESSIS** Roger
Ultrasound Imaging Specialist. — Prenatal Diagnosis consulting. — Research worker at Baudelocque-Port-Royal unit.

Centre d'échographie de l'Odéon
122, boulevard Saint-Germain
75006 Paris (France)
✆ (33) 1 - 43 26 54 63

■ AFFILIATIONS: commission nationale pour la médecine et la biologie de la reproduction, Paris, France (*Member*).
■ WORKING ON: human person [B12], professional competence [B52], medical education [B80], medical ethics [C52], advisory committees [D69], progress [E26], cognitive research [F7], physicians [G40], obstetrics and gynecology [G66], medical fees [H15], prenatal diagnosis [I3], amniocentesis [I4], mass screening [I6], referral and consultation [J24], human development [L12], embryos [M10], fetuses [M11], mother fetus relationship [M43], pregnancy [M44], selective abortion [M49], therapeutic abortion [M50], congenital defects [N23], genetic defects and hereditary diseases [Q2], diagnostic imaging [E32].
■ INTERESTED IN: human person [B12], professional competence [B52], health and biology mediatisation [B85], Bioethics bill, 1992 (France) [B98], medical ethics [C52], epidemiology [E9], diagnostic imaging [E32], obstetrics and gynecology [G66], health care costs [H9], costs and benefits [H10], prenatal diagnosis [I3], mass screening [I6], malpractice [K19], embryos [M10], fetuses [M11], therapeutic abortion [M50], fetal therapy [M57], congenital defects [N23], genetic defects and hereditary diseases [Q2], genetic screening [Q37], eugenics [Q39].

F 28 **BIOT** Christian
Hospital Chaplain (Catholic Centre). — Lecturer in Moral Ethics in a pastoral training centre of Lyon.

Centre Léon-Bérard
28, rue Laënnec
69008 Lyon (France)
✆ (33) 78 78 28 28

■ AFFILIATIONS: Santé, Ethique et Libertés — SEL, Lyon, France (*Member*).
■ TEACHING: Participates in the training units for health care personnel (units organised by the SEL).
■ PUBLICATIONS: yes.
■ WORKING ON: dignity [B5], CCPPRB — French committees for the protection of human subjects of biomedical research [B32], nursing education [B79], CCNE advice (France) [B91], scriptural interpretation [C40], nursing ethics [C53], mental institutions [G29], patient information [J4], consent to treatment [J8], treatment refusal [J15], quality adjusted life years [O14].

F 29 **BIOULAC** Bernard
Professor of Medicine (Neurophysiology). — M.D. Clinician at the CHU (University hospital) of Bordeaux. — M.P. of Dordogne. — Chairperson of the Parliamentary Working Group on Bioethics. — Former Chairman of the Conseil général of the Dordogne (1982-1992).

Université de Bordeaux-II
Faculté de médecine
146, rue Léo Saignat
33076 Bordeaux (France)

■ AFFILIATIONS: comité consultatif national d'éthique — CCNE, Paris, France (*Member*).
■ TEACHING: Université de Bordeaux-II (Faculté de médecine). — Information conferences.
■ PUBLICATIONS: yes.

F 30 **BLANC** Patrick
Cardiologist. — Chairman of the CCPPRB.

CCPPRB Limousin
Hôpital Jean-Rebeyrol
Avenue du Buisson
87042 Limoges cedex (France)
✆ (33) 55 05 65 40

■ WORKING ON: review [A2], fundamental rights of the individuals [B2], protection of rights — involved institutions [B16], ethical rules and principles [B47], biomedical ethics education [B78], bioethics [B89], medical ethics [C52], biomedical research and experimentation [F1], biological specimens procurement, blood transfusion, organ transplantation [I8], confidentiality [J20], jurisprudence — accountability [K8].
■ INTERESTED IN: fundamental rights of the individuals [B2], protection of rights — involved institutions [B16], CCPPRB — French committees for the protection of human subjects of biomedical research [B32], ethical rules and principles [B47], CCNE advice (France) [B91], medical ethics [C52], advisory committees [D69], clinical trials [F3], therapeutic research [F6], human experimentation [F11], methodology of research and experimentation [F15], biological specimens procurement, blood transfusion, organ transplantation [I8], donors [I14], blood transfusions [I19], spousal consent [J12], physicians and researchers accountability [J23], legal personality [K2], expert evaluation [K15], disease [N1].

F 31 **BOUCAN** Marie-Hélène
Head of the Department of Functional Re-education and Re-adaptation for adults.

Hôpital René-Sabran
GIENS
83406 Hyères Cedex (France)

■ AFFILIATIONS: Centre de bioéthique (Lyon), Lyon Cedex 02, France (*Participates in a research team on persistent vegetative states*).
■ TEACHING: organizing of a national colloquium to be held on June 11, 12, 1993.: "Re-Educational Ethics", Renée-Sabran Hospital.
■ PUBLICATIONS: yes.
■ WORKING ON: human person [B12], respect [B63], communication [B69], medical ethics [C52],

nursing ethics [C53], handicapped [D43], physically handicapped [D44], technology assessment [E65], human experimentation [F11], accidents [G6], traffic accidents [G7], health care costs [H9], patient information [J4], consent to treatment [J8], patient participation [J16], legal guardians [K7], brain [L6], procreation [M15], sexuality [M20], coma [N41], persistent vegetative state [N42], brain diseases [N44], injuries [N49], brain death [O6], euthanasia [O15], withholding treatment [O10], central nervous system diseases [N43].

■ INTERESTED IN: human person [B12], respect [B63], communication [B69], medical ethics [C52], nursing ethics [C53], handicapped [D43], physically handicapped [D44], technology assessment [E65], human experimentation [F11], accidents [G6], traffic accidents [G7], health care costs [H9], patient information [J4], consent to treatment [J8], patient participation [J16], legal guardians [K7], brain [L6], procreation [M15], sexuality [M20], coma [N41], persistent vegetative state [N42], central nervous system diseases [N43], brain diseases [N44], injuries [N49], brain death [O6], euthanasia [O15], withholding treatment [O10].

F 32 **BOUCAUD** Pascale

Expert for the Council of Europe. — Expert for the UN's Human Rights Centre. — Professor of law. — Director of the Institut des Droits de l'Homme of Lyon.

Institut des Droits de l'Homme de Lyon
10-12, rue Alphonse-Fochier
69002 Lyon (France)
✆ (33) 72 32 50 50 Fax: (33) 72 32 50 19

■ TEACHING: Postgraduate lectures, DEA level (Human Rights). — Master's training (Human Rights Masters). — Lecturing for diverse associations or training organisms (nursing schools, universities).

■ PUBLICATIONS: yes.

■ WORKING ON: respect of human dignity and human rights [B1], protection of rights — involved institutions [B16], ethical rules and principles [B47], history of the 20th century [C60], philosophy [C61], determinism [C62], sociobiology [C63], existentialism [C64], hedonism [C65], humanism [C66], Marxism [C67], psychology [C68], mental processes [C69], comprehension [C70], judgement [C71], imprisonment [D11], prisoners [D12], legislation [D13], model legislation [D14], capital punishment [D15], regulation [D16], government regulation [D17], aliens [D23], family members [D24], single persons [D25], couple [D26], cohabitation [D27], married persons [D28], marital relationship [D29], ethnic groups [D36], whites [D37], Hispanic Americans [D38], American Indians [D39], Jews [D40], Blacks [D41], minority groups [D42], handicapped [D43], physically handicapped [D44], males [D45], disadvantaged [D46], associations [D47], indigents [D48], social impact [D49], social interaction [D50], cultural pluralism [D51], political issues [D56], violence [D85], science, technology, methods [E1], biomedical research and experimentation [F1], health care, medical acts [I1], diagnosis [I2], intracerebral fetal tissue transplantation [I28], law, legislation and jurisprudence [K1].

■ INTERESTED IN: health care, medical acts [I1], sperm banks [I11], tissue banking [I12], tissue donation [I13], donors [I14], organ donors [I15], donor cards [I16], anonymous donation [I17], directed donation [I18].

F 33 **BOUÉ** André

Director of Unit 73 of INSERM. — Professor of Medical Genetics, University of Paris-V.

INSERM U 73 — Génétique et pathologie foetale
Château de Longchamp
Carrefour Longchamp
75016 Paris (France)
✆ (33) 1 - 42 88 53 17 Fax: (33) 1 - 46 47 95 01

■ AFFILIATIONS: CCNE, France (*Member*).

■ WORKING ON: human body commercialization [B13], medical ethics [C52], genetics [E6], data protection [E52], diagnosis [I2], reproductive technologies [M26], abortion [M47], genetics and applications — biotechnology [Q1].

F 34 **BOURGE** M.

Honorary Chairperson of the Tribunal de Grande Instance of Lyon.

CCPPRB- Lyon C
Hôpital de l'Hôtel-Dieu
1, place de l'Hôpital
69288 Lyon Cedex 02 (France)
✆ (33) 78 92 20 00

■ WORKING ON: justice [B9], CCPPRB — French committees for the protection of human subjects of biomedical research [B32].

F 34 b **BRAIBANT** Guy
Chairman of Section of the Conseil d'Etat. — Vice-chairman of the Commission supérieure de codofication. — Chairman of the Institut international des sciences administratives.

Commission supérieure de codification
72, rue de Varennes
75700 Paris Cedex (France)
✆ (33) 42 75 85 83 Fax: (33) 42 75 72 06

F 35 **BRETEAU** Michel
Head of the Laboratory of Clinical Pharmacology and Toxicology. — Chairperson of the CCPPRB of Tours.

CCPPRB Centre hospitalier régional universistaire
2, boulevard Tonnelle
37044 Tours Cedex (France)
✆ (33) 47 47 60 11

■ WORKING ON: CCPPRB — French committees for the protection of human subjects of biomedical research [B32], biomedical research and experimentation [F1], biomedical research [F2], experimentation [F9].
■ INTERESTED IN: *idem.*

F 36 **BRIARD** Marie-Louise
Research Director at INSERM.

Hôpital Necker-Enfants-Malades
149, rue de Sèvres
75043 Paris Cedex 15 (France)
✆ (33) 1 - 42 73 84 58 Fax: (33) 1 - 47 34 85 14

■ TEACHING: postgraduate DEA Ethics, Necker Faculty.
■ WORKING ON: health services research [F4], human genome project [F8], diagnosis [I2], tissue donation [I13], donors [I14], organ donors [I15], donor cards [I16], anonymous donation [I17], sexuality and procreation — unborn child [M1], genetic defects and hereditary diseases [Q2], medical genetics [Q35].
■ INTERESTED IN: medical ethics [C52], biological containment [E39], health services research [F4], molecular biology [E8], diagnosis [I2], prenatal diagnosis [I3], amniocentesis [I4], chorionic villus sampling [I5], mass screening [I6], mandatory screening [I7], required request [I10], sperm banks [I11], donors [I14], health care professionals, researchers and patient relationship [J1], law, legislation and jurisprudence [K1], sexuality and procreation — unborn child [M1], congenital defects [N23], eye diseases [N38], genetics and applications — biotechnology [Q1].

F 37 **BROYER** Michel
Head of the Department of Paediatric Nephrology at the Enfants Malades-Necker hospital. — Director of the INSERM research Unit 192. — Vice-Chairperson of France Transplant.

Hôpital Necker-Enfants-Malades
149, rue de Sèvres
76015 Paris (France)
✆ (33) 1 - 45 66 01 39 Fax: (33) 1 - 42 73 84 51

■ AFFILIATIONS: comité d'éthique hospitalier Necker, Paris, France (*Chairperson*). — Groupe de réflexion sur l'éthique des transplantations — GRET, Paris (*Member*).
■ TEACHING: contributes to the teaching of the postgraduate DEA on Ethics at the Necker Hospital.
■ PUBLICATIONS: yes.
■ WORKING ON: required request [I10], donors [I14], organ donors [I15], donor cards [I16], anonymous donation [I17], directed donation [I18], transplantation [I22], transplant recipients [I23], fetal tissue transplantation [I27], intracerebral fetal tissue transplantation [I28].
■ INTERESTED IN: human body commercialization [B13], traffic of organs [B15], Council of Europe [B25], psychology [C68], profit [H17], prenatal diagnosis [I3], biological specimens procurement, blood transfusion, organ transplantation [I8], treatment [I30], anesthesia [I31], renal dialysis [I32], child's interest [K24], somatotropin [L11], fetal therapy [M57], kidney diseases [N39], attitudes to death [O20].

F 38 **BUFFARD** Simone

Psychologist and Lecturer (Lyon Prisons, Edouard-Herriot hospital of Lyon, Criminology Institute). — Retired.

307, rue Fontanières
69350 La Mulanière (France)
✆ (33) 78 51 39 60

■ AFFILIATIONS: Santé, Ethique et liberté — SEL, Lyon Cedex 08, France
■ PUBLICATIONS: yes.

F 39 **BYK** Christian

Special Adviser on Bioethics for the Secretary General of the Council of Europe. — Secretary General of the Association internationale Droit, éthique et science.

Conseil de l'Europe
BP 431 - R6
67006 Strasbourg Cedex (France)
✆ (33) 88 41 28 20 Fax: (33) 88 41 27 99 et 88 41 27 94

■ AFFILIATIONS: Association Internationale Droit, Ethique et Science (groupe de Milazzo), Strasbourg, France (*Secretary General*).
■ TEACHING: Medical Network of Human Rights, Adult Education, René-Descartes University (Paris-IV). — DES in Medical Law, universities of Poitiers and Tours.
■ PUBLICATIONS: yes.
■ WORKING ON: literature [A7], human person [B12], ethics committees [B22], codes of ethics [B49], obligations of society [B53], interdisciplinary communication [B70], bioethical issues [B96], quality of life [C16], religious beliefs [C32], biology and human future [C43], medical ethics [C52], humanities [C55], socioeconomic factors [D18], community services [D78], biological warfare [D88], biology [E5], epidemiology [E9], progress [E26], biomedical technologies [E28], [E56], self regulation [E62], social control of science [E64], human genome project [F8], human experimentation [F11], research subjects [F12], selection of subjects [F19], health services [G25], health personnel [G37], medical staff [G41], medical records [G50], selection for treatment [G52], medicine [G63], life insurance [H8], mandatory screening [I7], required request [I10], physical restraint [I40], treatment refusal [J15], medical secrecy [J22], physician's role [J26], investigator subject relationship [J29], technical expertise [K16], accountability [K29], health legislation [K52], human body [L3], sexual behavior [M21], transmission [N15], HIV seropositivity [N36], withholding treatment [O10], attitudes to death [O20], sex linked defects [Q19], biotechnology [Q21], eugenics [Q39].

F 40 **CADORÉ** Bruno

Field of research, teaching and documentation.

Centre d'éthique médicale
60, boulevard Vauban
BP 109
59016 Lille Cedex (France)
✆ (33) 20 42 09 12 Fax: (33) Fax: 20 78 26 45

■ AFFILIATIONS: Association européenne des centres d'éthique médicale — AECEM, Brussels, Belgium (*Member*). — Centre d'éthique contemporaine de l'Institut catholique de Lille, Lille Cedex, France (*Member*).
■ TEACHING: yes.
■ PUBLICATIONS: yes.
■ WORKING ON: information sources, data bases [A1], data bases [A3], bibliography [A4], information centers [A5], documentation [A8], dignity [B5], human person [B12], ethical rules and principles [B47], ethics [B48], codes of ethics [B49], moral obligations [B50], obligations of society [B53], obligations to society [B54], moral policy [B55], deontology [B59], morality [B61], respect [B63], virtues [B64], solidarity [B65], biomedical ethics education [B78], European Convention of Bioethics [B97], Bioethics bill, 1992 (France) [B98], philosophical ethics [C2], morals [C12], conscience [C13], moral development [C14], values [C15], quality of life [C16], social worth [C17], value of life [C18], religion and religious ethics [C19], clergy [C20], religious ethics [C21], Christian ethics [C22], Roman catholic ethics [C23], Roman catholicism [C31], religious beliefs [C32], religion [C36], Christian science [C37], religious sciences [C39], scriptural interpretation [C40], theology [C41], professional ethics [C51], medical ethics [C52], nursing ethics [C53], philosophy [C61], pastoral care [C88], teaching methods [E59], social control of science [E64], health care costs [H9], prenatal diagnosis [I3], inter-personal relationship [J27], nurse patient relationship [J30], physician nurse relationship

[J31], physician patient relationship [J32], professional patient relationship [J33], newborns [L29], prematurity [L30], in vitro fertilization [M29], FIV center [M30], suffering [N11], pain [N12], congenital defects [N23], neural tube defects [N24], anencephaly [N25], spina bifida [N26], death and resuscitation [O1], terminal care [O7], euthanasia [O15], attitudes to death [O20], genetic defects and hereditary diseases [Q2], genetic defects [Q3], down's syndrome [Q6], hereditary diseases [Q18], medical genetics [Q35], genetic counseling [Q36], genetic screening [Q37], preimplantation genetic diagnosis [Q38], eugenics [Q39], negative eugenics [Q40], positive eugenics [Q41].
■ INTERESTED IN: *idem.*

F 41 **CADOUX** Louise
Deputy Chairman of CNIL.

Commission nationale de l'informatique et des libertés (CNIL)
21, rue Saint-Guillaume
75007 Paris (France)
✆ (33) 1 - 45 49 04 55

■ WORKING ON: human rights [B3], self determination [B4], freedom [B10], privacy [B11], CNIL — Commission Nationale Informatique et Libertés (France) [B19].
■ INTERESTED IN: Bioethics Bill, 1992 (France) [B98].

F 42 **CAILLAVET** Henri
Chairperson of the Association pour le droit de mourir dans la dignité (ADM.D.). — Honorary barrister-at-law of Cour d'Appel de Paris. — Honorary M.P. — Former Minister. — Former French M.P., Chambre des Députés. — Member of the Comité consultatif national d'éthique CCNE. — Member of the Commission nationale de l'informatique et des libertés (CNIL). — Member of the Commission d'évaluation de la lutte contre la drogue.

Association pour le Droit de mourir dans la dignité (ADMD)
103, rue Lafayette
75010 Paris (France)
✆ (33) 1 - 42 85 12 22 Fax: (33) 1 - 45 96 00 50

■ INTERESTED IN: euthanasia [O15].

F 43 **CHANGEUX** Jean-Pierre
Professor of the Collège de France and at Institut Pasteur. — Chairman of Comité consultatif national d'éthique (CCNE).

Institut Pasteur
25, rue du Docteur-Roux
75015 Paris (France)
✆ (33) 1 - 45 68 88 05 Fax: (33) 1 - 45 68 88 35

■ PUBLICATIONS: yes.
■ WORKING ON: information sources, data bases [A1], review [A2], respect of human dignity and human rights [B1], human rights [B3], CCNE — Comité Consultatif National d'Ethique (France) [B23], Declaration on Human and Citizen Rights [B38], Universal Declaration of Human Rights [B39], ethics [B48], codes of ethics [B49], moral obligations [B50], CCNE advice (France) [B91], specific approaches to ethics [C1], philosophy of biology [C42], ethics and humanities [C54], humanities [C55], history [C56], historical aspects [C57], history before the 20th century [C58], ancient history [C59], history of the 20th century [C60], philosophy [C61], determinism [C62], sociobiology [C63], existentialism [C64], hedonism [C65], humanism [C66], Marxism [C67], psychology [C68], mental processes [C69], comprehension [C70], judgement [C71], recall [C72], behavior [C73], attitudes [C74], intention [C75], motivation [C76], incentives [C77], normality [C78], emotions [C79], love [C80], trust [C81], personality [C82], self concept [C83], egoism [C84], intelligence [C85], applied psychology [C86], social sciences [C89], science, technology, methods [E1], science [E2], history of science [E3], life sciences [E4], biology [E5], genetics [E6], molecular biology [E8], research [E14], investigators [E15], neurosciences — psychiatry [P1], biotechnology — genetic engineering [Q20].
■ INTERESTED IN: *idem.*

F 44 **CHERMANN** Jean-Claude
Director of the Research Unit INSERM U323 (retrovirus and associated diseases). — M.P., Assemblée nationale, for the Bouches-du-Rhône.

Unité de recherches INSERM U 322 (rétrovirus et maladies associées)
Campus universitaire de Luminy

BP 33
13273 Marseille Cedex 9 (France)
✆ (33) 91 41 32 32 Fax: (33) 91 41 92 50

■ AFFILIATIONS: Commission sur la bioéthique, Paris 07 SP, France (*Member*).
■ WORKING ON: CCPPRB — French committees for the protection of human subjects of biomedical research [B32], International Charter of Human Rights [B34], Declaration on Human and Citizen Rights [B38], Bioethics bill, 1992 (France) [B98], political issues [D56], political activity [D57], EC — European Communities [D62], French National Assembly [D68], research [E14], investigators [E15], research team [E16], research institutes [E17], research policy [E18], pilot projects [E20], industrial research [E22], international research [E23], biomedical technologies [E28], containment [E38], biological containment [E39], safety [E51], biomedical research [F2], clinical trials [F3], therapeutic research [F6], human experimentation [F11], research subjects [F12], WHO — World Health Organization [G36], biological substances contamination [I9], blood transfusions [I19], blood donation [I20], blood substitutes [I21], transplantation [I22], immunotherapy [I41], vaccination [I43], consent to treatment [J8], consent forms [J14], confidentiality [J20], medical secrecy [J22], patents [K47], bill [K60], mother-child transmission [N16], acquired immunodeficiency syndrome [N35], HIV seropositivity [N36], biotechnology — genetic engineering [Q20], biotechnology [Q21], genetic intervention [Q22], genome mapping [Q23], cloning [Q24], clones [Q25], recombinant DNA research [Q31], DNA linkage [Q33], gene therapy [Q42], human genome [Q47], life sciences [E4], biology [E5], genetics [E6], mortality [E13], microbiology [E7], molecular biology [E8], epidemiology [E9].
■ INTERESTED IN: review [A2], human rights [B3], human person [B12], human body commercialization [B13], protection of rights — involved institutions [B16], European Convention on Human Rights [B36], patient advocacy [B42], ethical rules and principles [B47], solidarity [B65], humanitarian organizations [B66], codes of biomedical ethics [B90], biology and human future [C43], future generations [C44], consequences [C45], humanities [C55], biological warfare [D88], nuclear warfare [D89], history of science [E3], life sciences [E4], quality of environment [E41].

F 45 **CHOUTET** Patrick

M.D. Clinician, Head of University Department — Professor, School of Medicine, Tours (Tropical and Infectious Diseases).

CHU Bretonneau
Service des maladies infectueuses
37044 Tours Cedex (France)
✆ (33) 47 47 37 14 Fax: (33) 47 47 37 31

■ TEACHING: coordinates the teaching of medical deontology, DCEM 1, School of Medicine, Tours (within natural science curriculum).
■ WORKING ON: deontology [B59], education [B76], biomedical ethics education [B78], medical education [B80], students [B82], universities [B84], medical ethics [C52], HIV seropositivity [N36], religious ethics [C21], deontological ethics [C6], ethical analysis [C3].
■ INTERESTED IN: protection of rights — involved institutions [B16], ethical rules and principles [B47], education [B76], biomedical ethics education [B78], medical education [B80], students [B82], universities [B84], medical ethics [C52], history of science [E3], ethical analysis [C3], health care professionals, researchers and patient relationship [J1], conflict of interest [K9], therapeutic risk [K32], accountability [K29], conscience clause [K63], communicable diseases [N27], vaccination [I43], medical devices [G23], history [C56], deontological ethics [C6], religious ethics [C21].

F 46 **CLAVEL**

Professor of Cancerology. — Coordinator of the Département de Médecine of the Centre Régional de Lutte contre le cancer.

Centre anticancéreux régional Léon-Bérard
28, rue Laënnec
69008 Lyon (France)
✆ (33) 78 78 26 43 Fax: (33) 78 78 26 15

■ TEACHING: AEU (Philosophy and Ethics, Lyon).
■ WORKING ON: morals [C12], value of life [C18].
■ INTERESTED IN: biomedical research [F2], clinical trials [F3], genetic counseling [Q36], patient information [J4].

F 47 **CLAVERT** André

University Lecturer. — Director of CECOS Alsace. — M.D. Clinician.

CECOS Alsace
Hôpitaux Universitaires de Strasbourg
67091 Strasbourg (France)
✆ (33) 88 35 61 30

■ AFFILIATIONS: Comité d'éthique des Facultés de médecine ondotologie et pharmacie of Strasbourg, Strasbourg, France (*Member*). — Groupe de réflexion de la fédération des CECOS, France (*Member*). — Groupe de recherche éthique des cliniques Saint-Vincent, Strasbourg, France (*Member*).
■ TEACHING: teaching ethical problems arising from medical reproduction: "Certificat nursing" (M.A. level degree in Human Science), midwife school.
■ PUBLICATIONS: yes.
■ WORKING ON: fathers [D34], biology [E5], cryonic suspension [E31], nurse midwives [G43], sperm banks [I11], anonymous donation [I17], semen donors [M28], artificial insemination [M31], AID [M32].
■ INTERESTED IN: fathers [D34], teaching methods [E59], physician nurse relationship [J31], potentiality of personhood [K5], child donation [K22], beginning of life [L24], children's rights [L25].

F 48 **CLÉMENT** Jean-Loup

Clinical Psychologist, CECOS Lyon. — Lecturer, Lumière University (Lyon-II) and Claude-Bernard University (Lyon-I).

51, rue Président Herriot
69002 Lyon (France)
✆ (33) 78 37 37 06

■ AFFILIATIONS: CCPPRB Centre Léon-Bérard, Lyon Cedex 08, France (*qualified member for Ethics*).
■ TEACHING: responsible for teaching "Introduction to Bioethics" course, Science DEUG undergraduate degree, Claude-Bernard University (Lyon-I).
■ PUBLICATIONS: yes.
■ WORKING ON: human person [B12], human body commercialization [B13], CCPPRB — French committees for the protection of human subjects of biomedical research [B32], ethics [B48], codes of ethics [B49], deontology [B59], solidarity [B65], biomedical ethics education [B78], students [B82], universities [B84], health and biology mediatisation [B85], bioethical issues [B96], Bioethics bill, 1992 (France) [B98], social sciences [C89], research subjects [F12], healthy volunteers [F13], sperm banks [I11], donors [I14], aid children [L34], volontary sterilization [M8], procreation [M15], wish of children [M16], reproduction [M17], fertility [M18], infertility [M19], ovum donors [M27], semen donors [M28], in vitro fertilization [M29], FIV center [M30], artificial insemination [M31], AID [M32], AIH [M33], excess embryos [M34], host mothers [M35], embryo transfer [M36], embryo donation [M38].
■ INTERESTED IN: information centers [A5], privacy [B11], ethics committees [B22], CCNE — Comité Consultatif National d'Ethique (France) [B23], patient advocacy [B42], medical ethics [C52], applied psychology [C86], genetics [E6], epidemiology [E9], computerized file [E35], social control of science [E64], consent to treatment [J8], informed consent [J9], privileged communication [J21], medical secrecy [J22], genetic screening [Q37], preimplantation genetic diagnosis [Q38], eugenics [Q39], negative eugenics [Q40], positive eugenics [Q41], gene therapy [Q42].

F 49 **COEN** Abram

Hospital Psychiatrist (Grussel Hospital Complex, Saint-Denis). — M.D., Head of the Infant and Juvenile Psychiatric Ward. — Psychoanalyst. — Lecturer.

Hôpital Casanova
11, rue Danielle-Casanova
93200 Saint-Denis (France)
✆ (33) 1 - 42 43 59 80 Fax: (33) 1 - 42 35 61 84

■ AFFILIATIONS: Comité d'éthique du CHG of Saint-Denis, Saint-Denis Cedex, France (*Responsible for the research group "Applied social Ethics"*).
■ TEACHING: Postgraduate DEA level medical studies, UFR-Bobigny, Paris-XIII. — Educational Science, Saint-Denis University (Paris-VIII).
■ PUBLICATIONS: yes.
■ WORKING ON: respect of human dignity and human rights [B1], fundamental rights of the individuals [B2], protection of rights — involved institutions [B16], ethical rules and principles [B47], information — communication — media [B68], bioethics [B89], specific approaches to ethics [C1], philosophical ethics [C2], religion and religious ethics [C19], philosophy of biology [C42], professional ethics [C51], ethics and humanities [C54], social and political issues [D1], social issues [D2], stigmatization [D54], violence [D85], science, technology, methods [E1], research [E14], research institutes [E17], containment [E38], methods [E53], evaluation [E61], biomedical research and experimentation [F1],

biomedical research [F2], experimentation [F9], methodology of research and experimentation [F15], organization of health care; facilities, manpower and services; health occupations [G1], health [G2], public health [G5], health policy [G13], health facilities [G22], organization of health care [G35], health care [G45], convenience medicine [G57], family practice [G60], health occupations [G61], health economics, population characteristics [H1], population [H19], biological specimens procurement, blood transfusion, organ transplantation [I8], health care professionals, researchers and patient relationship [J1], confidentiality [J20], physicians and researchers accountability [J23], interpersonal relationship [J27], law, legislation and jurisprudence [K1], jurisprudence — accountability [K8], legislation and law [K34], anatomy, physiology, development [L2], aged — problems related with aging [L45], sexuality and procreation — unborn child [M1], reproductive technologies [M26], abortion [M47], fetal development [M54], pain [N12], congenital defects [N23], cancer [N17], nervous system diseases [N40], injuries, occupational diseases, intoxications [N46], determination of death [O2], terminal care [O7], euthanasia [O15], suicide [O19], neurosciences — psychiatry [P1], outpatient commitment [P18], genetics and applications — biotechnology [Q1], biotechnology — genetic engineering [Q20], behavioral genetics [Q34], population genetics [Q43].
■ INTERESTED IN: *idem.*

F 50 **CONGOURDEAU** Marie-Hélène
Research Fellow at the CNRS (History) URA 186.

Société française de réflexion bioéthique
30, rue d'Auteuil
75016 Paris (France)
✆ (33) 1 - 45 25 33 46

■ AFFILIATIONS: Société française de réflexion bioéthique — SFRB, Paris, France (*Chairperson*).
■ WORKING ON: human person [B12], information dissemination [B71], history of biomedical ethics [B94], Christian ethics [C22], Roman catholic ethics [C23], eastern orthodox ethics [C26], Protestant ethics [C27], Islamic ethics [C28], Jewish ethics [C29], history [C56], history before the 20th century [C58], ancient history [C59], history of science [E3], investigators [E15], human experimentation [F11], prenatal diagnosis [I3], beginning of life [L24], newborns [L29], aid children [L34], unwanted children [L35], handicapped children [L38], infanticide [L42], embryos [M10], fetuses [M11], procreation [M15], excess embryos [M34], embryo transfer [M36], embryo donation [M38], pregnant women [M42], mother fetus relationship [M43], illegal abortion [M48], selective abortion [M49], therapeutic abortion [M50], aborted fetuses [M51], abortion on demand [M52], fetal therapy [M57], determination of death [O2], autopsies [O3], preimplantation genetic diagnosis [Q38], eugenics [Q39], negative eugenics [Q40], positive eugenics [Q41].
■ INTERESTED IN: *idem.*

F 51 **CÜER** Pierre
Honorary Professor, University of Louis-Pasteur, Strasbourg. — Professor at the ethopsychology medical centre, Pierre-et-Marie-Curie University (Paris-VI). — Coordinates the teaching of biomedical ethics for the European Network of Scientific Cooperation "Medicine and Human Rights", of the Council of Europe, Strasbourg.

Réseau européen de coopération scientifique "Médecine et Droits de l'Homme"
7, rue des Tilleuls
67800 Hoenheim (France)
✆ (33) 88 33 44 65 Fax: (33) 88 41 27 87 c/o Dr. J.-P. Massue

■ AFFILIATIONS: Réseau européen de coopération scientifique "Médecine et Droits de l'Homme", Hoenheim, France (*Coordinator for the teaching of Medical Bioethics*).
■ TEACHING: Co-organizer and teacher for postgraduate "Medicine, Ethics and Human Rights" at Pierre et Marie Curie University (Paris VI). Co-organizer and teacher of Bioethics B.A. and M.A. levels, at the school of Pharmacology, René Descartes University (Paris V).
■ PUBLICATIONS: yes.
■ WORKING ON: Council of Europe [B25], ethics [B48], codes of ethics [B49], codes of biomedical ethics [B90], religious ethics [C21], history of science [E3], presumed consent [J10], brain [L6], host mothers [M35].
■ INTERESTED IN: ethics [B48], codes of ethics [B49], religious ethics [C21], history of science [E3], presumed consent [J10], brain [L6], sexuality and procreation [M2], reproductive technologies [M26], aborted fetuses [M51], acquired immunodeficiency syndrome [N35], biotechnology [Q21], codes of biomedical ethics [B90].

F 52 **D'HAUTEFEUILLE**
Head of the Cardiology Department.

CCPPRB Centre hospitalier Docteur-Schaffner
99, route La Bassée
62307 Lens Cedex (France)
✆ (33) 21 69 12 34

■ AFFILIATIONS: CCPPRB Centre hospitalier Docteur-Schaffner, Lens Cedex, France (*Chairperson*).
■ WORKING ON: CCPPRB — French committees for the protection of human subjects of biomedical research [B32], biomedical research [F2], diagnosis [I2], cardiovascular diseases [N19].
■ INTERESTED IN: human body commercialization [B13], traffic of organs [B15], Amnesty International [B18], ethics committees [B22], CCPPRB — French committees for the protection of human subjects of biomedical research [B32], religious ethics [C21], biomedical research [F2], medical ethics [C52], health care costs [H9], nursing ethics [C53], patients' rights [J2], North-South relationships [D60], patients' rights [J2], terminal care [O7], cardiovascular diseases [N19], euthanasia [O15], genetics and applications — biotechnology [Q1].

F 53 **DADOUNE** J.P.
Head of the Biology Department at the Hôtel-Dieu University (Paris). — Head of the Cytogenetics Department at Broussais-Hôtel-Dieu hospital,University of Paris-VI .

CCPPRB Hôtel-Dieu
1, place du Parvis Notre-Dame
75004 Paris (France)
✆ (33) 1 - 42 34 80 30 Fax: (33) 1 - 40 51 01 97

■ AFFILIATIONS: CCPPRB Hôtel-Dieu, Paris, France (*Chairperson*).
■ WORKING ON: documentation [A8], human person [B12], CCPPRB — French committees for the protection of human subjects of biomedical research [B32], ethics [B48], information dissemination [B71], CCNE advice (France) [B91], biology and human future [C43], medical ethics [C52], biology [E5], genetics [E6], molecular biology [E8], research team [E16], freezing [E29], clinical trials [F3], human experimentation [F11], research subjects [F12], healthy volunteers [F13], prenatal diagnosis [I3], sperm banks [I11], donors [I14], informed consent [J9], spousal consent [J12], parental consent [J13], consent forms [J14], medical secrecy [J22], referral and consultation [J24], physician patient relationship [J32], drugs [I34], contraception [M3], ovum [M12], sperm [M14], procreation [M15], wish of children [M16], reproduction [M17], fertility [M18], infertility [M19], semen donors [M28], in vitro fertilization [M29], FIV center [M30], artificial insemination [M31].
■ INTERESTED IN: documentation [A8], human person [B12], CCPPRB — French committees for the protection of human subjects of biomedical research [B32], information dissemination [B71], CCNE advice (France) [B91], biology and human future [C43], medical ethics [C52], freezing [E29], clinical trials [F3], prenatal diagnosis [I3], sperm banks [I11], aid children [L34], fertility [M18], infertility [M19], in vitro fertilization [M29], artificial insemination [M31], congenital defects [N23], genetic defects and hereditary diseases [Q2], biotechnology — genetic engineering [Q20], genetic counseling [Q36].

F 54 **DANCHIN** Antoine
Head of the Department of Regulation of Genetic Expression (Pasteur Institute). — Director of the Research Group on data processing and genome (CNRS).

CNRS Institut Pasteur
Département de biologie et génétique moléculaire
28, rue du Docteur-Roux
75724 Paris Cedex 15 (France)
✆ (33) 1 - 45 68 84 42 Fax: (33) 1 - 45 68 89 48 et 43 06 98 35

■ TEACHING: DESS Paris-I. — Ecole nationale de la Magistrature.
■ PUBLICATIONS: yes.
■ WORKING ON: book review [A6], editorial policies [A10], interdisciplinary communication [B70], information dissemination [B71], public debates [B75], philosophy of biology [C42], biology and human future [C43], consequences [C45], speciesism [C48], biological life [C49], historical aspects [C57], ancient history [C59], philosophy [C61], determinism [C62], mental processes [C69], recall [C72], love [C80], intelligence [C85], science [E2], history of science [E3], life sciences [E4], genetics [E6], microbiology [E7], molecular biology [E8], research [E14], research team [E16], research policy [E18], research design [E19], pilot projects [E20], industrial research [E22], scientific misconduct [E25], progress [E26], computers [E33], drug industry [E37], biological containment [E39], data protection [E52], human genome project [F8], biotechnology — genetic engineering [Q20], biotechnology [Q21], genetic intervention [Q22], genome mapping [Q23], cloning [Q24], genetically modified organisms [Q28], recombinant DNA research [Q31], DNA linkage [Q33], eugenics [Q39], gene pool [Q45].

■ INTERESTED IN: *idem, plus:* human rights [B3], human equality [B7], human body commercialization [B13], ethics committees [B22], CCNE — Comité Consultatif National d'Ethique (France) [B23], humanity heritage [B46], philosophical ethics [C2].

F 55 **DANTZER** Robert

Research Director for the Institut national de la recherche agronomique (INRA). — Interim Director of the INSERM U176 research unit.

INRA
INSERM U 176
rue Camille-Saint-Saëns
33077 Bordeaux cedex (France)
✆ (33) 56 00 02 50 Fax: (33) 56 98 90 29

■ AFFILIATIONS: Comité scientifique vétérinaire (*Member*).
■ TEACHING: teaching Animal Protection at Ecole nationale vétérinaire and at Ecole nationale supérieure d'agronomie.
■ PUBLICATIONS: yes.
■ WORKING ON: animal care committees [B21], ECC — European Community Commission [D63], animal experimentation [F10].
■ INTERESTED IN: philosophical ethics [C2], philosophy of biology [C42], biological life [C49], psychology [C68], emotions [C79], behavioral research [E21].

F 56 **DAUPHIN** Alain

Head of the Pharmacology Department of the Bichat-Claude-Bernard Hospital.

CCPPRB Groupe Hospitalier de Paris-Bichat/Claude-Bernard
Service de Pharmacie
46, rue Henri Huchard
75018 Paris (France)
✆ (33) 1 - 40 25 80 80

■ AFFILIATIONS: CCPPRB de Paris-Bichat/Claude-Bernard, Paris, France (*Chairperson*).
■ WORKING ON: fundamental rights of the individuals [B2], integrity [B8], freedom [B10], privacy [B11], human person [B12], human body commercialization [B13], protection of rights — involved institutions [B16], institutions involved in the protection of rights [B17], CNIL — Commission Nationale Informatique et Libertés (France) [B19], CCPPRB — French committees for the protection of human subjects of biomedical research [B32], patient advocacy [B42], state interest [B44], ethics [B48], codes of ethics [B49], competence [B51], bioethics [B89], codes of biomedical ethics [B90], CCNE advice (France) [B91], Bioethics bill, 1992 (France) [B98], professional ethics [C51], drug industry [E37], biomedical research [F2], clinical trials [F3], health services research [F4], nontherapeutic research [F5], therapeutic research [F6], cognitive research [F7], human genome project [F8], human experimentation [F11], research subjects [F12], healthy volunteers [F13], methodology of research and experimentation [F15], debriefing [F16], control groups [F17], random selection [F18], selection of subjects [F19], group of vulnerable subjects [F20], medical informatics [F21], patient information [J4], patient access [J5], parental notification [J6], disclosure [J7], consent to treatment [J8], informed consent [J9], presumed consent [J10], third party consent [J11], spousal consent [J12], parental consent [J13], consent forms [J14], investigator subject relationship [J29].

F 57 **DAUSSET** Jean

Chairperson of the Centre d'étude du polymorphisme humain (CEPH). — Chairperson of the Mouvement universel de la responsabilité scientifique (MURS).

Centre d'étude du polymorphisme humain (CEPH)
27, rue Juliette-Dodu
75010 Paris (France)
✆ (33) 1 - 42 49 98 50 Fax: (33) 1 - 42 06 16 19

■ AFFILIATIONS: Mouvement universel de la responsabilité scientifique — MURS, Paris, France (*Chairperson*).
■ PUBLICATIONS: yes.
■ WORKING ON: fundamental rights of the individuals [B2], human rights [B3], dignity [B5], integrity [B8], human person [B12], human body commercialization [B13], traffic of organs [B15], protection of rights — involved institutions [B16], ethics committees [B22], manifests and declarations concerning human rights [B33], Declaration on Human and Citizen Rights [B38], Universal Declaration of Human Rights [B39], humanity heritage [B46], philosophy of biology [C42], biology and human future [C43], future generations [C44], consequences [C45], speciesism [C48], professional ethics [C51], medical

ethics [C52], preventive medicine [G58], population genetics [Q43], gene pool [Q45], human genome [Q47], biotechnology — genetic engineering [Q20], genome mapping [Q23], DNA linkage [Q33].
■ INTERESTED IN: iatrogenic disease [N3], transmission [N15], diabetes [N22], genetic defects and hereditary diseases [Q2], genetic defects [Q3], dominant genetic conditions [Q8], Huntington's chorea [Q9], recessive genetic conditions [Q10], Duchenne muscular dystrophy [Q12], hereditary diseases [Q18], sex linked defects [Q19], behavioral genetics [Q34], eugenics [Q39], gene therapy [Q42].

F 58 **DAVID** Anne-Marie
General Supervisor, lung specialist, Pr.Chain's department (Pitié-Salpétrière hospital).

Hôpital de la Salpétrière Service de Neurologie
47, boulevard de l'Hôpital
75013 Paris (France)
✆ (33) 1 - 45 70 27 55

■ INTERESTED IN: human person [B12], European Convention of Bioethics [B97], patient care [G53], negligence [K20], central nervous system diseases [N43], brain diseases [N44], euthanasia [O15], active euthanasia [O16], Huntington's chorea [Q9].

F 59 **DAYDÉ** Marie-Claude
Independent private nurse.

Syndicat national des infirmiers et infirmières libéraux
15, rue Bayard
31000 Toulouse (France)
✆ (33) 61 62 12 12 Fax: (33) 61 63 76 11

■ TEACHING: teaches palliative care on a private basis within the Association soins palliatifs Midi-Pyrénées.
■ WORKING ON: human rights [B3], dignity [B5], Amnesty International [B18], CCPPRB — French committees for the protection of human subjects of biomedical research [B32], deontology [B59], humanitarian organizations [B66], nursing education [B79], nursing ethics [C53], prisoners [D12], torture [D94], home care [G55], nurse patient relationship [J30], palliative care [O8].
■ INTERESTED IN: patient advocacy [B42], ethics [B48], obligations of society [B53], obligations to society [B54], deontology [B59], code of deontology [B60], solidarity [B65], information dissemination [B71], nursing education [B79], European Convention of Bioethics [B97], social worth [C17], nursing ethics [C53], exclusion [D9], scarcity [D21], North-South relationships [D60], health services research [F4], nurses [G39], home care [G55], elderly [H23], consent to treatment [J8], physician nurse relationship [J31], suffering [N11], pain [N12].

F 60 **DE DINECHIN** Olivier
Delegate of the French Episcopate concerning moral questions . — Member of CCNE. — Member of the Conseil national du Sida (France).

Département d'éthique biomédicale du Centre Sèvres
12, rue d'Assas
75006 Paris (France)
✆ (33) 1 -44 39 48 33 Fax: (33) 1 - 40 49 01 92

■ AFFILIATIONS: Département d'éthique biomédicale du Centre Sèvres, Paris (*Member*).
■ TEACHING: Teaching at Centre Sèvres, Paris, département d'éthique biomédicale.
■ PUBLICATIONS: yes.
■ WORKING ON: Roman Catholic ethics [C23], theology [C41], legislation [D13], family members [D24], marital relationship [D29], family relationship [D31], advisory committees [D69], human genome project [F8], family planning [G8], population control [H22], prenatal diagnosis [I3], legitimacy [K21], child and family [L31], unwanted children [L35], handicapped children [L38], sexuality and procreation [M2], reproductive technologies [M26], abortion [M47], fetal therapy [M57], acquired immunodeficiency syndrome [N35], HIV seropositivity [N36], suicide [O19], medical genetics [Q35], population genetics [Q43].
■ INTERESTED IN: religion and religious ethics [C19], social issues [D2], political issues [D56], population [H19], sexuality and procreation — unborn child [M1], genetics and applications — biotechnology [Q1].

F 61 **DE ROSNAY** Joël
Director for the Advancement of International Relations of the Cité des Sciences et de l'Industrie.

Cité des Sciences et de l'Industrie
30, avenue Corentin-Cariou
75930 Paris Cedex 19 (France)
✆ (33) 1 - 40 05 73 38 et 72 68 Fax: (33) 1 - 40 05 73 44

■ AFFILIATIONS: Comité consultatif national d'éthique — CCNE, France (*Collaboration Bureau of the Cité des Sciences et de l'Industrie*).
■ WORKING ON: information sources, data bases [A1], respect of human dignity and human rights [B1], specific approaches to ethics [C1], social and political issues [D1], science, technology, methods [E1].

F 62 **DEFERT** Daniel

Sociology lecturer, Paris-VIII University. — Founder-Chairman of AIDES. — Member of sida. — Member of Commission mondiale du sida de l'OMS.

AIDES
247, rue de Belleville
75019 Paris (France)
✆ (33) 1 - 44 52 00 00 Fax: (33) 1 - 44 52 02 01

■ AFFILIATIONS: Conseil national du sida, Paris, France (*Member*).
■ TEACHING: teaching introductory courses of Medical Ethics and Bioethics at the University of Paris-VIII.
■ PUBLICATIONS: yes.
■ WORKING ON: human rights [B3], privacy [B11], patient advocacy [B42], humanitarian organizations [B66], health education [B77], counseling [C87], cohabitation [D27], ethnic groups [D36], minority groups [D42], stigmatization [D54], data protection [E52], clinical trials [F3], home care [G55], preventive medicine [G58], insurance [H5], mass screening [I6], mandatory screening [I7], patient association [J3], patient information [J4], consent forms [J14], patient participation [J16], living wills [J19], medical secrecy [J22], sexual behavior [M21], homosexuals [M23], acquired immunodeficiency syndrome [N35], HIV seropositivity [N36], palliative care [O8], withholding treatment [O10], allowing to die [O11], quality adjusted life years [O14], voluntary euthanasia [O18], drug abuse [P6], psychological stress [P17].
■ INTERESTED IN: patient advocacy [B42], health education [B77], codes of biomedical ethics [B90], ethicists [B92], bioethics movement [B95], Bioethics bill, 1992 (France) [B98], quality of life [C16], nursing ethics [C53], self concept [C83], counseling [C87], cohabitation [D27], minority groups [D42], stigmatization [D54], sociology of medicine [D55], research policy [E18], data protection [E52], clinical trials [F3], contact tracing [G12], health legislation [G14], preventive medicine [G58], life insurance [H8], health care costs [H9], compensation [H12], required request [I10], placebos [I36], immunotherapy [I41], patient association [J3], investigator subject relationship [J29], physician patient relationship [J32], health legislation [K52], therapeutic injunction [K64], condom [M4], wish of children [M16], sexual behavior [M21], homosexuals [M23], acquired immunodeficiency syndrome [N35], HIV seropositivity [N36], withholding treatment [O10], voluntary euthanasia [O18], drug abuse [P6], psychological stress [P17].

F 63 **DEHAN** M.

Head of the Neonatal Reanimation Department, Antoine-Béclère Hospital. — Professor of Paediatrics, Kremlin-Bicêtre University (Paris-XI), School of Medicine.

Service de réanimation néonatale
Hôpital Antoine-Béclère
157, rue de la Porte de Trivaux
92141 Clamart (France)
✆ (33) 1 - 45 37 46 14 Fax: (33) 1 - 45 37 49 25

■ AFFILIATIONS: Groupe d'étude en néonatologie et urgences pédiatriques de la région parisienne — GENEUP-RP, Paris, France (*Coordinator*).
■ TEACHING: Teaching Medicine to professionals, continuous training.
■ PUBLICATIONS: yes.
■ WORKING ON: terminal care [O7], palliative care [O8], life-sustaining treatment [O9], withholding treatment [O10], euthanasia [O15], resuscitation [O21], disease [N1], case studies [N4], patients [N5], physically [N8], suffering [N11], prognosis [N14], fetuses [M11], FIV center [M30], pregnancy and childbirth [M39], therapeutic abortion [M50], fetal development [M54], fetal therapy [M57], newborns [L29], prematurity [L30], Bioethics bill, 1992 (France) [B98].
■ INTERESTED IN: terminal care [O7], palliative care [O8], life-sustaining treatment [O9], withholding treatment [O10], euthanasia [O15], active euthanasia [O16], involuntary euthanasia [O17], resuscitation [O21], disease [N1], case studies [N4], patients [N5], physically [N8], suffering [N11], prognosis [N14], fetuses [M11], FIV center [M30], pregnancy and childbirth [M39], therapeutic

abortion [M50], fetal development [M54], fetal therapy [M57], newborns [L29], prematurity [L30], Bioethics bill, 1992 (France) [B98].

F 64 **DELAISI DE PARSEVAL** Geneviève
Psychoanalyste. — Bioethics Consultant at the Catholic University of Brussels (UCL).

118-122, rue de Vaugirard
75015 Paris (France)
✆ (33) 1 - 42 22 30 74

■ AFFILIATIONS: Groupe Franco-Québécois de Recherches en Bioéthique — B.O.F, Saint-Brice, France (*Chairperson*).
■ TEACHING: Centre for Human Bioethics (Melbourne) Australia, Centre de bioéthique de Montréal, Centre d'études bioéthiques de l'UCL (Brussels), Centre Sèvres (Paris).
■ PUBLICATIONS: yes.
■ WORKING ON: fundamental rights of the individuals [B2], human person [B12], Universal Declaration of Human Rights [B39], behavioral and mental disorders [P2], psychoactive drugs [P33], psychosurgery [P38], group therapy [P32], aggression [P14], dangerousness [P15], hyperkinesis [P16], volontary admission [P19], duration of commitment [P20], positive reinforcement [P28].
■ INTERESTED IN: human rights [B3], justice [B9], freedom [B10], codes of ethics [B49], moral policy [B55], uncontrolled information [B88], CCNE advice (France) [B91], quality of life [C16], social worth [C17], value of life [C18], future generations [C44], evolution [C47], psychology [C68], counseling [C87], family members [D24], couple [D26], sibling [D30], whites [D37], prenatal diagnosis [I3], amniocentesis [I4], sperm banks [I11], tissue banking [I12], donors [I14], anonymous donation [I17], directed donation [I18], blood donation [I20], bone marrow transplantation [I29], patient access [J5], patient access [J5], disclosure [J7], child development [L23], children's rights [L25], adoption [L32], aid children [L34], unwanted children [L35], brain [L6], livers [L7], bone marrow [L8], adolescents [L14], adults [L15], nutrition [L16], food [L17], animal organs [L18], mind [L20], infants [L28], newborns [L29], adoption [L32], abandoned (child) [L33], aid children [L34], disease and aged [L46].

F 65 **DELMAS-MARTY** Mireille
Professor, Panthéon- Sorbonne University (Paris I).

Institut de droit comparé de Paris
28, rue Saint-Guillaume
75007 Paris (France)
✆ (33) 1 - 42 22 35 90 Fax: (33) 1 - 42 84 03 27

■ PUBLICATIONS: yes.
■ WORKING ON: legislation and law [K34], rights [K42], constitutional law [K48], European law [K49], international law [K50], criminal law [K53], criminal code [K54].
■ INTERESTED IN: judicial action [K35], constitutional amendments [K36], authorization [K37], prohibition [K38], court decision [K40], guidelines [K55], law [K56], law enforcement [K57].

F 66 **DEMAISON** Michel
Teacher of Moral Theology and Biomedical Ethics, Catholic University of Lyon.

Centre de bioéthique
Université catholique de Lyon
25, rue du Plat
69288 Lyon Cedex 02 (France)
✆ (33) 72 32 50 22 Fax: (33) 72 32 50 19

■ AFFILIATIONS: Centre de bioéthique, Lyon Cedex 02, France (*Member*).
■ TEACHING: Teacher at the Faculty of Theology of the Catholic University of Lyon, Centre of Bioethics.
■ PUBLICATIONS: yes.
■ WORKING ON: human body commercialization [B13], ethical rules and principles [B47], codes of biomedical ethics [B90], philosophical ethics [C2], religion and religious ethics [C19], philosophy [C61], killing [D90], justifiable killing [D91], decision making [E54], evaluation [E61], human experimentation [F11], health care costs [H9], prenatal diagnosis [I3], biological specimens procurement, blood transfusion, organ transplantation [I8], consent to treatment [J8], wrongful life [K33], conscience clause [K63], child and family [L31], aid children [L34], sexuality and procreation [M2], reproductive technologies [M26], abortion [M47], determination of death [O2], terminal care [O7], euthanasia [O15], suicide [O19], attitudes to death [O20], biotechnology — genetic engineering [Q20], sex determination [Q26], sex preselection [Q30], genetic counseling [Q36], preimplantation genetic diagnosis [Q38], eugenics [Q39].

■ INTERESTED IN: bibliography [A4], documentation [A8], CCNE — Comité Consultatif National d'Ethique (France) [B23], CCNE advice (France) [B91], data protection [E52], social control of science [E64], human genome project [F8], population control [H22], physician's role [J26], suffering [N11], pain [N12], acquired immunodeficiency syndrome [N35].

F 67 **DESCAMPS-LATSCHA** Béatrice

Research Director at the INSERM. — Director of the INSERM Laboratory, U 25 (Necker Hospital).

INSERM U25
Hôpital Necker
161, rue de Sévres
75015 Paris (France)
✆ (33) 1- 42 73 89 29 Fax: (33) 1 - 43 06 23 88

■ AFFILIATIONS: Comité consultatif national d'éthique — CCNE, Paris, France (*Member*).
■ TEACHING: Coordinator of the postgraduate DEA Diploma of Medical Ethics (Necker). — Participates in the annual meetings of the Comité consultatif national d'éthique.
■ PUBLICATIONS: yes.
■ WORKING ON: CCNE — Comité Consultatif National d'Ethique (France) [B23], self determination [B4], dignity [B5], investigators [E15], selection of subjects [F19], donors [I14], organ donors [I15], transplantation [I22], transplant recipients [I23], organ transplantation [I24], organ donation [I25], organ procurement [I26], renal dialysis [I32], kidneys [L9], HIV seropositivity [N36], kidney diseases [N39].
■ INTERESTED IN: human body commercialization [B13], CCNE — Comité Consultatif National d'Ethique (France) [B23], dignity [B5], investigators [E15], cognitive research [F7], human genome project [F8], experimentation [F9], animal experimentation [F10], human experimentation [F11], medicine [G63], donors [I14], organ donors [I15], transplantation [I22], transplant recipients [I23], organ transplantation [I24], organ donation [I25], organ procurement [I26], renal dialysis [I32], kidneys [L9], acquired immunodeficiency syndrome [N35], HIV seropositivity [N36], kidney diseases [N39], terminal care [O7], euthanasia [O15], neurosciences — psychiatry [P1].

F 68 **DESNOS** Jacques

Professor Emeritus of the School of Medicine (Angers). — Former Head of the Otorhinolaryngology Department, and of Cervicofacial surgery, CHU (University hospital) of Angers.

Faculté de Médecine
Rue Haute-de-Reculée
49100 Angers (France)

■ AFFILIATIONS: Comité éthique, Angers, France (*Secretary*). — Comité de protection des personnes (CCPPRB), Angers, France (*Member*).
■ TEACHING: yes.
■ WORKING ON: philosophical ethics [C2], medical ethics [C52], medicine [G63], childhood difficulties, diseases, protection [L21].

F 69 **DIDIER** Elisabeth

M.D., Coordinator of the COMEDE (Medical Committee for Exiles).

Comité médical pour les exilés (COMEDE)
Hôpital Kremlin-Bicêtre
BP 31
94272 Le Kremlin Bicêtre Cedex (France)
✆ (33) 1 - 45 21 38 40 ou 45 21 38 41

■ AFFILIATIONS: Groupe franco-québecois de recherche en bioéthique, Saint-Brice/Forêt, France (*Member*).
■ PUBLICATIONS: yes.
■ WORKING ON: human rights [B3], human person [B12], humanitarian organizations [B66], exclusion [D9], aliens [D23], social problems [D52], social discrimination [D53], torture [D94], patient compliance [G49], mass screening [I6], mandatory screening [I7], law enforcement [K57], informed consent [J9], presumed consent [J10], patient participation [J16], expert testimony [K18], acquired immunodeficiency syndrome [N35], HIV seropositivity [N36], group therapy [P32].
■ INTERESTED IN: human rights [B3], human person [B12], humanitarian organizations [B66], exclusion [D9], aliens [D23], social problems [D52], social discrimination [D53], torture [D94], patient compliance [G49], mass screening [I6], mandatory screening [I7], law enforcement [K57], informed consent [J9], presumed consent [J10], patient participation [J16], expert testimony [K18], acquired immunodeficiency syndrome [N35], HIV seropositivity [N36], group therapy [P32], obligations of

society [B53], ethical review [B93], European Convention of Bioethics [B97], institutions involved in the protection of rights [B17], due process [B40], public advocacy [B41], morals [C12], moral development [C14], epidemiology [E9], health services research [F4], state medicine [G9], sexuality and procreation [M2], contraception [M3], procreation [M15], wish of children [M16], sexuality [M20], abortion [M47], abortion on demand [M52], terminal care [O7].

F 70 **DORAY** Bernard
Chargé de Mission (Chief State Officer) at the MIRE. — Psychiatrist public sector.

MIRE
1, Place Fontenoy
75007 Paris (France)

■ AFFILIATIONS: Association pour une instance des psychanalystes — APUI, Paris, France (*Member of the Board of Administration*).
■ PUBLICATIONS: yes.
■ WORKING ON: fundamental rights of the individuals [B2], human rights [B3], dignity [B5], human person [B12], reification [B14], traffic of organs [B15], ethics [B48], codes of ethics [B49], philosophy of biology [C42], professional ethics [C51], ethics and humanities [C54], biomedical research [F2], experimentation [F9], biological specimens procurement, blood transfusion, organ transplantation [I8], patients' rights [J2], patient association [J3], patient information [J4], informed consent [J9], presumed consent [J10], third party consent [J11], spousal consent [J12], treatment refusal [J15], inter-personal relationship [J27], legal personality [K2], personhood [K3], legally incompetent person [K4], neurosciences — psychiatry [P1], psychiatric technics [P24].
■ INTERESTED IN: ethics [B48], codes of ethics [B49], philosophy of biology [C42], professional ethics [C51], ethics and humanities [C54], biomedical research [F2], experimentation [F9], biological specimens procurement, blood transfusion, organ transplantation [I8], sexuality and procreation [M2], reproductive technologies [M26].

F 71 **DROUET** Nicolas
M.D. Clinician in Anaesthesiology-Reanimation.

CCPPRB Grenoble-I
Centre hospitalier universitaire
BP 217
38043 Grenoble Cedex 9 (France)
✆ (33) 76 76 57 83 Fax: (33) 76 76 51 77

■ AFFILIATIONS: Espace de réflexion éthique du diocèse de Grenoble (pastorale de la santé), Meylan, France (*Chairperson*).
■ TEACHING: initiation in ethical questions to nursing students in anaesthesiology (Ecole des infirmières anesthésistes), CHU (University hospital) of Grenoble.
■ WORKING ON: CCPPRB — French committees for the protection of human subjects of biomedical research [B32], interdisciplinary communication [B70], professional ethics [C51], medical ethics [C52], nursing ethics [C53], blood transfusions [I19], anesthesia [I31], surgery [I37], extraordinary treatment [I39], health care professionals, researchers and patient relationship [J1], physicians and researchers accountability [J23], inter-personal relationship [J27].
■ INTERESTED IN: bioethics [B89], situational ethics [C9], religious ethics [C21], professional ethics [C51], medical ethics [C52], nursing ethics [C53], biomedical research and experimentation [F1], organization of health care; facilities, manpower and services; health occupations [G1], health care [G45], donors [I14], blood transfusions [I19], transplantation [I22], anesthesia [I31], surgery [I37], extraordinary treatment [I39], health care professionals, researchers and patient relationship [J1], confidentiality [J20], physicians and researchers accountability [J23], inter-personal relationship [J27], accountability [K29], legal liability [K30], ghost surgery [K31], therapeutic risk [K32], death and resuscitation [O1], terminal care [O7], resuscitation [O21], blood donation [I20], blood substitutes [I21].

F 72 **DUBAS** Frédéric
Professor of Neurology. — Head of the Department of Neurology A.

CHU d'Angers
Service de neurologie A
4, rue Larrey
49033 Angers Cedex 01 (France)
✆ (33) 41 35 35 91

■ AFFILIATIONS: CCPPRB Pays-de-Loire, Angers Cedex, France (*Member*).
■ TEACHING: ethics of applied medicine, M.A. level School of Medicine, Angers.
■ WORKING ON: medical education [B80], internship and residency [B81], students [B82], curriculum [B83], universities [B84], medical ethics [C52], decision making [E54], decision analysis [E55], [E56], goals [E57], survey [E58], teaching methods [E59], standards [E60].
■ INTERESTED IN: medical education [B80], internship and residency [B81], students [B82], curriculum [B83], universities [B84], ethical review [B93], medical ethics [C52], nursing ethics [C53], central nervous system diseases [N43], involuntary euthanasia [O17], review [A2], book review [A6], publications [A9], human experimentation [F11], research subjects [F12], healthy volunteers [F13], clinical trials [F3], nontherapeutic research [F5], therapeutic research [F6], physician patient relationship [J32], physician nurse relationship [J31], nurse patient relationship [J30], investigator subject relationship [J29], patient information [J4], consent to treatment [J8], toxicity [E50].

F 73 **DUBOUIS** Louis
Professor of Law and Political Science, University of Aix-Marseille-III.

Université de droit, économie, sciences
Aix-Marseille-III
3, avenue Robert-Schumann
13626 Aix-en-Provence (France)
✆ (33) 42 17 28 00 et 42 17 29 72 Fax: (33) 42 64 03 96

■ AFFILIATIONS: Association française de droit de la santé — AFDS, Pessac Cedex, France (*Vice-Chairperson*). — CCPPRB Aix — CCPPRB, Aix-en-Provence, France (*Vice-Chairperson*).
■ TEACHING: Responsable et enseignant de l'option droit de la santé du DEA de droit social, faculté de droit de l'université d'Aix-Marseille-III.
■ PUBLICATIONS: yes.
■ WORKING ON: manifests and declarations concerning human rights [B33], European Convention on Human Rights [B36], medical ethics [C52], patients' rights [J2], jurisprudence — accountability [K8], European law [K49], international law [K50], medical law [K51], health legislation [K52].
■ INTERESTED IN: College of Physicians [B27], CCPPRB — French committees for the protection of human subjects of biomedical research [B32], manifests and declarations concerning human rights [B33], European Convention on Human Rights [B36], codes of ethics [B49], medical ethics [C52], French Council of State [D70], health policy [G13], organization of health care [G35], patients' rights [J2], confidentiality [J20], physicians and researchers accountability [J23], jurisprudence — accountability [K8], legislation and law [K34].

F 74 **DUMAS**
M.D. Clinician, Pharmacist.

CCPPRB, CHRU hôpital Pasteur
30, avenue de la Voie-Romaine
BP 069
06031 Nice Cedex (France)
✆ (33) 92 03 81 46

■ WORKING ON: CCPPRB — French committees for the protection of human subjects of biomedical research [B32], human person [B12], fundamental rights of the individuals [B2], patient advocacy [B42], codes of ethics [B49], moral obligations [B50], obligations of society [B53], respect [B63], clinical trials [F3], therapeutic research [F6], cognitive research [F7], hospital [G28], physicians [G40], medical staff [G41], pharmacists [G42], patient information [J4], parental notification [J6], informed consent [J9], consent forms [J14], patient participation [J16], right to treatment [J17].
■ INTERESTED IN: *idem*.

F 75 **DUPRET** Daniel
Scientific Director of Appligène.

Société Appligène
Parc d'Innovation
BP 72
67402 Illkirch cedex (France)
✆ (33) 88 67 22 67 Fax: (33) 88 67 19 45

■ PUBLICATIONS: yes.
■ WORKING ON: genetics [E6], molecular biology [E8], research policy [E18], industrial research [E22], human genome project [F8], prenatal diagnosis [I3], referral and consultation [J24], genetic defects and hereditary diseases [Q2], biotechnology — genetic engineering [Q20], biotechnology [Q21],

genetic intervention [Q22], genome mapping [Q23], sex determination [Q26], sex preselection [Q30], DNA fingerprinting [Q32], DNA linkage [Q33], genetic identity [Q44], human genome [Q47].
■ INTERESTED IN: genetics [E6], molecular biology [E8], research policy [E18], industrial research [E22], human genome project [F8], prenatal diagnosis [I3], referral and consultation [J24], genetic defects and hereditary diseases [Q2], biotechnology — genetic engineering [Q20], biotechnology [Q21], genetic intervention [Q22], genome mapping [Q23], sex determination [Q26], sex preselection [Q30], DNA fingerprinting [Q32], DNA linkage [Q33], genetic identity [Q44], human genome [Q47].

F 76 **DUPUY** Caroline

Fondation Marcel-Mérieux
17, rue Bourgelat
69002 Lyon (France)
✆ (33) 72 73 79 68 Fax: (33) 72 73 79 93

■ AFFILIATIONS: Centre de bioéthique, Institut Catholique (Lyon), Lyon, France (*Member of the Scientific Board*).
■ TEACHING: Organisation de colloques en bioéthique (Organisation of symposiums on bioethics).

F 77 **DURAND** Bernard

District Head of the Infant and Juvenile Psychiatric, Centre hospitalier intercommunal de Créteil.

Centre hospitalier intercommunal de Créteil (CHIC)
40, avenue de Verdun
94000 Créteil (France)
✆ (33) 1 - 48 98 77 64 Fax: (33) 1 - 48 99 53 98

■ PUBLICATIONS: yes.
■ WORKING ON: handicapped [D43], social interaction [D50], teaching methods [E59], mental health [G4], psychiatry [G69], rehabilitation [I42], parental notification [J6], presumed consent [J10], parental consent [J13], treatment refusal [J15], parent child relationship [J34], minors [K6], child's interest [K24], abandoned (child) [L33], handicapped children [L38], ill-treat children [L39], child abuse [L41], selective abortion [M49], therapeutic abortion [M50], mentally handicapped [P7], mentally retarded [P8], mentally ill [P9], hyperkinesis [P16].
■ INTERESTED IN: European Convention of Bioethics [B97], Bioethics bill, 1992 (France) [B98], future generations [C44], sibling [D30], family relationship [D31], stigmatization [D54], public policy [D77], community services [D78], handicapped [D43], social interaction [D50], teaching methods [E59], maternal health [G3], mental health [G4], health services [G25], residential facilities [G26], mental institutions [G29], medical evaluation [G51], psychiatry [G69], health care costs [H9], intracerebral fetal tissue transplantation [I28], rehabilitation [I42], parental notification [J6], presumed consent [J10], parental consent [J13], treatment refusal [J15], parent child relationship [J34], minors [K6], child's interest [K24], human body [L3], handicapped children [L38], ill-treat children [L39], child abuse [L41], selective abortion [M49], therapeutic abortion [M50], mentally handicapped [P7], mentally retarded [P8], mentally ill [P9], hyperkinesis [P16], behavior control [P25], eugenics [Q39].

F 78 **DUROUX** Pierre

University Professor (Paris-Sud Faculty). — Head of Clinical Department, M.D. Clinician, Paris Hospital.

CCPPRB Centre hospitalier de Bicêtre
Cour de Sibérie
78, rue du Général-Leclerc
94270 Le Kremlin-Bicêtre (France)
✆ (33) 1 - 45 21 28 46 Fax: (33) 1 - 45 21 21 45

■ AFFILIATIONS: CCPPRB Bicêtre, France (*Chairperson*).
■ WORKING ON: respect of human dignity and human rights [B1], biomedical research and experimentation [F1], organization of health care; facilities, manpower and services; health occupations [G1], health care, medical acts [I1], sexuality and procreation — unborn child [M1], disease [N1], neurosciences — psychiatry [P1], genetics and applications — biotechnology [Q1].

F 79 **DUYME** Michel

Research Director at the CNRS. — Member of the Conseil supérieur de l'adoption auprès du ministère de la Santé. — Member of the Comité opérationel d'éthique auprès des sciences de la vie, CNRS.

CNRS
UFR Biomédicale
45, rue des Saints-Pères
75270 Paris (France)
✆ (33) 1 - 42 86 21 48 Fax: (33) 1 - 42 86 21 50

■ AFFILIATIONS: Centre national de la recherche scientifique — CNRS, Paris, France (*Member*). — Sciences de la vie — SDV, Paris, France (*Member*).
■ PUBLICATIONS: yes.
■ WORKING ON: ethics committees [B22], ethical review [B93], sociobiology [C63], psychology [C68], behavior [C73], intelligence [C85], genetics [E6], epidemiology [E9], mortality [E13], behavioral research [E21], nontherapeutic research [F5], mental health [G4], public heaith [G5], third party consent [J11], professional deontology [J25], human development [L12], mind [L20], child development [L23], abandoned (child) [L33], aid children [L34], child abuse [L41], fetuses [M11], sperm [M14], procreation [M15], wish of children [M16], semen donors [M28], artificial insemination [M31], AID [M32], multiple pregnancy [M46], twinning [M55], central nervous system diseases [N43], alcohol abuse [P4], mentally handicapped [P7], behavior disorders [P13], hereditary diseases [Q18], biotechnology [Q21], DNA fingerprinting [Q32], genetic screening [Q37], eugenics [Q39], genetic identity [Q44], human genome [Q47].
■ INTERESTED IN: *idem*.

F 80 **ECHARD** Nicole

Ethnologist. — Research Director at the CNRS.

CNRS – Laboratoire de sociologie et de géographie africaines
EHESS
54, boulevard Raspail
75006 Paris (France)

■ AFFILIATIONS: Comité consultatif national d'éthique pour les sciences de la vie et de la santé — CCNE, Paris Cedex 13, France (*Member*).
■ WORKING ON: human person [B12], religious beliefs [C32], religious sciences [C39], social control [D3], social groups [D22], females [D35], ethnic groups [D36], cultural pluralism [D51], North-South relationships [D60], developing countries [H21], procreation [M15], reproduction [M17], fertility [M18].
■ INTERESTED IN: human rights [B3], women's rights [B6], Marxism [C67], sex offenses [D92], rape [D93], force feeding [D95], social control of science [E64], maternal welfare [G10], maternal life [G11], nurse midwives [G43], obstetrics and gynecology [G66], population growth [H20], population control [H22], elderly [H23], biological specimens procurement, blood transfusion, organ transplantation [I8], inter-personal relationship [J27], aged — problems related with aging [L45], reproductive technologies [M26], pregnancy and childbirth [M39], abortion [M47], fetal development [M54], venereal diseases [N31], death and resuscitation [O1], sex linked defects [Q19], sex preselection [Q30].

F 81 **EMILE** Jean

Professor of Neurology, Head of the Neurology B Department, CHU (University hospital) of Angers.

CCPPRB
Neurologie B, CHU d'Angers
49033 Angers Cedex 01 (France)
✆ (33) 41 35 44 25 Fax: (33) 41 35 35 89

■ AFFILIATIONS: CCPPRB-I des Pays de la Loire, Angers Cedex 01, France (*Chairperson*).
■ TEACHING: university level degree in Methodology and Clinical Pharmacology.
■ WORKING ON: inter-personal relationship [J27], human person [B12], traffic of organs [B15], ethics committees [B22], CCNE — Comité Consultatif National d'Ethique (France) [B23], office of science and technology assessment [B28], CCPPRB — French committees for the protection of human subjects of biomedical research [B32], Helsinki Declaration [B37], due process [B40], patient advocacy [B42], equal protection [B43], codes of ethics [B49], deontology [B59], medical education [B80], health and biology mediatisation [B85], codes of biomedical ethics [B90], bioethical issues [B96], quality of life [C16], social worth [C17], value of life [C18], utilitarianism [C8], philosophy of biology [C42], professional ethics [C51], medical ethics [C52], applied psychology [C86], counseling [C87], public opinion [D7], formal social control [D10], imprisonment [D11], prisoners [D12], research [E14], progress [E26], technology [E27], computerized file [E35], containment [E38], data protection [E52], biomedical research and experimentation [F1], medical informatics [F21], patients' rights [J2], informed consent [J9], parent child relationship [J34].
■ INTERESTED IN: bioethics [B89].

F 82 **FABREGUETTES**
Head of Pharmacology Department of the CHR regional Robert-Ballanger Hospital.

CCPPRB CHR Robert-Ballanger
Boulevard Robert-Ballanger
93602 Aulnay-sous-Bois (France)
✆ (33) 1 - 49 36 73 57

■ WORKING ON: medical ethics [C52], patient information [J4], parental notification [J6], parental consent [J13], medical secrecy [J22], health care costs [H9], price [H16].
■ INTERESTED IN: patient advocacy [B42], clinical trials [F3], religious ethics [C21], medical ethics [C52].

F 83 **FAGOT-LARGEAULT** Anne
Professor of the University of Paris-X (Philosophy). — Specialised M.D. at the Henri-Mondor Hospital, Créteil (Psychiatry).

Université Paris-X
Département de philosophie
92001 Nanterre Cedex (France)
✆ (33) 1 - 40 97 75 17

■ AFFILIATIONS: Comité consultatif national d'éthique — CCNE, Paris Cedex 13, France (*Member*).
■ TEACHING: University of Paris-X, at the Ecole doctorale de philosophie and at the Ecole normale supérieure (Biology Doctorate).
■ PUBLICATIONS: yes.
■ WORKING ON: human person [B12], human body commercialization [B13], ethical rules and principles [B47], philosophical ethics [C2], quality of life [C16], normality [C78], epidemiology [E9], statistics [E10], biomedical research and experimentation [F1], psychiatry [G69], biological specimens procurement, blood transfusion, organ transplantation [I8], sexuality and procreation — unborn child [M1], determination of death [O2], neurosciences — psychiatry [P1], medical genetics [Q35], gene pool [Q45].

F 84 **FARISY** Jacques
Chairman of the Conseil régional de l'Ordre des médecins de Poitou-Charentes.

Conseil régional de l'Ordre des médecins de Poitou-Charentes
17, boulevard Pont-Achard
86000 Poitiers (France)
✆ (33) 49 37 15 77 Fax: (33) 49 37 18 15

■ WORKING ON: deontological ethics [C6], medical ethics [C52].
■ INTERESTED IN: nursing ethics [C53].

F 85 **FAUCHER** Jacques
Catholic priest, Students' chaplain. — G.P.

Comité d'éthique pour les techniques de procréation artificielle
4, rue Mably
33000 Bordeaux (France)
✆ (33) 56 81 35 90

■ AFFILIATIONS: Comité d'éthique pour les techniques de procréation artificielle, Bordeaux, France (*Chairperson*).
■ TEACHING: Participates in the teaching at CHR hospital nursing schools. D.U. (university degree) of palliative care (University of Bordeaux-II). — Teacher at the Institut catholique of Toulouse (IERP).
■ PUBLICATIONS: yes.
■ WORKING ON: fundamental rights of the individuals [B2], protection of rights — involved institutions [B16], ethical rules and principles [B47], information — communication — media [B68], bioethics [B89], philosophical ethics [C2], religion and religious ethics [C19], philosophy of biology [C42], professional ethics [C51], ethics and humanities [C54], political issues [D56], methods [E53], evaluation [E61], organization of health care [G35], health care [G45], reproductive technologies [M26], death and resuscitation [O1].
■ INTERESTED IN: political issues [D56], protection of rights — involved institutions [B16], bioethics [B89], specific approaches to ethics [C1].

F 86 **FEDIDA** Pierre

University Professor of Psychopathology and Psychoanalysis. — Director of the doctorate education "Dysfunctioning Process". — Director of the Laboratory of Fundamental Psychopathology. — Member of the Association psychanalytique de France (*Former Chairperson*).

Laboratoire de psychopathologie fondamentale
UFR de sciences humaines cliniques
13, rue de Santeuil
75231 Paris Cedex 05 (France)
✆ (33) 1 - 45 87 41 02

■ AFFILIATIONS: Institut des sciences du vivant/ sciences de l'homme, Paris Cedex 05, France (*Co-director*).
■ TEACHING: M.A. level. — Postgraduate DEA Seminary.
■ WORKING ON: bibliography [A4], information centers [A5], book review [A6], human rights [B3], justice [B9], privacy [B11], human person [B12], human body commercialization [B13], reification [B14], traffic of organs [B15], Amnesty International [B18], ethics committees [B22], Council of Europe [B25], CCPPRB — French committees for the protection of human subjects of biomedical research [B32], Nuremberg Code [B35], Helsinki Declaration [B37], humanity heritage [B46], codes of ethics [B49], deontology [B59], education [B76], biomedical ethics education [B78], universities [B84], health and biology mediatisation [B85], bioethics [B89], philosophical ethics [C2], religion and religious ethics [C19], philosophy of biology [C42], professional ethics [C51], ethics and humanities [C54], social and political issues [D1], science [E2], life sciences [E4], research [E14], technology [E27], containment [E38], methods [E53], biomedical research [F2], experimentation [F9], methodology of research and experimentation [F15], organization of health care; facilities, manpower and services; health occupations [G1], health care [G45], health occupations [G61], diagnosis [I2], biological specimens procurement, blood transfusion, organ transplantation [I8], treatment [I30], health care professionals, researchers and patient relationship [J1], law, legislation and jurisprudence [K1], jurisprudence — accountability [K8], legislation and law [K34], stages of life — problems proper to childhood and elderly [L1], sexuality and procreation — unborn child [M1], disease [N1], neurosciences — psychiatry [P1], genetics and applications — biotechnology [Q1].
■ INTERESTED IN: health economics, population characteristics [H1], stages of life — problems proper to childhood and elderly [L1], sexuality and procreation — unborn child [M1], death and resuscitation [O1], genetics and applications — biotechnology [Q1].

F 87 **FELLOUS** Marc

Professor of Human Genetics at the University of Paris-VII. — Head of the Human Genetics Department at the Institut Pasteur (Paris). — Treasurer of the Société française de génétique.

Institut Pasteur
25, rue du Docteur-Roux
75724 Paris Cedex 15 (France)
✆ (33) 1 - 45 68 85 37 Fax: (33) 1 - 45 68 86 39

■ PUBLICATIONS: yes.
■ WORKING ON: reproductive organs and embryonic structures [M9], embryos [M10], fetuses [M11], reproduction [M17], infertility [M19], sexuality [M20], genetic defects and hereditary diseases [Q2], genetic defects [Q3], XYY karyotype [Q5], sex linked defects [Q19], transgenic animals [Q29], sex determination [Q26], sex preselection [Q30], preimplantation genetic diagnosis [Q38].
■ INTERESTED IN: information — communication — media [B68], universities [B84], evolution [C47], sociobiology [C63], diagnosis [I2], prenatal diagnosis [I3], mass screening [I6], mandatory screening [I7], negative eugenics [Q40], positive eugenics [Q41], gene therapy [Q42], genetic identity [Q44], human genome [Q47].

F 88 **FOLSCHEID** Dominique

Professor of Philosophy, University of Rennes-I.

Université de Rennes-I
UFR de philosophie
Campus de Beaulieu
35042 Rennes Cedex (France)
✆ (33) 99 28 63 02

■ AFFILIATIONS: Société française de réflexion bioéthique — SFRB, Paris, France (*Vice-Chairman and co-founder*).
■ TEACHING: Postgraduate DEA seminar at the University. — Participating in colloquiums and training

courses (within the university in general).
■ PUBLICATIONS: yes.
■ WORKING ON: respect of human dignity and human rights [B1], ethical rules and principles [B47], ethics [B48], codes of ethics [B49], moral obligations [B50], competence [B51], professional competence [B52], obligations of society [B53], obligations to society [B54], moral policy [B55], altruism [B56], beneficence [B57], authoritarianism [B58], deontology [B59], code of deontology [B60], morality [B61], paternalism [B62], respect [B63], virtues [B64], solidarity [B65], biomedical ethics education [B78], ethicists [B92], ethical review [B93], history of biomedical ethics [B94], bioethics movement [B95], bioethical issues [B96], European Convention of Bioethics [B97], Bioethics bill, 1992 (France) [B98], philosophical ethics [C2], normative ethics [C5], deontological ethics [C6], teleological ethics [C7], morals [C12], conscience [C13], moral development [C14], values [C15], quality of life [C16], social worth [C17], value of life [C18], Christian ethics [C22], natural law [C24], Islamic ethics [C28], Roman catholicism [C31], religion [C36], theology [C41], philosophy of biology [C42], biology and human future [C43], future generations [C44], consequences [C45], double effect [C46], evolution [C47], speciesism [C48], biological life [C49], medical ethics [C52], humanities [C55], Marxism [C67], science, technology, methods [E1], investigators [E15], scientific misconduct [E25], progress [E26], technology [E27], containment [E38], evaluation [E61], biological specimens procurement, blood transfusion, organ transplantation [I8], physicians and researchers accountability [J23], law, legislation and jurisprudence [K1], sexuality and procreation — unborn child [M1], reproductive technologies [M26], ovum donors [M27], semen donors [M28], in vitro fertilization [M29], FIV center [M30], artificial insemination [M31], AID [M32], AIH [M33], excess embryos [M34], host mothers [M35], embryo transfer [M36], GIFT (Gamete Intrafallopian Transfer) [M37], embryo donation [M38], fetal development [M54], terminal care [O7], euthanasia [O15], suicide [O19], attitudes to death [O20], genetics and applications — biotechnology [Q1], biotechnology — genetic engineering [Q20], medical genetics [Q35].
■ INTERESTED IN: *idem*.

F 89 **FRYDMAN** René
Head of the Gynaecology and Obstetrics Department, Antoine-Béclère Hospital.

Hôpital Antoine-Béclère
Service de gynécologie-obstétrique
157, rue de la Porte de Trivaux
92141 Clamart (France)
✆ (33) 1 - 46 31 38 30

■ WORKING ON: obstetrics and gynecology [G66], diagnosis [I2], prenatal diagnosis [I3], amniocentesis [I4], chorionic villus sampling [I5], mass screening [I6], immunologic contraception [M5], procreation [M15], ovum donors [M27], in vitro fertilization [M29], FIV center [M30], excess embryos [M34], embryo donation [M38], pregnancy and childbirth [M39], childbirth [M40], caesarean section [M41], pregnant women [M42], mother fetus relationship [M43], pregnancy [M44], pregnancy in adolescence [M45], multiple pregnancy [M46], selective abortion [M49], mifepristone [M53], fetal development [M54], twinning [M55], fetal therapy [M57], sex determination [Q26], sex preselection [Q30], preimplantation genetic diagnosis [Q38], eugenics [Q39], gene therapy [Q42], gene therapy [Q42].
■ INTERESTED IN: embryos [M10], wish of children [M16], twinning [M55], fetal therapy [M57], eugenics [Q39].

F 90 **GARIN** Jean-Paul
Member of the organisation which regulates university degree attestations (AEU) in Philosophy and Medical Ethics, School of Medicine, Claude-Bernard University, Lyon.

Faculté de médecine
Laboratoire de parasitologie
8, avenue Rockefeller
69008 Lyon (France)
✆ (33) 78 77 70 16

■ TEACHING: AEU (university attestation) in Philosophy and Medical Ethics, School of Medicine, Claude-Bernard University, Lyon.

F 91 **GASTARD** Joseph
Chairperson of CCPPRB, Rennes.

CCPPRB
2, rue de l'Hôtel-Dieu

35033 Rennes Cedex (France)
✆ (33) 99 28 41 35 Fax: (33) 99 28 41 36

■ AFFILIATIONS: CCPPRB de Rennes, Rennes Cédex, France (*Chairperson*).

F 92 **GERVAIS** Pierre

Full Professor of Forensic Medicine and Toxicology.

Hôpital Fernand-Vidal
200, rue du faubourg Saint-Denis
75475 Paris Cedex 10 (France)
✆ (33) 1 - 40 05 41 91 Fax: (33) 1 - 40 37 92 20

■ AFFILIATIONS: Comité d'éthique des hôpitaux AP-HP, Paris, France (*Chairperson*).
■ TEACHING: M.A. level teaching in Medical studies.
■ PUBLICATIONS: yes.
■ WORKING ON: human person [B12], human body commercialization [B13], professional deontology organs [B26], CCPPRB — French committees for the protection of human subjects of biomedical research [B32], professional competence [B52], deontology [B59], code of deontology [B60], education [B76], medical education [B80], medical ethics [C52], nursing ethics [C53], biology [E5], genetics [E6], preservation [E30], cryonic suspension [E31], drug industry [E37], quality of environment [E41], ecology [E43], occupational medicine [E46], pollution [E47], toxicity [E50], clinical trials [F3], nontherapeutic research [F5], therapeutic research [F6], human experimentation [F11], health care delivery [G48], preventive medicine [G58], medicine [G63], occupational medicine [G65], prenatal diagnosis [I3], biological specimens procurement, blood transfusion, organ transplantation [I8], immunotherapy [I41], patient information [J4], consent to treatment [J8], informed consent [J9], medical secrecy [J22], professional deontology [J25], legally incompetent person [K4], potentiality of personhood [K5], malpractice [K19], legislation and law [K34], ill-treat children [L39], abortion on demand [M52], contraception [M3], iatrogenic disease [N3], acquired immunodeficiency syndrome [N35], occupational diseases [N48], terminal care [O7], suicide [O19], substance dependence [P3], outpatient commitment [P18], hereditary diseases [Q18], genetic screening [Q37].
■ INTERESTED IN: human person [B12], human body commercialization [B13], professional deontology organs [B26], CCPPRB — French committees for the protection of human subjects of biomedical research [B32], professional competence [B52], deontology [B59], code of deontology [B60], education [B76], medical education [B80], medical ethics [C52], nursing ethics [C53], biology [E5], genetics [E6], preservation [E30], cryonic suspension [E31], drug industry [E37], quality of environment [E41], ecology [E43], occupational medicine [E46], pollution [E47], toxicity [E50], clinical trials [F3], nontherapeutic research [F5], therapeutic research [F6], human experimentation [F11], health care delivery [G48], preventive medicine [G58], medicine [G63], occupational medicine [G65], prenatal diagnosis [I3], biological specimens procurement, blood transfusion, organ transplantation [I8], immunotherapy [I41], patient information [J4], consent to treatment [J8], informed consent [J9], medical secrecy [J22], professional deontology [J25], legally incompetent person [K4], potentiality of personhood [K5], malpractice [K19], legislation and law [K34], ill-treat children [L39], abortion on demand [M52], contraception [M3], iatrogenic disease [N3], acquired immunodeficiency syndrome [N35], occupational diseases [N48], terminal care [O7], suicide [O19], substance dependence [P3], outpatient commitment [P18], hereditary diseases [Q18], genetic screening [Q37].

F 93 **GIL** Roger

Professor of Neurology, Head of the Neurology Department of the CHU (University hospital) of Poitiers.

CHU La Milétrie
Service de neurologie
86021 Poitiers (France)
✆ (33) 49 44 44 46 Fax: (33) 49 44 39 75

■ AFFILIATIONS: Comité d'éthique du CHU de Poitiers, Poitiers Cedex, France (*Chairperson*).
■ TEACHING: responsible for the organisation of an optional Bioethics certificate delivered by the School of Medicine of Poitiers (for M.A. level medical students, DC2-DC3).
■ PUBLICATIONS: yes.
■ WORKING ON: human person [B12], protection of rights — involved institutions [B16], bioethics [B89].
■ INTERESTED IN: human person [B12], protection of rights — involved institutions [B16], ethical rules and principles [B47], bioethics [B89], specific approaches to ethics [C1], philosophical ethics [C2], religion and religious ethics [C19], political issues [D56], biomedical research and experimentation

[F1], legislation and law [K34], neurosciences — psychiatry [P1], genetics and applications — biotechnology [Q1].

F 94 **GILGENKRANTZ** Simone

Professor of Medical Genetics, School of Medicine of Nancy. — Head of the Medical Genetics Department at the CHRU Transfusion Hospital.

Laboratoire de génétique
CRTS
Avenue de Bourgogne
54511 Vandœuvre-les-Nancy Cedex (France)
✆ (33) 83 44 62 62 Fax: (33) 83 44 60 46

■ TEACHING: genetics from training in ethical questions related to prenatal diagnosis, biology and human reproduction.
■ INTERESTED IN: genetics and applications — biotechnology [Q1], abortion [M47], research [E14], patients' rights [J2].

F 95 **GIRAUD** Francis

University Professor of Genetics. — M.D. Clinician, Head of Paediatrics and Medical Genetics. — Director of the Centre for Teaching and Research in Medical Genetics and of the INSERM, research unit 242. — Chairperson of the Association régionale d'étude et de dépistage des encéphalopathies, malformations et affections génétiques.

CCPPRB
Hôpital d'adultes de la Timone
13385 Marseille Cedex 5 (France)
✆ (33) 91 47 71 00 Fax: (33) 91 47 71 94

■ AFFILIATIONS: Comité d'éthique de l'Association française de dépistage et de prévention des handicaps de l'enfant, Marseille Cedex 5, France (*Member*).
■ TEACHING: Diplôme universitaire d'éthique médicale (faculté de médecine de Marseille).
■ PUBLICATIONS: yes.
■ WORKING ON: science [E2], life sciences [E4], biomedical research [F2], public health [G5], preventive medicine [G58], family practice [G60], diagnosis [I2], treatment [I30], confidentiality [J20], sexuality and procreation — unborn child [M1], congenital defects [N23], genetics and applications — biotechnology [Q1].
■ INTERESTED IN: bioethics [B89], law, legislation and jurisprudence [K1].

F 96 **GOOSSENS** Michel

Head of the Laboratory of Antenatal Diagnosis and of the Laboratory of Biochemistry (Henri-Mondor Hospital). — Professor of Human Genetics.

Hôpital Henri-Mondor
51, avenue du Maréchal-de-Lattre-de-Tassigny
94010 Créteil (France)
✆ (33) 1 - 49 81 28 61 Fax: (33) 1 - 49 81 28 42

■ TEACHING: Medical Genetics (University of Paris-XII).
■ WORKING ON: medical education [B80], philosophy of biology [C42], genetics [E6], molecular biology [E8], research [E14], biomedical technologies [E28], records [E34], human genome project [F8], diagnosis [I2], patient association [J3], genetic defects and hereditary diseases [Q2], genetic defects [Q3], sickle cell anemia [Q11], hemophilia [Q13], cystic fibrosis [Q15], thalassemia [Q17], hereditary diseases [Q18], sex linked defects [Q19], genetic intervention [Q22], genome mapping [Q23], cloning [Q24], clones [Q25], sex determination [Q26], recombinant DNA research [Q31], DNA linkage [Q33], genetic counseling [Q36], genetic screening [Q37], gene therapy [Q42], carriers [Q46], human genome [Q47].
■ INTERESTED IN: human genome project [F8], mass screening [I6], genetics and applications — biotechnology [Q1].

F 97 **GOUYON** Pierre- Henri

Professor at the University of Paris-Sud. — Professor-Consultant for the Institut national agronomique, Paris-Grignon. — Head of the ESV Laboratory (URA 1492 CNRS).

Université de Paris-Sud
Bât 362

91405 Orsay Cedex (France)
✆ (33) (1) 69 41 72 22 Fax: (33) (1) 69 41 72 38

■ AFFILIATIONS: Comité opérationnel d'éthique des sciences de la vie, CNRS, Paris (*Member*).
■ TEACHING: Man-nature relationship course at the University of Paris-Sud, Agro (INAPG). — Concept of race, eugenism. — Ecology. — Risks associated with biotechnologies. — Diverse adult education programmes.
■ PUBLICATIONS: yes.
■ WORKING ON: ethics committees [B22], humanity heritage [B46], education [B76], universities [B84], health and biology mediatisation [B85], quality of life [C16], evolution [C47], speciesism [C48], biological life [C49], history [C56], historical aspects [C57], history before the 20th century [C58], determinism [C62], sociobiology [C63], behavior [C73], history of science [E3], life sciences [E4], quality of environment [E41], ecology [E43], genetically modified organisms [Q28], population genetics [Q43], genetics [E6], epidemiology [E9].
■ INTERESTED IN: deontology [B59], eugenics [Q39], sex determination [Q26], ecology [E43].

F 98 GRIMALDI Pierre

Otorhinolaryngologist (retired). — Regional Deputy-Counsellor in medicine.

Conseil départemental de l'Ordre des médecins des Pyrénées-Atlantiques
Complexe de la République
Rue Carnot
64000 Pau (France)
✆ (33) 59 27 85 65 Fax: (33) 59 83 79 88

■ AFFILIATIONS: Groupe d'études biomédicales des Pyrénées-Atlantiques, Conseil départemental de l'Ordre des médecins des Pyrénées-Atlantiques, Pau, France (*Co-founder, Chairperson*).
■ TEACHING: University conferences. — Training physicians. — Teaching in nursing schools.
■ PUBLICATIONS: yes.
■ WORKING ON: human person [B12], human body commercialization [B13], traffic of organs [B15], ethics committees [B22], CCNE — Comité Consultatif National d'Ethique (France) [B23], professional deontology organs [B26], College of Physicians [B27], office of science and technology assessment [B28], deontology [B59], education [B76], biomedical ethics education [B78], nursing education [B79], codes of biomedical ethics [B90], history of biomedical ethics [B94], bioethical issues [B96], philosophical ethics [C2], religious ethics [C21], Christian ethics [C22], Protestant ethics [C27], Islamic ethics [C28], Jewish ethics [C29], scriptural interpretation [C40], future generations [C44], medical ethics [C52], genetics [E6], computerized file [E35], biological containment [E39], biomedical research [F2], health occupations [G61], medicine [G63], medical secrecy [J22], physicians and researchers accountability [J23], reproductive organs and embryonic structures [M9], procreation [M15], wish of children [M16], reproduction [M17], reproductive technologies [M26], death [O5], brain death [O6], palliative care [O8], life-sustaining treatment [O9], withholding treatment [O10], allowing to die [O11], terminally ill [O12], prolongation of life [O13], quality adjusted life years [O14], euthanasia [O15], suicide [O19], attitudes to death [O20].
■ INTERESTED IN: review [A2], fundamental rights of the individuals [B2], CNIL — Commission Nationale Informatique et Libertés (France) [B19], International Charter of Human Rights [B34], European Convention on Human Rights [B36], humanity heritage [B46], press conference [B73], CCNE advice (France) [B91], ethical review [B93], European Convention of Bioethics [B97], Christian science [C37], Jehovah's witnesses [C38], biology and human future [C43], philosophy [C61], public policy [D77], justifiable killing [D91], sex offenses [D92], rape [D93], torture [D94], force feeding [D95], history of science [E3], review committees [E42], biomedical research [F2], experimentation [F9], methodology of research and experimentation [F15], health occupations [G61], medicine [G63], diagnosis [I2], biological specimens procurement, blood transfusion, organ transplantation [I8], patients' rights [J2], medical secrecy [J22], physicians and researchers accountability [J23], aged — problems related with aging [L45], reproductive technologies [M26], death [O5], brain death [O6], palliative care [O8], withholding treatment [O10], allowing to die [O11], terminally ill [O12], prolongation of life [O13], quality adjusted life years [O14], euthanasia [O15], suicide [O19], attitudes to death [O20], biotechnology — genetic engineering [Q20], behavioral genetics [Q34], medical genetics [Q35], population genetics [Q43].

F 99 GROS François

Permanent Secretary of the Académie des sciences. — Professor at the Collège de France and Institut Pasteur.

Comité consultatif national d'éthique (CCNE)
INSERM
101, rue de tolbiac
75000 Paris (France)

■ AFFILIATIONS: Comité Consultatif national d'éthique — CCNE, Paris, France (*Member*).
■ TEACHING: Conferences on bioethics (Association Descartes, MURS, etc.).
■ PUBLICATIONS: yes.
■ WORKING ON: philosophy of biology [C42], biomedical research and experimentation [F1], genetics and applications — biotechnology [Q1].
■ INTERESTED IN: specific approaches to ethics [C1].

F 100 **GUYOTAT** Jean

Professor Emeritus of Psychiatry, Department UFR Alexis-Carrel, Claude-Bernard University (Lyon).

Université Claude-Bernard Lyon-I
8, avenue Rockefeller
69008 Lyon (France)
✆ (33) 78 25 23 51

■ PUBLICATIONS: yes.
■ WORKING ON: nursing education [B79], medical education [B80], Roman catholicism [C31], mental institutions [G29], mind [L20], adoption [L32], aid children [L34], infertility [M19], AID [M32].
■ INTERESTED IN: placebos [I36], Roman catholicism [C31], terminal care [O7], infertility [M19], DNA fingerprinting [Q32].

F 101 **HATRON**

Professor of Internal Medicine.

CCPPRB Hôpital Huriez
Département de pharmacologie
2, avenue Oscar-Lambret
59037 Lille cedex (France)
✆ (33) 20 44 59 62

■ AFFILIATIONS: CCPPRB Hôpital Huriez, Lille Cédex, France (*Chairperson*).

F 102 **HAZEBROUCQ** Georges

Professor of Clinical Pharmacology, Paris-Sud University, Chatenay-Malabry Centre. — Chief-Pharmacist of the Cochin Hospital Complex.

CCPPRB Paris-Cochin
27, rue du faubourg Saint-Jacques
75014 Paris (France)
✆ (33) 1 - 42 34 16 09 and 44 41 23 11 Fax: (33) 1 - 40 51 86 11 and 44 41 23 12

■ WORKING ON: CCPPRB — French committees for the protection of human subjects of biomedical research [B32], education [B76], drug industry [E37], clinical trials [F3], healthy volunteers [F13], methodology of research and experimentation [F15], patient compliance [G49], life-sustaining treatment [O9].
■ INTERESTED IN: human genome project [F8], withholding treatment [O10], attitudes to death [O20], quality adjusted life years [O14].

F 103 **HÉRITIER-AUGÉ** Françoise

Professor at the Collège de France. — Director of studies, EHESS, Director of the Laboratory of social anthropology. — Chairperson of the Conseil national du sida.

Conseil national du sida
7, rue d'Anjou
75008 Paris (France)
✆ (33) 1 - 40 07 01 06 Fax: (33) 1 - 40 07 01 12

■ TEACHING: Course Director. — Head of a training programme (the laboratory of social anthropology is used as a research laboratory by postgraduate DEA students in medical bioethics supervised by Pr. C. Hervé, Paris V University).
■ PUBLICATIONS: yes.
■ WORKING ON: respect of human dignity and human rights [B1], fundamental rights of the individuals [B2], human rights [B3], dignity [B5], privacy [B11], human person [B12], ethics committees [B22], CCNE — Comité Consultatif National d'Ethique (France) [B23], professional deontology organs [B26], College of Physicians [B27], ethical rules and principles [B47], ethics [B48], codes of ethics [B49], deontology [B59], solidarity [B65], information — communication — media [B68], health and

biology mediatisation [B85], uncontrolled information [B88], CCNE advice (France) [B91], social and political issues [D1], social impact [D49], social discrimination [D53], stigmatization [D54], health care, medical acts [I1], mass screening [I6], mandatory screening [I7], health care professionals, researchers and patient relationship [J1], patients' rights [J2], consent to treatment [J8], informed consent [J9], confidentiality [J20], physicians and researchers accountability [J23], professional deontology [J25], communicable diseases [N27], acquired immunodeficiency syndrome [N35], HIV seropositivity [N36].

■ INTERESTED IN: respect of human dignity and human rights [B1], fundamental rights of the individuals [B2], human rights [B3], dignity [B5], privacy [B11], human person [B12], ethics committees [B22], CCNE — Comité Consultatif National d'Ethique (France) [B23], professional deontology organs [B26], College of Physicians [B27], ethical rules and principles [B47], ethics [B48], codes of ethics [B49], deontology [B59], solidarity [B65], information — communication — media [B68], health and biology mediatisation [B85], uncontrolled information [B88], CCNE advice (France) [B91], social and political issues [D1], social impact [D49], social discrimination [D53], stigmatization [D54], health care, medical acts [I1], mass screening [I6], mandatory screening [I7], health care professionals, researchers and patient relationship [J1], patients' rights [J2], consent to treatment [J8], informed consent [J9], confidentiality [J20], physicians and researchers accountability [J23], professional deontology [J25], communicable diseases [N27], acquired immunodeficiency syndrome [N35], HIV seropositivity [N36].

F 104 **HEROLD** Monique

Biologist, Doctor of Science. — Investigation Director (in a private centre) on new pharmaceutical product licences. — Toxicology and Pharmacology Expert, ministère de la Santé.

Commission santé/bioéthique
Ligue des droits de l'homme
27, rue Jean-Dolent
75014 Paris (France)
✆ (33) 1 - 47 07 56 35

■ TEACHING: within the Ligue des droits de l'homme.
■ WORKING ON: fundamental rights of the individuals [B2], human rights [B3], medical ethics [C52], social issues [D2], social control [D3].
■ INTERESTED IN: European Convention of Bioethics [B97], biology and human future [C43], behavioral research [E21], computerized file [E35], biological containment [E39], data protection [E52], clinical trials [F3], cognitive research [F7].

F 105 **HERVÉ** Christian

Director of the University Laboratory of Medical Ethics, Necker School of Medicine (René-Descartes University). — Co-director of the postgraduate DEA course in Medical Ethics.

Faculté de médecine Necker
Laboratoire d'éthique médicale
156, rue de Vaugirard
75730 Paris Cedex (France)
✆ (33) 1 - 40 61 56 52 Fax: (33) 1 - 40 61 55 88

■ AFFILIATIONS: Comité consultatif de protection des personnes dans la recherche biomédicale — CCPPRB, Paris Cedex, France (*Permanent Expert*). — Comité d'Ethique (*Scientific Executive Secretary*).
■ TEACHING: Lectures and training course for undergraduate medical students, Necker School of Medicine. — Interdisciplinary conferences, courses and interviews (postgraduate level).
■ PUBLICATIONS: yes.

F 106 **HIRSCH** Emmanuel

Director of Studies.

Institut de formation et de recherche en éthique médicale
65, boulevard du Château
92200 Neuilly (France)
✆ (33) 1 - 47 47 24 97

■ TEACHING: Annual seminar on medical ethics.
■ PUBLICATIONS: yes.
■ WORKING ON: respect of human dignity and human rights [B1], protection of rights — involved institutions [B16], ethical rules and principles [B47], universities [B84], specific approaches to ethics [C1], philosophy of biology [C42], professional ethics [C51], ethics and humanities [C54], political

issues [D56], biomedical research and experimentation [F1], organization of health care; facilities, manpower and services; health occupations [G1], health care professionals, researchers and patient relationship [J1], stages of life — problems proper to childhood and elderly [L1], death and resuscitation [O1], genetics and applications — biotechnology [Q1].
■ INTERESTED IN: *idem.*

F 107 **HOERNI** Bernard
Head of the Department of Medicine, Anti-Cancer Centre of Bordeaux.

CCPPRB de Bordeaux-A
Centre de pharmacovigilance
CHR, Bat. 1A, Hôpital Pellegrin
33076 Bordeaux Cedex (France)
✆ (33) 57 81 76 07

■ TEACHING: monthly debates on ethics, University of Bordeaux-II.
■ PUBLICATIONS: yes.

F 108 **HUBER** Gérard
Head of the Department "Sciences et Technologies du vivant", Association Descartes.
Member of the Comité opérationnel d'éthique des Sciences de la vie at the CNRS.

Association Descartes
1, rue Descartes
75231 Paris Cedex 05 (France)
✆ (33) 1 - 46 34 32 96 Fax: (33) 1 - 46 34 33 40

■ AFFILIATIONS: Association Descartes, Paris, France (*Head of Department*).
■ TEACHING: University of Jussieu (Paris-VII). — Necker School of Medicine.
■ PUBLICATIONS: yes.
■ WORKING ON: ethical rules and principles [B47], bioethics [B89], information — communication — media [B68], ethics and humanities [C54], biomedical research [F2], sexuality and procreation [M2], behavioral and mental disorders [P2].
■ INTERESTED IN: dignity [B5], public advocacy [B41], patient advocacy [B42], common good [B45], respect [B63], metaethics [C11], future generations [C44], ethics and humanities [C54], family relationship [D31], biological containment [E39], research subjects [F12], healthy volunteers [F13], population control [H22], patients' rights [J2], legal personality [K2], legislation and law [K34], mind [L20], childhood difficulties, diseases, protection [L21], life extension [L47], sexual behavior [M21], fetal therapy [M57], brain death [O6], attitudes to death [O20], behavioral and mental disorders [P2], sex preselection [Q30], DNA linkage [Q33], gene therapy [Q42], human genome [Q47].

F 109 **HURIET** Claude
Head of the Nephrology Department , CHU (University hospital) of Nancy. — School of Medicine of Nancy. — Senator of Meurthe-et-Moselle, Deputy Chairperson of the Conseil général. — Mayor of Vroncourt.

Palais du Luxembourg
15, rue de Vaugirard
75006 Paris (France)
✆ (33) 1 - 42 34 24 87 Fax: (33) 1 - 42 34 39 57

■ AFFILIATIONS: Commission des affaires sociales du Sénat, Paris, France (*Vice Président*).
■ TEACHING: yes.
■ PUBLICATIONS: yes.

F 110 **JALBERT** Pierre
Professor of Medical Genetics. — Head of the Department of Genetics (Michallon hospital, Grenoble).

CHU Hôpital Albert-Michallon
Laboratoire de cytogénétique
BP 217
38043 Grenoble (France)
✆ (33) 76 76 75 75

■ AFFILIATIONS: Fédération française des CECOS — CECOS, La Tronche Cedex, France (*Deputy-Chairperson, Director of the CECOS-Alpes*).

■ TEACHING: Postgraduate DEA Diploma, Paris.
■ PUBLICATIONS: yes.
■ WORKING ON: physician's role [J26], semen donors [M28], AID [M32], genetic defects [Q3], genetic counseling [Q36], genetic screening [Q37], eugenics [Q39], carriers [Q46].
■ INTERESTED IN: CCNE advice (France) [B91], medical ethics [C52], parental consent [J13], physician's role [J26], therapeutic abortion [M50], chromosomal disorders [Q4].

F 111 JASMIN Claude

Professor of Cancerology, University of Paris-XI. — Head of the Department of Haematology, Immunology and Tumour Biology, Paul-Brousse Hospital. — Director of the INSERM U 268 research unit (Applied Oncology).

Hôpital Paul-Brousse
12-14, avenue Paul-Vaillant-Couturier
94800 Villejuif (France)
✆ (33) 1 - 45 59 36 62 Fax: (33) 1 - 45 59 38 60

■ AFFILIATIONS: Comité d'éthique du Consistoire israélite de Paris, France (*Member*).
■ TEACHING: annual medical meeting of the Centre Rachi (Paris).
■ PUBLICATIONS: yes.
■ WORKING ON: human body commercialization [B13], CCPPRB — French committees for the protection of human subjects of biomedical research [B32], humanity heritage [B46], ethical rules and principles [B47], information — communication — media [B68], bioethics [B89], specific approaches to ethics [C1], science, technology, methods [E1], biomedical research and experimentation [F1], organization of health care; facilities, manpower and services; health occupations [G1], health economics, population characteristics [H1], treatment [I30], health care professionals, researchers and patient relationship [J1], legislation and law [K34], aged — problems related with aging [L45], fetal therapy [M57], cancer [N17], leukemia [N18], terminal care [O7], euthanasia [O15], attitudes to death [O20], medical genetics [Q35], population genetics [Q43].
■ INTERESTED IN: *idem*.

F 112 KAHN Axel

Director of the INSERM U 129 research unit (genetics and molecular pathology). — Chairperson of the Commission du génie biomoléculaire. — Member of the Commission du génie génétique. — Deputy-chairperson of the Ligue nationale française contre le cancer. — Chief-editor of the Médecine/sciences journal. — Chairperson of the technical section of the Comité consultatif national d'éthique pour les sciences de la vie et de la santé (CCNE).

Institut Cochin de génétique moléculaire
Unité 129 INSERM
24, rue du faubourg Saint-Jacques
75014 Paris (France)
✆ (33) 1 - 44 41 24 24 Fax: (33) 1 - 44 41 24 21

■ TEACHING: Debates and lectures at the UNAF and UDAF (Union nationale et Union départementale des associations familiales). — Ethics, CHU (University hospital) of Cochin. — Conferences on ethics for the INSERM-Jeune Clubs.
■ PUBLICATIONS: yes.
■ WORKING ON: philosophy of biology [C42], biology and human future [C43], evolution [C47], biology [E5], genetics [E6], molecular biology [E8], research [E14], containment [E38], biological containment [E39], biomedical research [F2], human genome project [F8], reproductive technologies [M26], genetic defects and hereditary diseases [Q2], biotechnology — genetic engineering [Q20], medical genetics [Q35].

F 113 KAMIONER Didier

CCPPRB Ile-de-France Paris-Saint-Antoine
184, rue du faubourg Saint-Antoine
75012 Paris (France)
✆ (33) 1 - 49 28 20 00

■ AFFILIATIONS: CCPPRB Ile-de-France Paris-Saint-Antoine, Paris, France (*Chairperson*).
■ TEACHING: EPU. — Colloquiums.
■ PUBLICATIONS: yes.
■ WORKING ON: human rights [B3], human person [B12], ethics committees [B22], CCPPRB — French committees for the protection of human subjects of biomedical research [B32], deontology [B59], codes of biomedical ethics [B90], medical ethics [C52], epidemiology [E9], computers [E33], safety

[E51], clinical trials [F3], patient information [J4], cancer [N17].
■ INTERESTED IN: patient information [J4], patient access [J5], parental notification [J6], disclosure [J7], consent to treatment [J8], cancer [N17], virus diseases [N37].

F 114 **KRESS** Jean-Jacques

Professor of adult psychiatry. — Head of Department. — Chairperson of the Association française de psychiatrie.

Clinique psychiatrique CHU
Hôpital de Brest-Bohars
29820 Bohars (France)
✆ (33) 98 22 38 35 Fax: (33) 98 22 38 35

■ AFFILIATIONS: Séminaire Psychanalyse et Bioéthique, Association Descartes, Paris, France (*Member of the scientific committee*).
■ TEACHING: Medical psychology, School of medicine, Brest.
■ PUBLICATIONS: yes.
■ WORKING ON: medical education [B80], healthy volunteers [F13], debriefing [F16], mental health [G4], mental institutions [G29], public hospitals [G30], psychiatry [G69], consent to treatment [J8], informed consent [J9], physician patient relationship [J32], neurosciences — psychiatry [P1], behavioral and mental disorders [P2], outpatient commitment [P18], psychiatric technics [P24].
■ INTERESTED IN: medical education [B80], healthy volunteers [F13], debriefing [F16], mental health [G4], mental institutions [G29], public hospitals [G30], psychiatry [G69], consent to treatment [J8], informed consent [J9], physician patient relationship [J32], neurosciences — psychiatry [P1], behavioral and mental disorders [P2], outpatient commitment [P18], psychiatric technics [P24], ethics and humanities [C54], personality [C82], medical ethics [C52].

F 115 **LACHAUX** Bernard

M.D. Clinician, psychiatry. — Chairperson of the CCPPRB of Lyon-B.

CHS Saint-Jean-de-Dieu
Service du Dr. S. Jallade
290, route de Vienne
69373 Lyon cedex 08 (France)
✆ (33) 78 09 78 09 Fax: (33) 78 77 96 67

■ AFFILIATIONS: Service Droit et éthique de la santé, Bron Cedex, France (*Manager*).
■ TEACHING: Department of Health Law and Ethics (Dr. Nicole Lery).
■ PUBLICATIONS: yes.
■ WORKING ON: human body commercialization [B13], CCPPRB — French committees for the protection of human subjects of biomedical research [B32], ethics [B48], codes of ethics [B49], moral obligations [B50], competence [B51], professional competence [B52], moral policy [B55], deontology [B59], code of deontology [B60], interdisciplinary communication [B70], health and biology mediatisation [B85], biology and human future [C43], medical ethics [C52], life sciences [E4], epidemiology [E9], research [E14], research team [E16], biomedical research and experimentation [F1], biomedical research [F2], clinical trials [F3], health services research [F4], nontherapeutic research [F5], therapeutic research [F6], cognitive research [F7], human genome project [F8], experimentation [F9], animal experimentation [F10], human experimentation [F11], research subjects [F12], healthy volunteers [F13], animal testing alternatives [F14], group of vulnerable subjects [F20], health occupations [G61], psychiatry [G69], health care professionals, researchers and patient relationship [J1], patients' rights [J2], patient association [J3], patient information [J4], patient access [J5], parental notification [J6], disclosure [J7], consent to treatment [J8], informed consent [J9], presumed consent [J10], third party consent [J11], spousal consent [J12], parental consent [J13], consent forms [J14], treatment refusal [J15], neurosciences — psychiatry [P1], behavioral and mental disorders [P2], mentally handicapped [P7], mentally retarded [P8], mentally ill [P9], psychotic disorders [P10], dementia [P11], schizophrenia [P12], behavior disorders [P13], aggression [P14], dangerousness [P15], hyperkinesis [P16], psychological stress [P17], outpatient commitment [P18], volontary admission [P19], duration of commitment [P20], involontary commitment [P21], psychiatric technics [P24], behavior control [P25], operant conditioning [P26], negative reinforcement [P27], positive reinforcement [P28], psychiatric diagnosis [P29], psychotherapy [P30], hypnosis [P31], group therapy [P32], psychoactive drugs [P33].
■ INTERESTED IN: *idem*.

F 116 **LAMAU** Marie-Françoise

Research fellow, teacher.

Centre d'éthique médicale
60, boulevard Vauban
BP 109
59016 Lille Cedex (France)
✆ (33) 20 42 09 12 Fax: (33) Fax: 20 78 26 45

■ AFFILIATIONS: Association européenne des centres d'éthique médicale — AECEM, Brussels, Belgium (*Member*). — Centre d'éthique contemporaine de l'Institut catholique de Lille, Lille Cedex, France (*Member*).
■ TEACHING: yes.
■ PUBLICATIONS: yes.
■ WORKING ON: information sources, data bases [A1], data bases [A3], bibliography [A4], information centers [A5], documentation [A8], dignity [B5], human person [B12], ethical rules and principles [B47], ethics [B48], codes of ethics [B49], moral obligations [B50], obligations of society [B53], obligations to society [B54], moral policy [B55], deontology [B59], morality [B61], respect [B63], virtues [B64], solidarity [B65], biomedical ethics education [B78], European Convention of Bioethics [B97], Bioethics bill, 1992 (France) [B98], philosophical ethics [C2], morals [C12], conscience [C13], moral development [C14], values [C15], quality of life [C16], social worth [C17], value of life [C18], religion and religious ethics [C19], clergy [C20], religious ethics [C21], Christian ethics [C22], Roman catholic ethics [C23], Roman catholicism [C31], religious beliefs [C32], religion [C36], Christian science [C37], religious sciences [C39], scriptural interpretation [C40], theology [C41], professional ethics [C51], medical ethics [C52], nursing ethics [C53], philosophy [C61], pastoral care [C88], teaching methods [E59], social control of science [E64], health care costs [H9], prenatal diagnosis [I3], inter-personal relationship [J27], nurse patient relationship [J30], physician nurse relationship [J31], physician patient relationship [J32], professional patient relationship [J33], newborns [L29], prematurity [L30], in vitro fertilization [M29], FIV center [M30], suffering [N11], pain [N12], congenital defects [N23], neural tube defects [N24], anencephaly [N25], spina bifida [N26], death and resuscitation [O1], terminal care [O7], euthanasia [O15], attitudes to death [O20], genetic defects and hereditary diseases [Q2], genetic defects [Q3], down's syndrome [Q6], hereditary diseases [Q18], medical genetics [Q35], genetic counseling [Q36], genetic screening [Q37], preimplantation genetic diagnosis [Q38], eugenics [Q39], negative eugenics [Q40], positive eugenics [Q41].
■ INTERESTED IN: *idem.*

F 117 **LANGLOIS** Anne
Professor of Arts, Lycée Jean-Jaurès (Montreuil). — Investigator in bioethics, Department of philosophy, Paris-X University (Nanterre).

25, rue de la Paix
94300 Vincennes (France)
✆ (33) 1 - 43 74 69 43

■ AFFILIATIONS: Groupe de bioéthique franco-québécois — B.O.F, Saint-Brice/Forêt, France (*Member*).
■ PUBLICATIONS: yes.
■ WORKING ON: fundamental rights of the individuals [B2], dignity [B5], human body commercialization [B13], ethics committees [B22], CCPPRB — French committees for the protection of human subjects of biomedical research [B32], codes of ethics [B49], altruism [B56], solidarity [B65], ethical review [B93], morals [C12], social control [D3], standards [E60], biomedical research [F2], clinical trials [F3], cognitive research [F7], human experimentation [F11], research subjects [F12], healthy volunteers [F13], presumed consent [J10], third party consent [J11], patient participation [J16].
■ INTERESTED IN: human person [B12], reification [B14], public debates [B75], ethical analysis [C3], utilitarianism [C8], selection of subjects [F19], organ donors [I15], organ donation [I25].

F 118 **LANSAC** Jacques
Professor of gynaecology and obstetrics.

CHU Tours
Hôpital Bretonneau
37044 Tours (France)
✆ (33) 47 47 47 36 Fax: (33) 47 47 38 01

■ AFFILIATIONS: Centre d'études et de conservation des œufs et du sperme — CECOS, Bicêtre, France (*Chairperson*).
■ TEACHING: postgraduate DEA diploma in medical and biology ethics, René-Descartes University (Paris).
■ PUBLICATIONS: yes.
■ WORKING ON: reproductive technologies [M26], pregnancy and childbirth [M39], abortion [M47],

fetal therapy [M57].
■ INTERESTED IN: *idem*.

F 119 **LAPORTE** Yves
Former Professor of neurophysiology.

Collège de France
11, place Marcelin-Berthelot
75231 Paris Cedex 05 (France)
✆ (33) 1 - 44 27 12 11

■ AFFILIATIONS: CCNE, Paris, France (*Member*).
■ WORKING ON: bioethics [B89], CCNE advice (France) [B91].
■ INTERESTED IN: biomedical research [F2], clinical trials [F3], therapeutic research [F6], experimentation [F9], animal experimentation [F10], human experimentation [F11], fetal tissue transplantation [I27], intracerebral fetal tissue transplantation [I28].

F 120 **LAROQUE** Pierre
Chairperson of the honorary section of the Conseil d'Etat.

Ministère des affaires sociales et de l'intégration
8, avenue de Ségur
75007 Paris (France)
✆ (33) 1 - 40 56 78 84

■ AFFILIATIONS: CCNE, Paris, France (*Member*).
■ PUBLICATIONS: yes.
■ WORKING ON: respect of human dignity and human rights [B1], protection of rights — involved institutions [B16], institutions involved in the protection of rights [B17], CNIL — Commission Nationale Informatique et Libertés (France) [B19], Council of Europe [B25], European Convention on Human Rights [B36], Helsinki Declaration [B37], Declaration on Human and Citizen Rights [B38], Universal Declaration of Human Rights [B39], due process [B40], public advocacy [B41], state interest [B44], common good [B45], humanity heritage [B46], ethical rules and principles [B47], CCNE advice (France) [B91], ethicists [B92], ethical review [B93], history of biomedical ethics [B94], bioethics movement [B95], bioethical issues [B96], European Convention of Bioethics [B97], Bioethics bill, 1992 (France) [B98], values [C15], quality of life [C16], social worth [C17], value of life [C18], biology and human future [C43], future generations [C44], humanism [C66], social sciences [C89], legislation [D13], socioeconomic factors [D18], employment [D19], strikes [D20], family members [D24], married persons [D28], marital relationship [D29], family relationship [D31], social problems [D52], political activity [D57], international aspects [D59], advisory committees [D69], French Council of State [D70], democracy [D82], occupational medicine [E46], data protection [E52], evaluation [E61], public health [G5], state medicine [G9].

F 121 **LASSAUNIERE** Jean Michel
M.D. Clinician. — Head of the Palliative Care Centre, Hôtel-Dieu Hospital.

Centre de Soins Palliatifs à l'Hôtel-Dieu
1, place du Parvis-Notre-Dame
75181 Paris Cedex 04 (France)
✆ (33) 1 - 42 34 84 00 Fax: (33) 1 - 42 34 84 06

■ AFFILIATIONS: Centre Sèvres, France
■ TEACHING: Palliative care.
■ PUBLICATIONS: yes.
■ WORKING ON: palliative care [O8], life-sustaining treatment [O9], withholding treatment [O10], allowing to die [O11], terminally ill [O12], quality adjusted life years [O14], euthanasia [O15], prolongation of life [O13].
■ INTERESTED IN: quality adjusted life years [O14], withholding treatment [O10].

F 122 **LAVARENNE** Jeannine
Professor, School of Medicine. Head of the Department of toxicology and clinical pharmacology, CHU (University hospital) of Clermont-Ferrand.

CCPPRB CHRU de Clermont-Ferrand
Administration centrale
30, place Henri-Dunant

63003 Clermont-Ferrand Cedex (France)
✆ (33) 73 60 61 44

■ WORKING ON: CCPPRB — French committees for the protection of human subjects of biomedical research [B32], clinical trials [F3], animal experimentation [F10], human experimentation [F11], health care professionals, researchers and patient relationship [J1].
■ INTERESTED IN: *idem.*

F 123 **LE SOMMER-PÉRÉ** Myriam
M.D. Clinician, Head of the Department of medical information, Suburban Hospital of Bouscat. — Temporary investigator at the Laboratoire d'information médicale, Pr. Salamon's Department (University of Bordeaux-II). — Chairperson of the Prévention Action auprès des personnes âgées (local department of gerontology coordination).

29, rue de Canéjan
33700 Mérignac (France)
✆ (33) 56 97 06 46

■ AFFILIATIONS: Groupe franco-québécois de recherche en bioéthique — B.O.F, St-Brice, France (*Executive Secretary*).
■ TEACHING: Ecole des cadres des infirmiers et infirmières du CHR de Bordeaux.
■ PUBLICATIONS: yes.
■ WORKING ON: self determination [B4], dignity [B5], integrity [B8], justice [B9], freedom [B10], human person [B12], CNIL — Commission Nationale Informatique et Libertés (France) [B19], office of science and technology assessment [B28], patient advocacy [B42], beneficence [B57], paternalism [B62], interdisciplinary communication [B70], health education [B77], biomedical ethics education [B78], nursing education [B79], history of biomedical ethics [B94], nursing ethics [C53], physically handicapped [D44], force feeding [D95], data protection [E52], health services [G25], home care [G55], rehabilitation [I42], legal guardians [K7], nutrition [L16], health care services and aged [L48], aged [L49], aging [L50], sexuality and procreation — unborn child [M1], terminal care [O7], life-sustaining treatment [O9], withholding treatment [O10], euthanasia [O15], dementia [P11].
■ INTERESTED IN: respect of human dignity and human rights [B1], medical ethics [C52], social control [D3], exclusion [D9], socioeconomic factors [D18], physically handicapped [D44], sociology of medicine [D55], dehumanization [D86], decision making [E54], self regulation [E62], health services research [F4], residential facilities [G26], patient care team [G38], medical records [G50], medical evaluation [G51], preventive medicine [G58], family practice [G60], physical restraint [I40], rehabilitation [I42], patient information [J4], patient access [J5], consent to treatment [J8], patient participation [J16], right to treatment [J17], confidentiality [J20], nurse patient relationship [J30], physician nurse relationship [J31], physician patient relationship [J32], legally incompetent person [K4], legal guardians [K7], nutrition [L16], mind [L20], old person abuse [L51], death and resuscitation [O1], euthanasia [O15], suicide [O19], dementia [P11].

F 124 **LEBRUN** Thérèse
Head of the Department of health economics of the CRESGE. — Research Fellow, INSERM.

Centre de recherches économiques, sociologiques et de gestion (CRESGE)
BP 109
1, rue Norbert-Ségard
59016 Lille Cedex (France)
✆ (33) 20 57 11 77 Fax: (33) 20 78 26 45

■ AFFILIATIONS: Centre d'éthique médicale, Lille, France (*Member*).
■ TEACHING: Conferences on bioethics organised by the Centre d'éthique médicale (economic aspects).
■ PUBLICATIONS: yes.
■ WORKING ON: value of life [C18], social sciences [C89], socioeconomic factors [D18], technology assessment [E65], costs and benefits [H10], therapeutic abortion [M50], persistent vegetative state [N42], euthanasia [O15].
■ INTERESTED IN: *idem.*

F 125 **LEHN** Pierre
Lecturer (university of Paris-V). — M.D. Clinician.

Hôpital Saint-Louis
Département d'hématologie
1, avenue Claude-Vellefaux
75475 Paris Cedex 10 (France)
✆ (33) 1 - 42 49 96 39 Fax: (33) 16 1 - 42 49 96 32

■ AFFILIATIONS: Association internationale Droit, éthique et science (groupe de Milazzo), Strasbourg, France (*Member*).
■ PUBLICATIONS: yes.
■ WORKING ON: bioethics [B89], bioethical issues [B96], life sciences [E4], genetics [E6], molecular biology [E8], investigators [E15], biological containment [E39], biomedical research [F2], experimentation [F9], animal experimentation [F10], human experimentation [F11], bone marrow transplantation [I29], informed consent [J9], physicians and researchers accountability [J23], therapeutic risk [K32], bone marrow [L8], child diseases [L43], genetic defects and hereditary diseases [Q2], biotechnology — genetic engineering [Q20], medical genetics [Q35], gene therapy [Q42], fetal development [M54], fetal therapy [M57], cancer [N17], leukemia [N18], human genome [Q47], gene pool [Q45].
■ INTERESTED IN: CCNE — Comité Consultatif National d'Ethique (France) [B23], Council of Europe [B25], CCPPRB — French committees for the protection of human subjects of biomedical research [B32], biomedical ethics education [B78], health and biology mediatisation [B85], codes of biomedical ethics [B90], bioethical issues [B96], European Convention of Bioethics [B97], biology and human future [C43], medical ethics [C52], genetics [E6], molecular biology [E8], biological containment [E39], biomedical research [F2], experimentation [F9], methodology of research and experimentation [F15], fetal tissue transplantation [I27], bone marrow transplantation [I29], immunotherapy [I41], informed consent [J9], investigator subject relationship [J29], fetal therapy [M57], cancer [N17], leukemia [N18], genetic defects and hereditary diseases [Q2], biotechnology [Q21], genetically modified organisms [Q28], medical genetics [Q35], gene therapy [Q42], gene pool [Q45], human genome [Q47].

F 126 **LEMÉNAGER** Jacques

Honorary Professor, School of Medicine of Caen. — Former Head of the Department of Pneumology, CHU (University hospital) of Caen. — Correspondent of the Académie de médecine. — Chairperson of the CCPPRB of Basse-Normandie (since 1991). — Chairperson of the Ethics Committee of Caen (since 1987). — Chairperson of the Association d'aide aux insuffisants respiratoires (help to the respiratory impaired) de Basse-Normandie (since 1976).

CCPPRB CHRU Côte-de-Nacre
Avenue Côte-de-Nacre
14033 Caen Cedex (France)
✆ (33) 31 06 31 06

F 127 **LENOIR** Noëlle

Member of the Conseil constitutionnel. — Maître des requêtes au Conseil d'Etat (Chief officer reporting on the matters to be judged). — Conseil supérieur de la Recherche et de la Technologie (French ministry of Research and Space). — President (with Mr. François Guinot, General Director of Rhone-Poulenc) of the commission "Enjeux socio-économiques et internationaux de la Recherche".

Conseil constitutionnel
2, rue Montpensier
75001 Paris (France)
✆ (33) 1 - 40 15 30 08 Fax: (33) 1 - 42 60 17 41

■ AFFILIATIONS: European Communities Commission, Groupe des conseillers pour l'éthique de la biotechnologie, Brussels, Belgium (*Expert*). — High Council on bioethics UNESCO, Paris, France (*Chairperson*).
■ PUBLICATIONS: yes.

F 128 **LERAT** Marc F.

Professor. — Clinical Gynaecologist and Obstetrician.

Centre hospitalier universitaire
HME
BP 1005
44035 Nantes Cedex 01 (France)
✆ (33) 40 08 31 79 Fax: (33) 40 08 46 48

■ AFFILIATIONS: Conseil national de l'ordre des médecins de France, Paris (*Vice Président*).
■ TEACHING: bioethics and medical deontology to physicians, gynaecologists, obstetricians and midwives.
■ PUBLICATIONS: yes.

F 129 **LEROY** Jacques

Head of the Department of Intensive Medical Care, CHU (University hospital) of Rouen.

CCPPRB Hôpital Charles-Nicolle
1, rue de Germont
76031 Rouen Cedex (France)
✆ (33) 35 08 84 46 Fax: (33) 35 08 80 53

■ AFFILIATIONS: CCPPRB de Haute-Normandie, Rouen, France (*Chairperson*).
■ WORKING ON: CCPPRB — French committees for the protection of human subjects of biomedical research [B32], due process [B40], clinical trials [F3], research subjects [F12], healthy volunteers [F13], hospital [G28], hospitalisation [G46], medical evaluation [G51], emergency care [G54], organ donors [I15], transplantation [I22], organ transplantation [I24], organ procurement [I26], renal dialysis [I32], biological substances contamination [I9], iatrogenic disease [N3], coma [N41], persistent vegetative state [N42], brain death [O6], life-sustaining treatment [O9], withholding treatment [O10], quality adjusted life years [O14], suicide [O19], resuscitation [O21].
■ INTERESTED IN: review [A2], data bases [A3], traffic of organs [B15], institutions involved in the protection of rights [B17], College of Physicians [B27], biomedical ethics education [B78], medical ethics [C52], nursing ethics [C53], toxicity [E50], nontherapeutic research [F5], informed consent [J9], medical secrecy [J22].

F 130 **LÉRY** Louis

M.D. Clinician, hospita"s occupational medicine. — Head of the Department of preventive medicine and vaccines.

Institut Pasteur
Avenue Tony-Garnier
69007 Lyon (France)
✆ (33) 72 72 25 23

■ AFFILIATIONS: Santé, éthique et libertés — SEL, Lyon, France (*Member*).
■ TEACHING: AEU (university attestation) in philosophy and medical ethics, Claude-Bernard University. — Teaching to physicians and health care professionnals.
■ PUBLICATIONS: yes.
■ WORKING ON: preventive medicine [G58], immunization [G59], communicable diseases [N27], acquired immunodeficiency syndrome [N35], HIV seropositivity [N36], occupational medicine [G65], biological containment [E39], health services research [F4], human experimentation [F11], confidentiality [J20], physicians and researchers accountability [J23].
■ INTERESTED IN: *idem*.

F 131 **LÉRY** Nicole

Lecturer. — M.D. Clinician.

Droit et éthique de la santé
95, boulevard Pinel
69677 Bron cedex (France)
✆ (33) 72 35 87 50

■ AFFILIATIONS: Association Européenne des centres d'éthique médicale — AECEM, Belgium (*Former President*).
■ TEACHING: AEU (university attestation) in philosophy and medical ethics. — Training to health care professionnals. — Public conferences on specialised issues.
■ PUBLICATIONS: yes.
■ WORKING ON: respect of human dignity and human rights [B1], professional ethics [C51], medical ethics [C52], nursing ethics [C53], sex offenses [D92], rape [D93], torture [D94], force feeding [D95], records [E34], computerized file [E35], data protection [E52], drug industry [E37], decision making [E54], medical records [G50], human experimentation [F11], research subjects [F12], healthy volunteers [F13], emergency care [G54], immunization [G59], family practice [G60], radiology [G70], alternative therapies [I33], drugs [I34], vaccination [I43], patient access [J5], consent to treatment [J8], informed consent [J9], presumed consent [J10], treatment refusal [J15], medical secrecy [J22], referral and consultation [J24], physician's role [J26], ill-treat children [L39], child abuse [L41], old person abuse [L51], sports [L53], embryos [M10], cancer [N17], acquired immunodeficiency syndrome [N35], autopsies [O3], death [O5], withholding treatment [O10], allowing to die [O11], euthanasia [O15], suicide [O19], attitudes to death [O20].
■ INTERESTED IN: alcohol abuse [P4], psychoactive drugs [P33].

F 132 **LÉVY** J.-P.
Head of the Haematology Department at Cochin Hospital. — Professor at René-Descartes University. — Director of the Institut Cochin for Molecular Genetics. — Director of the National Agency for AIDS Research.

Hôpital Cochin
27, rue du Faubourg Saint-Jacques
75014 Paris (France)
✆ (33) 40 51 64 57 Fax: (33) 40 51 64 73

■ WORKING ON: *idem.*
■ INTERESTED IN: life sciences [E4], research [E14], biomedical research [F2], experimentation [F9], methodology of research and experimentation [F15], cancer [N17], communicable diseases [N27].

F 133 **LION** Antoine
Director of the Centre Thomas-More. — Chairperson of the Association Chrétiens et sida.

Centre Thomas-More
La Tourette
69210 L'Arbresle (France)
✆ (33) 74 01 01 03 Fax: (33) 74 01 47 27

■ AFFILIATIONS: Centre Thomas-More, L'Arbresle, France (*Director*).
■ TEACHING: colloquiums organised by the Centre Thomas-More.
■ PUBLICATIONS: yes.
■ WORKING ON: respect of human dignity and human rights [B1], religion and religious ethics [C19], acquired immunodeficiency syndrome [N35], HIV seropositivity [N36].
■ INTERESTED IN: respect of human dignity and human rights [B1], ethical rules and principles [B47], philosophical ethics [C2], religion and religious ethics [C19], ethics and humanities [C54], sexuality and procreation [M2], acquired immunodeficiency syndrome [N35], acquired immunodeficiency syndrome [N35].

F 134 **LLUIS** M.
Research fellow at the CNRS URA 147 research unit.

CNRS URA 147- PRII
Institut Gustave-Roussy
39, rue Camille-Desmoulins
94805 Villejuif (France)
✆ (33) 1 - 45 59 47 92 Fax: (33) 1 - 46 78 41 20

■ AFFILIATIONS: Comité opérationnel sur l'éthique dans les sciences de la vie, CNRS–département sciences de la vie, Paris, France (*Member*).
■ WORKING ON: science [E2], life sciences [E4], research [E14], technology [E27], biomedical research and experimentation [F1], cancer [N17].
■ INTERESTED IN: human body commercialization [B13], ethics committees [B22], CCNE — Comité Consultatif National d'Ethique (France) [B23], CCPPRB — French committees for the protection of human subjects of biomedical research [B32], quality of life [C16], philosophy of biology [C42], science [E2], life sciences [E4], research [E14], technology [E27], drug industry [E37], containment [E38], health hazards [E49], toxicity [E50], biomedical research [F2], experimentation [F9], methodology of research and experimentation [F15], anesthesia [I31], drugs [I34], insulin [I35], placebos [I36], extraordinary treatment [I39], immunotherapy [I41], vaccination [I43], patient information [J4], informed consent [J9], presumed consent [J10], physicians and researchers accountability [J23], self induced illness [N10], suffering [N11], pain [N12], follow-up studies [N13], prognosis [N14], cancer [N17], diabetes [N22], terminal care [O7], attitudes to death [O20].

F 135 **LOBEL** B.
Head of the Department of Urology, Rennes-I University. — Chairperson of the CECOS Ouest.

CHR Pontchaillou, service d'urologie
Rue Henri-le-Guilloux
35033 Rennes Cedex (France)
✆ (33) 99 28 42 69 and 99 28 42 Fax: (33) 99 28 41 13

■ TEACHING: secondary school and university teaching.
■ WORKING ON: CCNE advice (France) [B91], bioethical issues [B96], referral and consultation [J24],

contraception [M3], volontary sterilization [M8], procreation [M15], palliative care [O8], active euthanasia [O16], involuntary euthanasia [O17], voluntary euthanasia [O18].

F 136 **LOPES** Patrice

Head of the Department of Gynaecology and Obstetrics A, CHU (University hospital) of Nantes. — Professor. — Gynaecologist and Obstetrician. — Founding member of the European College of Obstetrics and Gynaecology. — Member of the Commission médicale d'établissement of the Nantes university hospital.

Centre Hospitalier Régional de Nantes, service de gynécologie obstétrique A
Centre clinique de procréation médicalement assistée
Hôtel Dieu
44035 Nantes (France)
✆ (33) 40 08 31 75 Fax: (33) 40 08 46 48

■ AFFILIATIONS: Commission Nationale de Médecine et biologie de la reproduction, sous-section PMA (*Member*).
■ WORKING ON: information — communication — media [B68], public debates [B75], education [B76], nursing education [B79], medical education [B80], internship and residency [B81], students [B82], sex offenses [D92], rape [D93], biomedical research [F2], clinical trials [F3], therapeutic research [F6], control groups [F17], random selection [F18], family planning [G8], maternal welfare [G10], maternal life [G11], medical devices [G23], WHO — World Health Organization [G36], nurses [G39], physicians [G40], medical staff [G41], prenatal diagnosis [I3], amniocentesis [I4], chorionic villus sampling [I5], mass screening [I6], mandatory screening [I7], sperm banks [I11], anonymous donation [I17], directed donation [I18], fetal tissue transplantation [I27], sexuality and procreation — unborn child [M1], sexuality and procreation [M2], reproductive technologies [M26], abortion [M47], fetal development [M54], fetal therapy [M57], expert evaluation [K15], child donation [K22], child's interest [K24], legal liability [K30], European law [K49], human development [L12], breast feeding [L22], beginning of life [L24], adoption [L32], aid children [L34], ill-treat children [L39], disease and aged [L46], life extension [L47], pain [N12], congenital defects [N23], neural tube defects [N24], anencephaly [N25], venereal diseases [N31], biotechnology [Q21], sex determination [Q26], sex preselection [Q30], genetic counseling [Q36], genetic screening [Q37], preimplantation genetic diagnosis [Q38], gene therapy [Q42].
■ INTERESTED IN: press conference [B73], health and biology mediatisation [B85], uncontrolled information [B88], European Convention of Bioethics [B97], Bioethics bill, 1992 (France) [B98], international aspects [D59], EC — European Communities [D62], decision making [E54], social control of science [E64], program descriptions [G18], patient association [J3], European law [K49].

F 137 **LUCAS** Philippe

Professor of sociology. — Director of the Académie de Bordeaux (Educational district of Bordeaux), Chancellor of the universities of Aquitaine.

Rectorat de Bordeaux
5, rue Joseph-de-Carayon-Latour
33060 Bordeaux Cedex (France)
✆ (33) 56 56 38 10 Fax: (33) 56 96 29 42

■ AFFILIATIONS: CCNE, Paris Cedex 13, France (*Member, Chairperson of the EducationCommission*).
■ TEACHING: Sociology at the University.
■ PUBLICATIONS: yes.
■ WORKING ON: *idem.*
■ INTERESTED IN: information — communication — media [B68], medical ethics [C52], social and political issues [D1].

F 138 **MANGIN** Patrice

Professor, Strasbourg School of Medicine (Forensic Medicine). — M.D. Clinician, Strasbourg hospitals. — Director of the Institute for forensic medicine of the Strasbourg University.

Institut de médecine légale
11, rue Humann
67085 Strasbourg Cedex (France)
✆ (33) 88 24 00 85

■ WORKING ON: religious beliefs [C32].
■ INTERESTED IN: clergy [C20].

F 139 **MARC-VERGNES** Jean-Pierre

Research Director of the INSERM U 230 research unit (medical imagery, neuropsychology, pharmacology for the aged). — Chairman at the CCPPRB of Toulouse-I. — Former founding-chairperson of the Regional Biomedical Ethics Committee Midi-Pyrénées.

INSERM U 230
CHU Purpan
Place Baylac
31059 Toulouse Cedex (France)
✆ (33) 61 49 89 26 Fax: (33) 61 49 95 24

■ AFFILIATIONS: Comité régional d'éthique biomédicale Midi-Pyrénées, Toulouse, France (*Former founding Chairperson*).
■ TEACHING: Member of the education committee for the postgraduate DEA diploma of biological and medical ethics, René-Descartes University (Paris).
■ PUBLICATIONS: yes.
■ WORKING ON: ethics committees [B22], CCPPRB — French committees for the protection of human subjects of biomedical research [B32], ethics [B48], competence [B51], diagnostic imaging [E32], decision making [E54], decision analysis [E55], biomedical research and experimentation [F1], biomedical research [F2], clinical trials [F3], nontherapeutic research [F5], therapeutic research [F6], cognitive research [F7], experimentation [F9], human experimentation [F11], research subjects [F12], healthy volunteers [F13], methodology of research and experimentation [F15], debriefing [F16], control groups [F17], random selection [F18], selection of subjects [F19], aged — problems related with aging [L45], disease and aged [L46], life extension [L47], aging [L50], hypertension [N21].
■ INTERESTED IN: economics [H2], resource allocation [H4], health insurance [H6], health care costs [H9], costs and benefits [H10], risks and benefices [H11], compensation [H12], remuneration [H14], medical fees [H15], price [H16], profit [H17], death and resuscitation [O1], death [O5], brain death [O6], terminal care [O7], euthanasia [O15], suicide [O19], attitudes to death [O20].

F 140 **MARGNON** Véronique

Chaplain of the Christian community of Tours. — Teaching in moral theology.

5, rue de la Barre
37000 Tours (France)
✆ (33) 47 66 20 84 Fax: (33) 47 64 86 20

■ AFFILIATIONS: Comité consultatif de protection des personnes dans la recherche biomédicale — CCPPRB, Tours, France (*Member*).
■ TEACHING: Moral Theology, general bioethics (Diocese of Tours) — Lecturer, moral theology, Faculty of theology, Institut catholique of Angers.
■ WORKING ON: philosophical ethics [C2], morals [C12], conscience [C13], religion and religious ethics [C19], clergy [C20], religious ethics [C21], Christian ethics [C22], Roman catholic ethics [C23], theology [C41].
■ INTERESTED IN: review [A2], book review [A6], publications [A9], fundamental rights of the individuals [B2], human person [B12], ethics committees [B22], CCNE — Comité Consultatif National d'Ethique (France) [B23], Council for International Organization of Medical Sciences [B24], Council of Europe [B25], office of science and technology assessment [B28], CCPPRB — French committees for the protection of human subjects of biomedical research [B32], patient advocacy [B42], ethical rules and principles [B47], morals [C12], conscience [C13], religion and religious ethics [C19], clergy [C20], religious ethics [C21], Christian ethics [C22], theology [C41], homosexuality [M22].

F 141 **MATRAY** Bernard

Researcher and teacher in biomedical Ethics.

Revue Laënnec
12, rue d'Assas
75006 Paris (France)
✆ (33) 1 - 45 44 18 99

■ AFFILIATIONS: Centre Sèvres, Paris, France (*Member, teacher*).
■ TEACHING: Centre Sèvres. — Nursing school. — Training centre for hospital staff.
■ PUBLICATIONS: yes.
■ WORKING ON: death and resuscitation [O1].
■ INTERESTED IN: religion and religious ethics [C19], sexuality and procreation — unborn child [M1], reproductive technologies [M26], abortion [M47], fetal development [M54], fetal therapy [M57].

F 142 **MATTEI** Jean-François

Professor of medical genetics, CHU (University hospital) of Marseille, Director of the Centre for Medical Genetics and the Centre of Prenatal Diagnosis. — Head of a research team on genetic diseases (INSERM). — Director of the medical ethics teaching programme (AEU, university attestation), Medical School of Marseille. — Chairperson of the Centre for Teaching and Research in Medical Ethics (CEREM). — M.P. for the Bouches-du-Rhône.

Centre de génétique médicale
Hôpital d'enfants de la Timone
13385 Marseille Cedex 5 (France)
✆ (33) 91 38 67 49 Fax: (33) 91 49 41 94

■ TEACHING: Director of the medical ethics teaching programme (AEU, university attestation), Medical School of Marseille.
■ PUBLICATIONS: yes.
■ WORKING ON: information sources, data bases [A1], editorial policies [A10], respect of human dignity and human rights [B1], human body commercialization [B13], reification [B14], traffic of organs [B15], ethics committees [B22], patient advocacy [B42], ethical rules and principles [B47], ethics [B48], codes of ethics [B49], professional competence [B52], obligations of society [B53], moral policy [B55], information dissemination [B71], public debates [B75], biomedical ethics education [B78], health and biology mediatisation [B85], bioethics [B89], codes of biomedical ethics [B90], Bioethics bill, 1992 (France) [B98], specific approaches to ethics [C1], values [C15], religion and religious ethics [C19], philosophy of biology [C42], professional ethics [C51], medical ethics [C52], social and political issues [D1], political issues [D56], political activity [D57], French National Assembly [D68], French Senate [D73], government agencies [D74], state responsibility [D75], politics [D76], community services [D78], political systems [D79], science, technology, methods [E1], health care professionals, researchers and patient relationship [J1], biomedical research and experimentation [F1], stages of life — problems proper to childhood and elderly [L1], childhood difficulties, diseases, protection [L21], children's rights [L25], organization of health care; facilities, manpower and services; health occupations [G1], sexuality and procreation — unborn child [M1], health economics, population characteristics [H1], disease [N1], health care, medical acts [I1], genetics and applications — biotechnology [Q1], medical genetics [Q35], human genome [Q47].

F 143 **MELLIER** Georges

Professor, M.D. Clinician.

CCPPRB Hôpital Edouard-Herriot
Pavillon K
Place d'Arsonval
69003 Lyon (France)
✆ (33) 72 34 48 59

■ AFFILIATIONS: Comité d'éthique, université Claude-Bernard (Lyon), Lyon, France (*Member*).
■ WORKING ON: pregnancy and childbirth [M39], foetal development [M54].
■ INTERESTED IN: pregnancy and childbirth [M39], abortion [M47], fetal development [M54], fetal therapy [M57], reproductive technologies [M26], sexuality and procreation [M2].

F 144 **METROT** Jacques

M.D. clinician, specialised in anaethesia and intensive care. — Vice-Chairperson of the Centre européen de medecine des catastrophes (Council of Europe). — Member of the Réseau européen Médecine, éthique et droits de l'homme. — Director of Education, University of Paris-VI.

Hôpital Broussais
96, rue Didot
75014 Paris (France)
✆ (33) 43 95 92 65

■ AFFILIATIONS: Council of Europe, Réseau européen Médecine et droits de l'homme, Strasbourg, France (*Member*).
■ TEACHING: medicine, ethics, human rights and international solidarity (University of Paris-VI).
■ PUBLICATIONS: yes.
■ WORKING ON: traffic of organs [B15], CNIL — Commission Nationale Informatique et Libertés (France) [B19], World Medical Assembly [B20], ethics [B48], education [B76], medical education [B80], European Convention of Bioethics [B97], Bioethics bill, 1992 (France) [B98], medical ethics [C52], social impact [D49], ECC — European Community Commission [D63], government agencies [D74], computers [E33], radiation [E48], required request [I10], organ transplantation [I24], organ

donation [I25], treatment [I30], patients' rights [J2], physician's role [J26], nervous system diseases [N40], health care [G45], death and resuscitation [O1].
■ INTERESTED IN: fundamental rights of the individuals [B2], bioethics [B89], professional ethics [C51], political issues [D56], biomedical research and experimentation [F1], health facilities [G22], organization of health care [G35], health care [G45], treatment [I30], patients' rights [J2], physicians and researchers accountability [J23], nervous system diseases [N40], injuries, occupational diseases, intoxications [N46], death and resuscitation [O1].

F 145 **MICHAUD** Jean
Counsellor at the Cour de Cassation (High Appeal Court).

Cour de cassation
Palais de Justice
5, quai de l'Horloge
75001 Paris (France)
✆ (33) 1 - 44 32 50 50

■ AFFILIATIONS: Comité consultatif national d'éthique — CCNE, Paris, France (*Vice-Chairperson*). — Commission nationale de médecine et de biologie de la reproduction (ministère de la Santé) (*Chairperson*). — Groupe de réflexion éthique sur les transplantations, France (*Chairperson*).
■ TEACHING: Medical Ethics, Necker School of Medicine Necker. — Medicine, Ethics and Human Rights (Pierre-et-Marie-Curie University, Paris). — AEU (university attestation) in Philosophy and Medical ethics, Claude-Bernard University (Lyon-I).
■ PUBLICATIONS: yes.
■ WORKING ON: fundamental rights of the individuals [B2], justice [B9], human body commercialization [B13], CCNE — Comité Consultatif National d'Ethique (France) [B23], Council of Europe [B25], ethical rules and principles [B47], public debates [B75], biomedical ethics education [B78], nursing education [B79], health and biology mediatisation [B85], CCNE advice (France) [B91], European Convention of Bioethics [B97], Bioethics bill, 1992 (France) [B98], legislation [D13], ECC — European Community Commission [D63], French Council of State [D70], research policy [E18], consent to treatment [J8], informed consent [J9], presumed consent [J10], medical secrecy [J22], potentiality of personhood [K5], rights [K42], sports [L53], embryos [M10], fetuses [M11], ovum [M12], sperm [M14], in vitro fertilization [M29], persistent vegetative state [N42], palliative care [O8], health personnel [G37], patient care team [G38], patient admission [G47].

F 146 **MISES** Roger
Professor, Department of Psychology and Infant and Juvenile Psychiatry, Paris-Sud University.

Centre public de psychiatrie de l'enfant et de l'adolescent
Fondation Vallée
7, rue Bensérade
94250 Gentilly (France)
✆ (33) 1 - 49 86 11 11

■ AFFILIATIONS: Comité d'éthique de l'UNAPEI, Paris, France (*Member*).
■ WORKING ON: sexuality and procreation — unborn child [M1].
■ INTERESTED IN: sexuality and procreation — unborn child [M1], organization of health care; facilities, manpower and services; health occupations [G1], psychiatry [G69].

F 147 **MOATTI** Jean-Paul
Research fellow, INSERM research unit for public health economics.

Unité INSERM 357 - Salle Déjerine
Hôpital de Bicêtre
78, rue du Général-Leclerc
94275 Le Kremlin-Bicêtre Cedex (France)
✆ (33) 1 - 46 71 32 72 Fax: (33) 1 - 46 71 32 70

■ TEACHING: medical ethics and public health economics. — Permanent training seminars. — INSERM Summer university course on public health (Bicêtre hospital).
■ PUBLICATIONS: yes.
■ WORKING ON: resource allocation [H4], health care costs [H9], costs and benefits [H10], risks and benefices [H11], prenatal diagnosis [I3], amniocentesis [I4], chorionic villus sampling [I5], mass screening [I6], mandatory screening [I7], sexuality [M20], sexual behavior [M21], selective abortion [M49], therapeutic abortion [M50], abortion on demand [M52], mother-child transmission [N16], medical genetics [Q35], genetic counseling [Q36], eugenics [Q39], gene therapy [Q42], evaluation [E61].

■ INTERESTED IN: social and political issues [D1], evaluation [E61], organization of health care [G35], health care professionals, researchers and patient relationship [J1].

F 148 **MOISAN** J.-P.

Head of the Laboratory of Molecular Genetics, CHRU (Regional University hospital) of Nantes. — M.D., docteur ès sciences. — Research Fellow, INSERM U 211, unit research (Nantes). — Forensic expert (DNA finger printing).

Laboratoire de génétique moléculaire
CHU Hôtel-Dieu
44035 Nantes Cedex (France)
✆ (33) 40 08 40 22 Fax: (33) 40 35 66 97

■ WORKING ON: sibling [D30], family relationship [D31], rape [D93], genetics [E6], molecular biology [E8], records [E34], human genome project [F8], prenatal diagnosis [I3], chorionic villus sampling [I5], mass screening [I6], parental notification [J6], referral and consultation [J24], expert evaluation [K15], technical expertise [K16], expert testimony [K18], legitimacy [K21], child's interest [K24], child and family [L31], child diseases and problems [L37], illegal abortion [M48], selective abortion [M49], therapeutic abortion [M50], twinning [M55], cadavers [O4], genetic defects [Q3], single gene defects [Q7], recessive genetic conditions [Q10], sickle cell anemia [Q11], Duchenne muscular dystrophy [Q12], hemophilia [Q13], cystic fibrosis [Q15], hereditary diseases [Q18], sex linked defects [Q19], genome mapping [Q23], cloning [Q24], sex determination [Q26], genetically modified organisms [Q28], DNA fingerprinting [Q32], DNA linkage [Q33], genetic counseling [Q36], genetic screening [Q37], genetic identity [Q44], carriers [Q46], human genome [Q47].
■ INTERESTED IN: CNIL — Commission Nationale Informatique et Libertés (France) [B19], ethics committees [B22], CCNE — Comité Consultatif National d'Ethique (France) [B23], codes of ethics [B49], biomedical ethics education [B78], codes of biomedical ethics [B90], bioethics movement [B95], European Convention of Bioethics [B97], Bioethics bill, 1992 (France) [B98], Christian ethics [C22], biology and human future [C43], family relationship [D31], associations [D47], records [E34], decision making [E54], WHO — World Health Organization [G36], life insurance [H8], patient access [J5], parental notification [J6], medical secrecy [J22], referral and consultation [J24], expert evaluation [K15], technical expertise [K16], expert testimony [K18], child's interest [K24], voluntary euthanasia [O18], alcohol abuse [P4], drug abuse [P6], aggression [P14], hypnosis [P31], genetically modified organisms [Q28], DNA fingerprinting [Q32], eugenics [Q39], negative eugenics [Q40], positive eugenics [Q41], genetic identity [Q44].

F 149 **MOISSELIN** Vincent

Assistant of Mr. Franck Sérusclat, Senator for the Rhône.

Palais du Luxembourg
15, rue de Vaugirard
75006 Paris (France)
✆ (33) 1 - 42 34 34 72

■ AFFILIATIONS: Ethique et Société (association), Paris, France (*Treasurer*).
■ WORKING ON: fundamental rights of the individuals [B2], ethical rules and principles [B47], biomedical research [F2], experimentation [F9], jurisprudence — accountability [K8], legislation and law [K34].

F 150 **MONTAGUT** Jacques

Biologist physician. — Chairperson of the Groupe d'étude de la fécondation in vitro en France (GEFF). — Member of the Commission nationale de médecine et de biologie de la reproduction. — Coordinator of the Institut francophone de recherche et d'études appliquées à la reproduction et à la sexologie (IFREARES). — Member of the CCPPRB of Toulouse-II.

Institut francophone de recherche et d'études appliquées à la reproduction et à la sexologie (IFREARES)
20, route de Revel
31400 Toulouse (France)
✆ (33) 61 34 80 31

■ AFFILIATIONS: Comité d'éthique en reproduction humaine — CERH, Toulouse, France (*Administrator*).
■ TEACHING: Colleges, Agriculture School.
■ PUBLICATIONS: yes.
■ WORKING ON: bibliography [A4], dignity [B5], integrity [B8], human person [B12], human body commercialization [B13], CCNE — Comité Consultatif National d'Ethique (France) [B23], professional deontology organs [B26], CCPPRB — French committees for the protection of human subjects of

biomedical research [B32], ethics [B48], codes of ethics [B49], professional competence [B52], obligations to society [B54], moral policy [B55], public debates [B75], education [B76], biomedical ethics education [B78], codes of biomedical ethics [B90], CCNE advice (France) [B91], European Convention of Bioethics [B97], Bioethics bill, 1992 (France) [B98], deontological ethics [C6], situational ethics [C9], value of life [C18], religion and religious ethics [C19], biology and human future [C43], consequences [C45], biological life [C49], medical ethics [C52], life sciences [E4], biology [E5], genetics [E6], microbiology [E7], molecular biology [E8], epidemiology [E9], incidence [E11], research institutes [E17], research institutes [E17], research design [E19], progress [E26], biomedical technologies [E28], freezing [E29], preservation [E30], cryonic suspension [E31], computerized file [E35], peer review [E63], social control of science [E64], technology assessment [E65], nontherapeutic research [F5], therapeutic research [F6], experimentation [F9], animal experimentation [F10], human experimentation [F11], research subjects [F12], healthy volunteers [F13], control groups [F17], random selection [F18], selection of subjects [F19], medical records [G50], medical evaluation [G51], medicine [G63], obstetrics and gynecology [G66], prenatal diagnosis [I3], amniocentesis [I4], chorionic villus sampling [I5], sperm banks [I11], donors [I14], anonymous donation [I17], consent to treatment [J8], informed consent [J9], spousal consent [J12], parental consent [J13], consent forms [J14], confidentiality [J20], medical secrecy [J22], referral and consultation [J24], physician's role [J26], medical etiquette [J28], physician patient relationship [J32], legal personality [K2], potentiality of personhood [K5], technical expertise [K16], expert testimony [K18], legitimacy [K21], child's interest [K24], moratory [K25], preconception injuries [K27], accountability [K29], guidelines [K55], bill [K60], human body [L3], body parts and fluids [L4], beginning of life [L24], children's rights [L25], aid children [L34], sexuality and procreation [M2], reproductive organs and embryonic structures [M9], embryos [M10], ovum [M12], sperm [M14], procreation [M15].

■ INTERESTED IN: bibliography [A4], information centers [A5], human rights [B3], human equality [B7], integrity [B8], justice [B9], human person [B12], human body commercialization [B13], traffic of organs [B15], institutions involved in the protection of rights [B17], World Medical Assembly [B20], ethics committees [B22], Council of Europe [B25], office of science and technology assessment [B28], United Nations [B29], International Charter of Human Rights [B34], European Convention on Human Rights [B36], public advocacy [B41], patient advocacy [B42], ethics [B48], obligations to society [B54], solidarity [B65], humanitarian organizations [B66], biomedical ethics education [B78], health and biology mediatisation [B85], codes of biomedical ethics [B90], CCNE advice (France) [B91], European Convention of Bioethics [B97], Bioethics bill, 1992 (France) [B98], situational ethics [C9], value of life [C18], religion and religious ethics [C19], biology and human future [C43], consequences [C45], biological life [C49], medical ethics [C52], existentialism [C64], social control [D3], regulation [D16], family members [D24], single persons [D25], couple [D26], cohabitation [D27], sibling [D30], family relationship [D31], parents [D32], ethnic groups [D36], cultural pluralism [D51], sociology of medicine [D55], international organizations [D61], EC — European Communities [D62], ECC — European Community Commission [D63], French National Assembly [D68], advisory committees [D69], French Council of State [D70], French Senate [D73], government agencies [D74], politics [D76], community services [D78], democracy [D82], socialism [D84], dehumanization [D86], life sciences [E4], biology [E5], genetics [E6], microbiology [E7], molecular biology [E8], epidemiology [E9], incidence [E11], research institutes [E17], research design [E19], progress [E26], biomedical technologies [E28], freezing [E29], preservation [E30], cryonic suspension [E31], computerized file [E35], biological containment [E39], health hazards [E49], toxicity [E50], data protection [E52], decision making [E54], decision analysis [E55], peer review [E63], social control of science [E64], technology assessment [E65], health services research [F4], human genome project [F8], experimentation [F9], animal experimentation [F10], human experimentation [F11], research subjects [F12], healthy volunteers [F13], debriefing [F16], control groups [F17], random selection [F18], selection of subjects [F19], state medicine [G9], health legislation [G14], institutional obligations [G16].

F 151 MORETTI Jean Marie

Honorary Professor of Biochemistery, Faculty of Science, Montpellier. — Former Research Director at the CNRS, School of Medicine, Paris-VI University. — Member of the Groupe de recherche éthique des transplantations (Ethical Research of Transplants Group, GRET).

42, rue de Grenelle
75007 Paris (France)
✆ (33) 1 - 44 39 46 63 Fax: (33) 1 - 44 39 46 93

■ AFFILIATIONS: Centre catholique des médecins français, Paris, France (*Member of the board of editors*).
■ TEACHING: public conferences on bioethics in various towns (audience composed of nurses, teachers individuals, secondary school students, member of cultural centres).
■ PUBLICATIONS: yes.
■ WORKING ON: human person [B12], human body commercialization [B13], ethics [B48], Christian ethics [C22], biology [E5], biological substances contamination [I9].

■ INTERESTED IN: human body commercialization [B13], codes of ethics [B49], codes of biomedical ethics [B90], CCNE advice (France) [B91], history of biomedical ethics [B94], Bioethics bill, 1992 (France) [B98], medical ethics [C52], genetics [E6], molecular biology [E8], prenatal diagnosis [I3], sexuality and procreation — unborn child [M1].

F 152 **MUZET** Alain

Director of the UMR 32 Laboratory, CNRS/INRS. — Research Director at the CNRS.

CNRS — LPPE
21, rue Becquerel
67087 Strasbourg Cedex (France)
✆ (33) 88 28 67 83 Fax: (33) 88 28 62 45

■ AFFILIATIONS: Comité opérationnel d'éthique des sciences de la vie, CNRS, Paris, France (*Member*).
■ WORKING ON: respect of human dignity and human rights [B1], CCPPRB — French committees for the protection of human subjects of biomedical research [B32], research [E14], containment [E38], human experimentation [F11], research subjects [F12], healthy volunteers [F13], methodology of research and experimentation [F15], sports [L53].
■ INTERESTED IN: ethics committees [B22], ethical rules and principles [B47], information — communication — media [B68], bioethics [B89], philosophy of biology [C42], physicians and researchers accountability [J23].

F 153 **NIHOUL-FEKETE** Claire

Head of the Department of Paediatric Visceral Surgery , Hôpital des Enfants-Malades de Paris. — Full Professor, School of Medicine, Paris V University. — Representative, European Union of specialist doctors (Paediatric surgery). — Chairperson of the Société française de chirurgie pédiatrique.

Hôpital des Enfants-Malades
Clinique chirurgicale infantile
149, rue de Sèvres
75015 Paris (France)
✆ (33) 1 - 42 73 83 93 Fax: (33) 1 - 42 73 83 92

■ AFFILIATIONS: Laboratoire d'éthique médicale, Paris, France (*Member*). — Groupe d'études en néonatologie et urgences pédiatriques de la région parisienne, Clamart, France (*Member*).
■ TEACHING: Ethics for postgraduate DEA students, Paris -V University. — Laboratory of Ethics, Faculty Necker-Enfants-Malades.
■ PUBLICATIONS: yes.
■ WORKING ON: treatment [I30], childhood difficulties, diseases, protection [L21], abortion [M47], fetal development [M54], fetal therapy [M57], congenital defects [N23], euthanasia [O15], genetic defects and hereditary diseases [Q2].
■ INTERESTED IN: treatment [I30], childhood difficulties, diseases, protection [L21], fetal development [M54], fetal therapy [M57], congenital defects [N23], euthanasia [O15].

F 154 **NIVELON-CHEVALIER** Annie

Head of the Genetics Centre, Dijon (Burgundy regional centre for genetics).

Centre de génétique
Hôpital d'enfants
10, boulevard du Maréchal de-Lattre-de-Tassigny
21034 Dijon Cedex (France)
✆ (33) 80 29 33 00

■ AFFILIATIONS: Commission d'éthique de l'Association française de dépistage et prévention du handicap, Lyon, France (*Member*). — Commission d'éthique du Service commun INSERM U12, France (*Member*).
■ TEACHING: School of Medicine. — Health Care Schools.
■ PUBLICATIONS: yes.
■ WORKING ON: prenatal diagnosis [I3], mass screening [I6], required request [I10], potentiality of personhood [K5], wrongful life [K33], sterilization [M6], wish of children [M16], embryo donation [M38], therapeutic abortion [M50], fetal therapy [M57], genetic defects and hereditary diseases [Q2], biotechnology — genetic engineering [Q20], medical genetics [Q35], population genetics [Q43].
■ INTERESTED IN: prenatal diagnosis [I3], mass screening [I6], required request [I10], potentiality of personhood [K5], wrongful life [K33], sterilization [M6], wish of children [M16], embryo donation [M38], therapeutic abortion [M50], fetal therapy [M57], genetic defects and hereditary diseases [Q2], biotechnology — genetic engineering [Q20], medical genetics [Q35], population genetics [Q43].

F 155 **NOVAES** Simone
Research Fellow (CNRS, URA 896) Centre of Sociology of Ethics.

Centre de Sociologie de l'Ethique (CNRS)
IRESCO 59-61, Rue POuchet
59-61, rue Pouchet
75849 Paris Cedex 17 (France)
✆ (33) 1 - 40 25 10 75 Fax: (33) 1 - 42 28 95 44

■ AFFILIATIONS: Centre de sociologie de l'éthique — CESE, Paris, France (*Research fellow*).
■ TEACHING: Conferences for postgraduate DEA students, René Descartes University (Paris V), Necker School of Medicine.
■ PUBLICATIONS: yes.
■ WORKING ON: human person [B12], ethics committees [B22], CCNE — Comité Consultatif National d'Ethique (France) [B23], altruism [B56], bioethics [B89], social sciences [C89], social issues [D2], regulation [D16], biomedical research [F2], research subjects [F12], sperm banks [I11], donors [I14], anonymous donation [I17], physicians and researchers accountability [J23], physician's role [J26], human body [L3], sexuality and procreation [M2], procreation [M15], semen donors [M28], artificial insemination [M31], abortion on demand [M52], medical genetics [Q35], genetic screening [Q37], eugenics [Q39].
■ INTERESTED IN: human person [B12], protection of rights — involved institutions [B16], ethics committees [B22], CCNE — Comité Consultatif National d'Ethique (France) [B23], manifests and declarations concerning human rights [B33], bioethics [B89], social sciences [C89], social issues [D2], biomedical research [F2], experimentation [F9], physicians and researchers accountability [J23], human body [L3], sexuality and procreation [M2], abortion [M47], medical genetics [Q35], eugenics [Q39].

F 156 **OURY** Jean-François
Gynaecologist-Obstetrician. — M.D. Clinician, assistant to the head of the Department.

Hôpital Robert-Debré
48, boulevard Sérurier
75019 Paris (France)
✆ (33) 1 - 40 03 21 70 Fax: (33) 1 - 40 03 20 20

■ AFFILIATIONS: GENEV/RP–UNAPEI, France (*Member of the Ethics Committee*).
■ WORKING ON: research [E14], evaluation [E61], health occupations [G61], diagnosis [I2], biological specimens procurement, blood transfusion, organ transplantation [I8], treatment [I30], pregnancy and childbirth [M39], abortion [M47], fetal development [M54], fetal therapy [M57], diabetes [N22], congenital defects [N23], genetic defects and hereditary diseases [Q2].
■ INTERESTED IN: diagnosis [I2], biological specimens procurement, blood transfusion, organ transplantation [I8], treatment [I30], abortion [M47], fetal development [M54], fetal therapy [M57].

F 157 **PALEY-VINCENT** Catherine
Barrister-at-Law (Paris).

8,rue Murillo
75008 Paris (France)
✆ (33) 1 - 47 54 92 92

■ AFFILIATIONS: Groupe de réflexion sur l'éthique biomédicale de Bicêtre (CHU Bicêtre) — GREBB, Le Kremlin-Bicêtre, France (*Member*).
■ PUBLICATIONS: yes.
■ INTERESTED IN: fundamental rights of the individuals [B2], legal personality [K2], jurisprudence — accountability [K8], legislation and law [K34], reproductive technologies [M26].

F 158 **PELICIER** Yves
Professor, Necker School of Medicine (Paris). — M.D. Clinician. — Director of the postgraduate DEA diploma of Medical Ethics and Biology.

Hôpital Necker
Service de psychiatrie
149, rue de Sèvres
75015 Paris (France)
✆ (33) 1 - 47 83 49 79 Fax: (33) 1 - 46 51 12 51

■ TEACHING: Etudes médicales, troisième cycle.
■ PUBLICATIONS: yes.
■ WORKING ON: psychiatry [G69], mentally handicapped [P7], quality of environment [E41].

F 159 **PERROTIN** Catherine
Head of the Centre for Bioethics.

Centre de bioéthique
Université catholique de Lyon
25, rue du Plat
69288 Lyon Cedex 02 (France)
✆ (33) 72 32 50 28 Fax: (33) Fax: 72 32 50 19

■ AFFILIATIONS: Association européenne des Centres d'éthique médicale — AECEM, Brussels, Belgium (*Member*).
■ TEACHING: Lectures and seminars, Catholic University of Lyon. — Conferences, adult education.
■ PUBLICATIONS: yes.
■ WORKING ON: ethics [B48], diagnosis [I2], utilitarianism [C8], conscience [C13], Christian ethics [C22], Roman catholicism [C31], theology [C41], mental institutions [G29], transplantation [I22], procreation [M15], reproductive technologies [M26], therapeutic abortion [M50], suicide [O19].
■ INTERESTED IN: dignity [B5], reification [B14], respect [B63], biomedical ethics education [B78], interdisciplinary communication [B70], counseling [C87], genetics [E6], decision making [E54], evaluation [E61], human genome project [F8], acquired immunodeficiency syndrome [N35], persistent vegetative state [N42], genetic intervention [Q22], genetic counseling [Q36], genetic screening [Q37], preimplantation genetic diagnosis [Q38], eugenics [Q39], negative eugenics [Q40], positive eugenics [Q41], gene therapy [Q42], human genome [Q47].

F 160 **PETON** Patrick
M.D. Clinician, Emergency Department. — Lecturer, School of Medicine, University of Nancy. — Chairperson of the CCPPRB of Nancy.

CCPPRB de Nancy
29, avenue du Maréchal-de-Lattre-de-Tassigny
54035 Nancy Cedex (France)

■ AFFILIATIONS: CCPPRB de Nancy, France (*Chairperson*).
■ TEACHING: Forensic Medicine and Medical Law, University degree in Forensic Medicine and Toxicology, post-university education in ethics and Medical Law (University of Forensic Medicine).
■ PUBLICATIONS: yes.
■ INTERESTED IN: protection of rights — involved institutions [B16], bioethics [B89], biomedical research and experimentation [F1], biological specimens procurement, blood transfusion, organ transplantation [I8], physicians and researchers accountability [J23], legislation and law [K34], determination of death [O2].

F 161 **PETTITI** Louis
Judge, European Court of Human Rights. — Barrister-at-Law, former President of the Bar (Ordre des avocats de Paris).

4, square La Bruyère
75009 Paris (France)
✆ (33) 1 - 42 80 49 54 Fax: (33) 1 - 48 74 15 00

■ AFFILIATIONS: Institut de formation en Droits de l'Homme du barreau de Paris, Paris, France (*Chairperson*). — Human Rights Committee, French Commission Unesco, France (*Chairperson*).
■ TEACHING: Human rights: Institut de formation en Droits de l'Homme du Barreau de Paris. — Centre de formation des avocats. — Ecole de police.
■ PUBLICATIONS: yes.
■ WORKING ON: respect of human dignity and human rights [B1], protection of rights — involved institutions [B16].
■ INTERESTED IN: sexuality and procreation [M2], reproductive technologies [M26], abortion [M47], fetal development [M54], fetal therapy [M57], genetic defects and hereditary diseases [Q2], biotechnology — genetic engineering [Q20], behavioral genetics [Q34], medical genetics [Q35], population genetics [Q43].

F 162 **POHIER** Jacques
National Chairperson of the Association pour le droit de mourir dans la dignité (Association for the right to die with dignity, ADM.D.).

Association pour le droit de mourir dans la dignité (ADMD)
103, rue Lafayette
75010 Paris (France)
✆ (33) 1 - 42 85 12 22 Fax: (33) 1 - 45 96 00 50

■ AFFILIATIONS: Association pour le droit de mourir dans la dignité — ADM.D., Paris, France (*Chairperson*).
■ TEACHING: Within the ADMD writing of documents, leaflets and newletters (25 000 readers). — Basic training to local representatives (80).
■ PUBLICATIONS: yes.
■ WORKING ON: self determination [B4], dignity [B5], due process [B40], public advocacy [B41], patient advocacy [B42], ethics [B48], moral policy [B55], morality [B61], European Convention of Bioethics [B97], Bioethics bill, 1992 (France) [B98], Roman catholic ethics [C23], Protestant ethics [C27], theology [C41], historical aspects [C57], legislation [D13], French National Assembly [D68], French Senate [D73], state responsibility [D75], elderly [H23], patient access [J5], disclosure [J7], consent to treatment [J8], informed consent [J9], presumed consent [J10], third party consent [J11], spousal consent [J12], consent forms [J14], treatment refusal [J15], patient participation [J16], right to treatment [J17], advance directives [J18], living wills [J19], nurse patient relationship [J30], physician patient relationship [J32], professional patient relationship [J33], economic value of life [K28], law enforcement [K57], bill [K60], aged — problems related with aging [L45], disease and aged [L46], life extension [L47], health care services and aged [L48], aged [L49], aging [L50], suffering [N11], pain [N12], persistent vegetative state [N42], brain death [O6], terminal care [O7], palliative care [O8].
■ INTERESTED IN: self determination [B4], dignity [B5], due process [B40], public advocacy [B41], patient advocacy [B42], ethics [B48], moral policy [B55], morality [B61], European Convention of Bioethics [B97], Bioethics bill, 1992 (France) [B98], Roman catholic ethics [C23], Protestant ethics [C27], theology [C41], historical aspects [C57], legislation [D13], French National Assembly [D68], French Senate [D73], state responsibility [D75], elderly [H23], patient access [J5], disclosure [J7], consent to treatment [J8], informed consent [J9], presumed consent [J10], third party consent [J11], spousal consent [J12], consent forms [J14], treatment refusal [J15], patient participation [J16], right to treatment [J17], advance directives [J18], living wills [J19], nurse patient relationship [J30], professional patient relationship [J33], economic value of life [K28], law enforcement [K57], bill [K60], aged — problems related with aging [L45], disease and aged [L46], life extension [L47], health care services and aged [L48], aged [L49], aging [L50], suffering [N11], pain [N12], persistent vegetative state [N42], brain death [O6], terminal care [O7], palliative care [O8], life-sustaining treatment [O9].

F 163 **POMPIDOU** Alain
Professor. — M.D. Clinician, Head of the Department (Histology, Embryology, Cytogenetics, Anatomopathology), Saint-Vincent-de-Paul Hospital. — European M.P.

Faculté de médecine Cochin
24, rue du faubourg Saint-Jacques
75014 Paris (France)
✆ (33) 1 - 44 41 23 50

■ PUBLICATIONS: yes.
■ WORKING ON: human person [B12], human body commercialization [B13], traffic of organs [B15], ethics [B48], political activity [D57], EC — European Communities [D62], biology [E5], genetics [E6], molecular biology [E8], research policy [E18], technology assessment [E65], human genome project [F8], embryos [M10], fetuses [M11], genetic defects and hereditary diseases [Q2], genetic defects [Q3], chromosomal disorders [Q4], XYY karyotype [Q5], down's syndrome [Q6], single gene defects [Q7], genetic counseling [Q36], genetic screening [Q37], human genome [Q47].
■ INTERESTED IN: ethics committees [B22], Council of Europe [B25], office of science and technology assessment [B28], public advocacy [B41], biomedical ethics education [B78], CCNE advice (France) [B91], European Convention of Bioethics [B97], cultural pluralism [D51], social discrimination [D53], French National Assembly [D68], French Council of State [D70], industrial research [E22], military research [E24], technology [E27], industry [E36], containment [E38], technology assessment [E65], human genome project [F8], fetal therapy [M57], death and resuscitation [O1], genetic defects and hereditary diseases [Q2], biotechnology — genetic engineering [Q20], medical genetics [Q35].

F 164 **QUERE** France
Research Fellow in theology.

3, rue Laplace
75005 Paris (France)
✆ (33) 1 - 43 26 61 12

■ AFFILIATIONS: CCNE, Paris, France (*Member*).
■ TEACHING: INSERM Clubs. — Meeting of the CNRS (Arc-et-Senans).

■ PUBLICATIONS: yes.
■ WORKING ON: child and family [L31], handicapped children [L38], psychiatric technics [P24], children's rights [L25], CCNE advice (France) [B91], fundamental rights of the individuals [B2].
■ INTERESTED IN: population growth [H20], population control [H22].

F 165 **RAPHAEL** Martine
Lecturer. — M.D. Clinician, Department of Haematology.

CCPPRB Groupe hospitalier Pitié-Salpétrière
47, boulevard de l'Hôpital
75651 Paris Cedex 13 (France)
✆ (33) 1 - 45 70 37 45 Fax: (33) 1 - 45 70 26 20

■ AFFILIATIONS: CCPPRB Pitié-Salpétrière, Paris Cedex 13, France (*Chairperson*).
■ WORKING ON: CCPPRB — French committees for the protection of human subjects of biomedical research [B32], molecular biology [E8], clinical trials [F3], human experimentation [F11], research subjects [F12], healthy volunteers [F13], control groups [F17], random selection [F18], hospital [G28], medical evaluation [G51], blood transfusions [I19], bone marrow transplantation [I29], informed consent [J9], consent forms [J14], bone marrow [L8], cancer [N17], leukemia [N18], acquired immunodeficiency syndrome [N35], HIV seropositivity [N36].
■ INTERESTED IN: integrity [B8], European Convention of Bioethics [B97], epidemiology [E9], clinical trials [F3], cognitive research [F7], human genome project [F8], intracerebral fetal tissue transplantation [I28], placebos [I36].

F 166 **RAPPAPORT** R.
Head of the Department of Paediatric Endocrinology. — Head of a INSERM research unit. — Coordinator of the Paediatrics Department, Enfants-Malades Hospital (Paris).

Höpital Necker-Enfants-Malades
149, rue de Sèvres
75743 Paris Cedex 15 (France)
✆ (33) 1 - 42 73 83 61 Fax: (33) 1 - 42 73 83 60

■ PUBLICATIONS: yes.
■ WORKING ON: biomedical research [F2], clinical trials [F3], therapeutic research [F6], health care [G45], diagnosis [I2], treatment [I30], health care professionals, researchers and patient relationship [J1].
■ INTERESTED IN: *idem.*

F 167 **RENARD** Jean-Paul
Research Director at the INRA, Head of the Developement Biology Department.

INRA
Unité de biologie du développement
78352 Jouy-en-Josas Cedex (France)
✆ (33) 1 - 34 65 25 94 Fax: (33) 1 - 34 65 22 73

■ AFFILIATIONS: CCNE, Paris, France
■ PUBLICATIONS: yes.
■ WORKING ON: biotechnology — genetic engineering [Q20], freezing [E29], preservation [E30], cryonic suspension [E31], biology [E5], genetics [E6].
■ INTERESTED IN: genetic defects and hereditary diseases [Q2], biotechnology — genetic engineering [Q20], animal experimentation [F10], excess embryos [M34], human person [B12].

F 168 **REVILLARD** Mariel
Lawyer.

CRIDON
59 bis, rue de Créqui
69006 Lyon (France)
✆ (33) 78 93 93 62 Fax: (33) 78 94 18 50

■ AFFILIATIONS: Association internationale de Droit, Ethique et Science, groupe de Milazzo, Strasbourg, France
■ PUBLICATIONS: yes.
■ WORKING ON: privacy [B11], human person [B12], ethics committees [B22], Council of Europe [B25],

ethics [B48], codes of ethics [B49], bioethical issues [B96], Bioethics bill, 1992 (France) [B98], legislation [D13], family relationship [D31], data protection [E52], sexuality and procreation [M2], embryos [M10], sperm [M14], procreation [M15], reproduction [M17], reproductive technologies [M26], genome mapping [Q23], sex preselection [Q30], diagnosis [I2], prenatal diagnosis [I3], biological specimens procurement, blood transfusion, organ transplantation [I8], sperm banks [I11], donors [I14], organ donors [I15], transplantation [I22], transplant recipients [I23], organ donation [I25], patients' rights [J2], law, legislation and jurisprudence [K1], legal personality [K2], legislation and law [K34].

■ INTERESTED IN: privacy [B11], human person [B12], ethics committees [B22], Council of Europe [B25], ethics [B48], codes of ethics [B49], bioethical issues [B96], Bioethics bill, 1992 (France) [B98], legislation [D13], family relationship [D31], data protection [E52], sexuality and procreation [M2], embryos [M10], sperm [M14], procreation [M15], reproduction [M17], reproductive technologies [M26], genome mapping [Q23], sex preselection [Q30], diagnosis [I2], prenatal diagnosis [I3], biological specimens procurement, blood transfusion, organ transplantation [I8], sperm banks [I11], donors [I14], organ donors [I15], transplantation [I22], transplant recipients [I23], organ donation [I25], patients' rights [J2], law, legislation and jurisprudence [K1], legal personality [K2], legislation and law [K34], preventive medicine [G58].

F 169 **ROCHE** Louis

Editions Lacassagne
162, avenue A.-Lacassagne
69003 Lyon (France)
✆ (33) 72 33 40 40 Fax: (33) 72 39 16 74

F 170 **ROLLIN** Francis
Teacher of Moral Theology (Catholic).

2, place Gailleton
69002 Lyon (France)
✆ (33) 78 37 49 82

■ AFFILIATIONS: Centre de bioéthique de l'Université catholique de Lyon, Lyon Cedex 02, France (*Member*).
■ TEACHING: yes.
■ PUBLICATIONS: yes.
■ WORKING ON: fundamental rights of the individuals [B2], human person [B12], ethics committees [B22], CCNE — Comité Consultatif National d'Ethique (France) [B23], ethical rules and principles [B47], ethics [B48], codes of ethics [B49], moral policy [B55], education [B76], CCNE advice (France) [B91], ethicists [B92], ethical review [B93], bioethical issues [B96], philosophical ethics [C2], teleological ethics [C7], morals [C12], values [C15], religion and religious ethics [C19], Christian ethics [C22], Roman catholic ethics [C23], natural law [C24], Roman catholicism [C31], scriptural interpretation [C40], theology [C41], family members [D24], couple [D26], marital relationship [D29], decision making [E54], family planning [G8], prenatal diagnosis [I3], potentiality of personhood [K5], conflict of interest [K9], wrongful life [K33], beginning of life [L24], sexuality and procreation [M2], contraception [M3], reproductive technologies [M26], abortion [M47], terminal care [O7], euthanasia [O15], eugenics [Q39], preimplantation genetic diagnosis [Q38].
■ INTERESTED IN: *idem, plus:* humanity heritage [B46], philosophy of biology [C42], legislation [D13], social impact [D49], cultural pluralism [D51], democracy [D82], dehumanization [D86], health care costs [H9], compensation [H12], public participation [H13], remuneration [H14], price [H16], profit [H17], financial support [H18], population [H19], conscience clause [K63], human development [L12], mind [L20], infanticide [L42], reproductive technologies [M26], abortion [M47], determination of death [O2], terminal care [O7], euthanasia [O15], suicide [O19], biotechnology — genetic engineering [Q20], behavioral genetics [Q34], medical genetics [Q35], population genetics [Q43]

F 171 **ROPERT** Jean-Claude
Head of the Paediatrics and Neonatalogy Department.

Centre Hospitalier
Service de Pédiatrie
36, boulevard du Général-Leclerc
92200 Neuilly-sur-Seine (France)
✆ (33) 1 - 47 47 11 44 Fax: (33) 1 - 46 24 16 06

■ AFFILIATIONS: Groupe d'études en néonatalogie et urgences pédiatriques de la région parisienne, Paris, France (*Member*).

■ TEACHING: Seminars.
■ PUBLICATIONS: yes.
■ WORKING ON: freedom [B10], human person [B12], ethics committees [B22], moral obligations [B50], professional competence [B52], obligations of society [B53], obligations to society [B54], respect [B63], health and biology mediatisation [B85], uncontrolled information [B88], values [C15], quality of life [C16], value of life [C18], medical ethics [C52], nursing ethics [C53], history [C56], parents [D32], family relationship [D31], handicapped [D43], decision analysis [E55], self regulation [E62], peer review [E63], public hospitals [G30], intensive care units [G33], patient care team [G38], physicians [G40], medical evaluation [G51], parental notification [J6], disclosure [J7], parental consent [J13], physician's role [J26], physician nurse relationship [J31], physician patient relationship [J32], professional patient relationship [J33], child's interest [K24], public policy [D77], decision making [E54], emergency care [G54], pediatrics [G68], brain [L6], beginning of life [L24], newborns [L29], prematurity [L30], mother fetus relationship [M43], viability [M56], fetal therapy [M57], physically [N8], brain diseases [N44], prenatal injuries [N50], death [O5], brain death [O6], withholding treatment [O10], prolongation of life [O13], quality adjusted life years [O14], resuscitation [O21], life-sustaining treatment [O9], euthanasia [O15].
■ INTERESTED IN: resuscitation [O21], withholding treatment [O10], prolongation of life [O13], viability [M56], life-sustaining treatment [O9], euthanasia [O15], quality adjusted life years [O14].

F 172 **ROUFFY** Jacques

Professor, UFR Villemein, Department of the Paris VII University (Medicine). — Head of the Internal Medicine Department, Saint-Louis Hospital (Paris).

CCPPRB-Hôpital Saint-Louis
1, avenue Claude-Vellefaux
75475 Paris Cedex 10 (France)
✆ (33) 1 - 42 49 97 80 Fax: (33) 1 - 42 49 97 80

■ AFFILIATIONS: CCPPRB Hôpital Saint-Louis, France (*Chairperson*).
■ WORKING ON: fundamental rights of the individuals [B2], ethical rules and principles [B47], bioethics [B89], medical ethics [C52], containment [E38], biomedical research and experimentation [F1], treatment [I30], health care professionals, researchers and patient relationship [J1], disease [N1].
■ INTERESTED IN: CCNE — Comité Consultatif National d'Ethique (France) [B23], CCPPRB — French committees for the protection of human subjects of biomedical research [B32], CCNE advice (France) [B91], medical ethics [C52], review committees [E42], clinical trials [F3], cognitive research [F7], patients' rights [J2].

F 173 **ROYER** Pierre

Chairperson of the Centre International de l'Enfance.

Centre International de l'Enfance
Château de Longchamp
Carrefour de Lonchamp
75016 Paris (France)
✆ (33) 1 - 45 20 79 92 Fax: (33) 1 - 45 25 73 67

■ AFFILIATIONS: CCNE, Paris, France (*Member*).
■ TEACHING: conferences.
■ PUBLICATIONS: yes.
■ WORKING ON: public health [G5], pediatrics [G68], organ transplantation [I24], childhood difficulties, diseases, protection [L21], viability [M56], fetal therapy [M57], medical genetics [Q35].

F 174 **RUBELLIN-DEVICHI** Jacqueline

Full Professor. — Director of the Centre de droit de la famille (Centre for Family Law). — URA CNRS.

Université Jean-Moulin
1, rue de l'Université
BP 0638
69239 Lyon Cedex 02 (France)
✆ (33) 72 72 20 63

■ TEACHING: Family Law for postgraduate DEA students.
■ PUBLICATIONS: yes.
■ WORKING ON: human rights [B3], traffic of organs [B15], European Convention on Human Rights [B36], Declaration on Human and Citizen Rights [B38], Universal Declaration of Human Rights [B39], CCNE advice (France) [B91], Bioethics bill, 1992 (France) [B98], physicians [G40], social workers [G44], tissue banking [I12], tissue donation [I13], anonymous donation [I17], blood donation [I20],

transplant recipients [I23], physicians and researchers accountability [J23], legal personality [K2], minors [K6], child donation [K22], stages of life — problems proper to childhood and elderly [L1], childhood difficulties, diseases, protection [L21], sexuality and procreation [M2], genetics and applications — biotechnology [Q1], genetic defects and hereditary diseases [Q2], genetic counseling [Q36], gene pool [Q45].
■ INTERESTED IN: *idem, plus:* aged — problems related with aging [L45]

F 175 **SAILLY** Jean-Claude
Director of the Centre de recherches économiques, sociologiques et de gestion (Centre for Economic, Sociological and Managerial Research, CRESGE).

Centre de recherches économiques, sociologiques et de gestion (CRESGE)
BP 109
1, rue Norbert-Ségard
59016 Lille Cedex (France)
✆ (33) 20 57 11 77 Fax: (33) 20 78 26 45

■ AFFILIATIONS: Centre d'éthique médicale, Lille, France (*Member*).
■ TEACHING: conferences on bioethics organised by the Centre d'éthique médicale (economic aspects).
■ PUBLICATIONS: yes.
■ WORKING ON: value of life [C18], social sciences [C89], socioeconomic factors [D18], technology assessment [E65], costs and benefits [H10], therapeutic abortion [M50], persistent vegetative state [N42], euthanasia [O15].
■ INTERESTED IN: *idem.*

F 176 **SANN** Léon
M.D. Clinician, Paediatrics.

Hôpital Debrousse
29, rue Soeur-Bouvier
69005 Lyon (France)
✆ (33) 78 25 16 65

■ AFFILIATIONS: Santé, Ethique, Libertés (Lyon) — SEL, France (*Member*).
■ TEACHING: AEU (university attestation) on medical ethics and philosophy, Faculty of Lyon-Nord.
■ PUBLICATIONS: yes.
■ WORKING ON: medical ethics [C52], biomedical ethics education [B78], children [L27].
■ INTERESTED IN: morals [C12], philosophy [C61].

F 177 **SAUTIER** Jean-René
Civil Administrator. — Honorary Chairperson of SANOFI (Chemistry company). — Former Chairperson of the Syndicat national de l'industrie pharmaceutique (Natonal pharmaceutical industry organisation).

SANOFI
40, avenue Georges V
75008 Paris (France)
✆ (33) (1) 40 73 40 73

■ AFFILIATIONS: CCNE, Paris, France (*Member*).
■ PUBLICATIONS: yes.

F 178 **SCHAERER** René
Professor, M.D. Clinician (cancerology).

CHU Grenoble
BP 217
38043 Grenoble (France)
✆ (33) 76 76 75 75 Fax: (33) 76 44 77 40

■ AFFILIATIONS: Fédération des Associations "Jusqu'à la mort accompagner la vie" (Stay with life till death) (JALMALV) et associées — JALMALV, Grenoble, France (*Chairperson*).
■ TEACHING: Joseph-Fourier University, Grenoble (cancerology, University diploma of palliative care).
■ PUBLICATIONS: yes.
■ WORKING ON: dignity [B5], human person [B12], patient advocacy [B42], ethics [B48], codes of ethics [B49], deontology [B59], biomedical ethics education [B78], nursing education [B79], medical

education [B80], students [B82], curriculum [B83], universities [B84], bioethics [B89], bioethical issues [B96].

F 179 **SEBAG-LANOE**
M.D. Clinician, Head of the Department of Gerontology and Palliative care, Paul-Brousse Hospital.

Groupe hospitalier Paul-Brousse
Service gérontologie et soins palliatifs
14, avenue Paul Vaillant-Couturier
BP 200
94804 Villejuif Cedex (France)
✆ (33) 1 - 45 59 30 63

■ AFFILIATIONS: Commission Droits et libertés des personnes âgées de la Fondation nationale de gérontologie, France (*Member*). — Comité d'éthique des Hôpitaux de Paris, Paris, France (*Member*).
■ TEACHING: University diploma of palliative care. — Numerous conferences on education (France and abroad).
■ PUBLICATIONS: yes.
■ WORKING ON: dignity [B5], ethics committees [B22], due process [B40], quality of life [C16], medical ethics [C52], elderly [H23], disease and aged [L46], life extension [L47], health care services and aged [L48], aged [L49], aging [L50], palliative care [O8], terminally ill [O12], attitudes to death [O20].
■ INTERESTED IN: *idem.*

F 180 **SEDAT** Jacques
Psychoanalyst.

36, rue Pierre-Sémard
75009 Paris (France)
✆ (33) 1- 42 80 22 86

■ AFFILIATIONS: Revue Esprit, Paris, France (*Member*).
■ TEACHING: Conferences (Association Descartes).
■ PUBLICATIONS: yes.

F 181 **SÉRIE** Charles
Chairperson of the ADAPEI Ariège. — Chairperson of the URAPEI Midi-Pyrénées. — Delegate of the UNAPEI at the Comité de réflexion d'éthique sur le handicap mental de l'UNAPEI (ethics committee on mental handicaps).— Honorary Professor, Institut Pasteur.

ADAPEI Ariège
BP 133
09104 Pamiers Cedex (France)
✆ (33) 61 60 89 32 Fax: (33) 61 60 08 56

■ WORKING ON: prenatal diagnosis [I3], amniocentesis [I4], sexuality and procreation [M2], contraception [M3], condom [M4], immunologic contraception [M5], sterilization [M6], involuntary sterilization [M7], volontary sterilization [M8], procreation [M15], sexual behavior [M21], homosexuality [M22], homosexuals [M23], genetic counseling [Q36].
■ INTERESTED IN: codes of biomedical ethics [B90], CCNE advice (France) [B91], prenatal diagnosis [I3], amniocentesis [I4], mass screening [I6], mandatory screening [I7], sexuality and procreation [M2], contraception [M3], condom [M4], immunologic contraception [M5], sterilization [M6], involuntary sterilization [M7], volontary sterilization [M8], procreation [M15], sexual behavior [M21], homosexuality [M22], homosexuals [M23], resuscitation [O21], mentally handicapped [P7], mentally ill [P9], behavior disorders [P13], genetic defects [Q3], down's syndrome [Q6], genetic counseling [Q36], genetic screening [Q37].

F 182 **SERUSCLAT** Franck
Senator of Rhône, Mayor of Saint-Fons. — Member of the Office parlementaire d'évaluation des choix scientifiques et technologiques (Parliamentary Office on the evaluation of scientific and technological choices).

Palais du Luxembourg
15, rue de Vaugirard
75006 Paris (France)
✆ (33) 1 - 42 34 30 24 Fax: (33) 1 - 42 34 34 72

■ AFFILIATIONS: Ethique et Société (association), Paris, France (*Chairperson*).
■ PUBLICATIONS: yes.
■ WORKING ON: fundamental rights of the individuals [B2], ethical rules and principles [B47], biomedical research [F2], experimentation [F9], sexuality and procreation [M2], reproductive technologies [M26], abortion [M47], genetic defects and hereditary diseases [Q2].

F 183 **SEVE** Lucien

Institut de recherches marxistes
64, boulevard Auguste-Blanqui
75013 Paris (France)
✆ (33) (1) 43 36 45 34

■ AFFILIATIONS: Comité consultatif national d'éthique — CCNE, France (*Member*).
■ PUBLICATIONS: yes.

F 184 **SIKSOU** Maryse
Lecturer.

Université Paris-VII
13, rue de Santeuil
75231 Paris Cedex 05 (France)

■ TEACHING: Clinical human science, Laboratory of fundamental psychology, Paris-VII.
■ WORKING ON: book review [A6], psychology [C68], research [E14], investigators [E15], research team [E16], research institutes [E17], research policy [E18], research design [E19], biomedical research and experimentation [F1], biomedical research [F2], human experimentation [F11], methodology of research and experimentation [F15], health care professionals, researchers and patient relationship [J1], aged — problems related with aging [L45], neurosciences — psychiatry [P1].
■ INTERESTED IN: documentation [A8], publications [A9], respect of human dignity and human rights [B1], philosophical ethics [C2], religion and religious ethics [C19], professional ethics [C51], psychology [C68], political issues [D56], research [E14], containment [E38], evaluation [E61], biomedical research [F2], human experimentation [F11], methodology of research and experimentation [F15], organization of health care; facilities, manpower and services; health occupations [G1], health care professionals, researchers and patient relationship [J1], law, legislation and jurisprudence [K1], aged — problems related with aging [L45], neurosciences — psychiatry [P1].

F 185 **SINET** Pierre-Marie
Research Director, CNRS. — Assistant, Department of Genetics, Necker Hospital. — Director of the URA CNRS 1335, Research Unit.

Laboratoire de biochimie génétique
Hopital Necker-Enfants-Malades
149, rue de Sèvres
75743 Paris (France)
✆ (33) 1 - 43 06 92 64 et 42 73 88 18 Fax: (33) 1 - 42 73 06 59

■ AFFILIATIONS: Comité opérationnel sur l'éthique dans les sciences de la vie au CNRS, Paris, France (*Member*).
■ WORKING ON: respect of human dignity and human rights [B1], science, technology, methods [E1], biomedical research and experimentation [F1], health care, medical acts [I1], genetics and applications — biotechnology [Q1].
■ INTERESTED IN: science, technology, methods [E1], biomedical research and experimentation [F1], health care, medical acts [I1], health care professionals, researchers and patient relationship [J1], handicapped children [L38], abortion [M47], mentally handicapped [P7], mentally retarded [P8], genetics and applications — biotechnology [Q1].

F 186 **TAILLEMITE** Jean-Louis
Professor of Genetics, Saint-Antoine School of Medicine. — Head of the Laboratory of Cytogenetics, Saint-Antoine Hospital.

Laboratoire de cytogénétique
Hôpital Saint-Antoine
184, rue du faubourg Saint-Antoine
75012 Paris (France)
✆ (33) 1 - 49 28 21 89

■ WORKING ON: social issues [D2], biomedical research [F2], public health [G5], health policy [G13], preventive medicine [G58], family practice [G60], childhood difficulties, diseases, protection [L21], abortion [M47], fetal development [M54], fetal therapy [M57], congenital defects [N23], nervous system diseases [N40], genetics and applications — biotechnology [Q1], genetic defects and hereditary diseases [Q2], genetic defects [Q3], chromosomal disorders [Q4], XYY karyotype [Q5], down's syndrome [Q6], single gene defects [Q7], dominant genetic conditions [Q8], Huntington's chorea [Q9], recessive genetic conditions [Q10], sickle cell anemia [Q11], Duchenne muscular dystrophy [Q12], hemophilia [Q13], positive eugenics [Q41], cystic fibrosis [Q15], phenylketonuria [Q16], thalassemia [Q17], hereditary diseases [Q18], Tay Sachs disease [Q14], biotechnology — genetic engineering [Q20], biotechnology [Q21], genetic intervention [Q22], genome mapping [Q23].
■ INTERESTED IN: biotechnology [Q21], genetic intervention [Q22], cloning [Q24], sex determination [Q26], genetically modified organisms [Q28], transgenic animals [Q29], sex preselection [Q30], DNA fingerprinting [Q32], genetic counseling [Q36], genetic screening [Q37], preimplantation genetic diagnosis [Q38], eugenics [Q39], negative eugenics [Q40], positive eugenics [Q41], gene therapy [Q42], genetic identity [Q44], gene pool [Q45], carriers [Q46], human genome [Q47].

F 187 **TASSEAU** François

M.D. Clinician, Head of Department. — Coordinator of neurological rehabilitation programmes (centre médical de l'Argentière).

Centre médical de l'Argentière-Aveize
69610 Sainte-Foy-l'Argentière (France)
✆ (33) 74 26 41 41 Fax: (33) 74 26 42 87

■ PUBLICATIONS: yes.
■ WORKING ON: suffering [N11], prognosis [N14], coma [N41], persistent vegetative state [N42], life-sustaining treatment [O9], withholding treatment [O10], allowing to die [O11], prolongation of life [O13], quality adjusted life years [O14].
■ INTERESTED IN: nontherapeutic research [F5], cognitive research [F7].

F 188 **TEMPE** Jean-Daniel

Head of the Intensive Care and Emergency Department (Hautepierre Hospital).

CCPPRB Alsace-I Strasbourg
1, place de l'Hôpital
67091 Strasbourg Cedex (France)
✆ (33) 88 16 13 57 Fax: (33) 88 16 13 58

■ AFFILIATIONS: Comité d'éthique des facultés de médecine, d'odontologie, de pharmacie et du CHRU de Strasbourg, Paris, France (*Member*). — Comité d'éthique de la Société de réanimation de langue française, Strasbourg, France (*Member*).
■ TEACHING: Pratical classes on medical ethics for undergraduates and nurses. — Conferences.
■ PUBLICATIONS: yes.
■ WORKING ON: specific approaches to ethics [C1], professional ethics [C51], medical ethics [C52], nursing ethics [C53], biomedical research and experimentation [F1], biomedical research [F2], clinical trials [F3], experimentation [F9], methodology of research and experimentation [F15], health care professionals, researchers and patient relationship [J1], informed consent [J9], presumed consent [J10], disease [N1], suffering [N11], persistent vegetative state [N42], death and resuscitation [O1], brain death [O6], life-sustaining treatment [O9], withholding treatment [O10], suicide [O19], resuscitation [O21].
■ INTERESTED IN: *idem.*

F 189 **TERRENOIRE** Gwen

Research Engineer, CNRS.

Centre de sociologie de l'éthique
59, rue Pouchet
75849 Paris Cedex 17 (France)
✆ (33) (1) 40 25 10 74 Fax: (33) (1) 42 28 95 44

■ AFFILIATIONS: Comité de réflexion d'éthique de l'UNAPEI — UNAPEI, Paris, France (*Member*).
■ PUBLICATIONS: yes.
■ WORKING ON: information centers [A5], documentation [A8], social control [D3], family members [D24], handicapped [D43], social problems [D52], research policy [E18], human genome project [F8], Huntington's chorea [Q9], genetic counseling [Q36], genetic screening [Q37], eugenics [Q39].
■ INTERESTED IN: information centers [A5], documentation [A8], social and political issues [D1], human

genome project [F8], Huntington's chorea [Q9], genetic counseling [Q36], genetic screening [Q37], eugenics [Q39].

F 190 **THÉPOT F.**
Head of the Department of Reproductive Biology CHU (University Hospital) of Amiens (France). — Professor of hystology, embryology, cytogenetic, Faculty of Medicine, Amiens. — Chairperson of the Genetics Commission de of the Federation of the CECOS.

Commission génétique de la Fédération des CECOS
124, rue Camille-Desmoulins
80000 Amiens (France)
✆ (33) 22 53 36 77 Fax: (33) 22 53 36 79

■ AFFILIATIONS: Commission de génétique de la Fédération des CECOS, Amiens, France (*Chairperson*). — Commission d'éthique de la Fédération des CECOS, Amiens, France (*Member*).
■ TEACHING: Post-university education. — Health care teaching.
■ PUBLICATIONS: yes.
■ WORKING ON: quality of life [C16], social worth [C17], value of life [C18], medical ethics [C52], couple [D26], cohabitation [D27], married persons [D28], marital relationship [D29], sibling [D30], family relationship [D31], genetics [E6], biological containment [E39], prenatal diagnosis [I3], amniocentesis [I4], chorionic villus sampling [I5], mass screening [I6], mandatory screening [I7], biological substances contamination [I9], required request [I10], sperm banks [I11], tissue banking [I12], tissue donation [I13], donors [I14], organ donors [I15], donor cards [I16], anonymous donation [I17], directed donation [I18], sexuality and procreation — unborn child [M1], procreation [M15], wish of children [M16], reproduction [M17], fertility [M18], infertility [M19], reproductive technologies [M26], ovum donors [M27], semen donors [M28], in vitro fertilization [M29], FIV center [M30], artificial insemination [M31], AID [M32], AIH [M33], excess embryos [M34], host mothers [M35], embryo transfer [M36], GIFT (Gamete Intrafallopian Transfer) [M37], embryo donation [M38], selective abortion [M49], therapeutic abortion [M50], fetal development [M54], twinning [M55], viability [M56], fetal therapy [M57], genetics and applications — biotechnology [Q1], genetic defects and hereditary diseases [Q2], sex linked defects [Q19], DNA linkage [Q33], medical genetics [Q35], gene therapy [Q42].
■ INTERESTED IN: biomedical ethics education [B78], quality of life [C16], social worth [C17], value of life [C18], medical ethics [C52], couple [D26], family relationship [D31], genetics [E6], biological containment [E39], prenatal diagnosis [I3], mandatory screening [I7], biological substances contamination [I9], directed donation [I18], sexuality and procreation — unborn child [M1], procreation [M15], wish of children [M16], reproduction [M17], fertility [M18], infertility [M19], reproductive technologies [M26], ovum donors [M27], semen donors [M28], in vitro fertilization [M29], FIV center [M30], artificial insemination [M31], AID [M32], AIH [M33], excess embryos [M34], host mothers [M35], embryo transfer [M36], GIFT (Gamete Intrafallopian Transfer) [M37], embryo donation [M38], selective abortion [M49], therapeutic abortion [M50], fetal development [M54], viability [M56], fetal therapy [M57], genetics and applications — biotechnology [Q1], genetic defects and hereditary diseases [Q2], biotechnology — genetic engineering [Q20], medical genetics [Q35].

F 191 **THÉVENOT** Xavier
Professor, Institut catholique de Paris.

393bis, rue des Pyrénées
75020 Paris (France)
✆ (33) 1 - 47 97 77 14

■ TEACHING: Courses within the Catholic theology curriculum (B.A. and M.A.), Institut catholique de Paris.
■ PUBLICATIONS: yes.
■ WORKING ON: respect of human dignity and human rights [B1], ethical rules and principles [B47], CCNE advice (France) [B91], specific approaches to ethics [C1], religion and religious ethics [C19], Roman catholic ethics [C23], psychology [C68], counseling [C87], sexuality and procreation — unborn child [M1], homosexuality [M22], reproductive technologies [M26], abortion [M47], psychotherapy [P30].
■ INTERESTED IN: *idem*.

F 192 **THIBAULT** Charles
Professor Emeritus, Paris-VI University (retired).

INRA
Physiologie animale
78352 Jouy-en-Josas Cedex (France)
✆ (33) 1 - 34 65 23 61 Fax: (33) 1 - 34 65 23 64

■ PUBLICATIONS: yes.
■ WORKING ON: embryos [M10], ovum [M12], sperm [M14], reproductive technologies [M26], fetal therapy [M57], animal experimentation [F10], human experimentation [F11], cloning [Q24], genetically modified organisms [Q28], transgenic animals [Q29], sex preselection [Q30].

F 193 **THOUVENIN** Dominique
Lecturer, Faculty of Law, Jean-Moulin University (Lyon-III).

Université Jean-Moulin
1, rue de l'Université
69239 Lyon Cedex 02 (France)
✆ (33) 72 72 45 55

■ TEACHING: yes.
■ PUBLICATIONS: yes.
■ WORKING ON: legal personality [K2], jurisprudence — accountability [K8], authorization [K37], rights [K42], medical law [K51], health legislation [K52], criminal law [K53], criminal code [K54], state courts [K62], conscience clause [K63], therapeutic injunction [K64].
■ INTERESTED IN: biomedical research [F2], experimentation [F9], methodology of research and experimentation [F15], diagnosis [I2], biological substances contamination [I9], treatment [I30], confidentiality [J20], acquired immunodeficiency syndrome [N35], HIV seropositivity [N36], sexuality and procreation [M2], medical genetics [Q35], population genetics [Q43].

F 194 **TOLEDANO** Claude
M.D. Clinician, Head of Department. — Chief-Physician of the 17th District of General Psychiatry of Hauts-de-Seine.

Hôpital psychiatrique Paul-Guiraud
54, avenue de la République
94800 Villejuif (France)
✆ (33) 1 - 45 59 57 44 Fax: (33) 1 - 46 77 20 26

■ AFFILIATIONS: Association scientifique d'éthique, recherche informatique et enseignement (ethics, computer research and teaching) — ERIE, Villejuif, France (*Scientific secretary*).
■ TEACHING: Organisation of symposiums for ERIE and the CHS Paul-Guiraud on ethics in psychiatry. — Conférences et tables rondes sur ce sujet.
■ WORKING ON: consent to treatment [J8], treatment refusal [J15], medical secrecy [J22], inter-personal relationship [J27], psychotic disorders [P10], outpatient commitment [P18], psychotherapy [P30].
■ INTERESTED IN: child's interest [K24], medical law [K51], conscience clause [K63], children's rights [L25].

F 195 **TOMKIEWICZ** Stanislas
Research Director at the INSERM. — Consultant in Psychiatry.

INSERM U 69
1, rue du 11-Novembre
92120 Montrouge (France)
✆ (33) 1 - 47 35 80 89

■ AFFILIATIONS: Comité européen Droit, éthique et psychiatrie, Paris, France (*Member*).
■ TEACHING: Occasional conferences.
■ PUBLICATIONS: yes.
■ WORKING ON: fundamental rights of the individuals [B2], dignity [B5], human equality [B7], justice [B9], freedom [B10], human person [B12], ethical rules and principles [B47], respect [B63], biology and human future [C43], professional ethics [C51], nursing ethics [C53], social control [D3], social dominance [D6], legislation [D13], family members [D24], family relationship [D31], community services [D78], war [D87], nuclear warfare [D89], sex offenses [D92], torture [D94], mental institutions [G29], patient care team [G38], social workers [G44].
■ INTERESTED IN: mental institutions [G29], public hospitals [G30], private hospitals [G31], religious hospitals [G32], intensive care units [G33], hospices [G34], drug abuse [P6], psychological stress [P17], behavior control [P25], positive reinforcement [P28], terminal care [O7], quality adjusted life years [O14], voluntary euthanasia [O18], suicide [O19], wish of children [M16], child and family

[L31], abandoned (child) [L33], unwanted children [L35], handicapped children [L38], ill-treat children [L39], child abuse [L41], pediatrics [G68], psychiatry [G69], rehabilitation [I42], parental notification [J6], disclosure [J7], consent to treatment [J8], confidentiality [J20], medical secrecy [J22], physician patient relationship [J32].

F 196 **TOUBON** Jacques

Paris M.P., Mayor of the XIIIth Arrondissement of Paris, Deputy-Mayor of Paris. — Member of the Laws Commission.

Assemblée nationale
126, rue de l'Université
75007 Paris (France)
✆ (33) 1 - 40 63 83 14 Fax: (33) 1 - 40 63 83 45

■ AFFILIATIONS: Commission spéciale chargée d'examiner les projets de loi sur la bioéthique (commission on bioethics bills) (Assemblée Nationale), France (*Member*).
■ PUBLICATIONS: yes.

F 197 **VEDRINNE** Jacques

Professor. — M.D. Clinician, Head of the Department of Medical and Psychiatric Emergency (Lyon-Sud Hospital). — Medical Expert at the Cour de Cassation (High Court of Appeal).

Laboratoire de médecine légale
Faculté de médecine Alexis-Carrel
Rue Guillaume-Paradin
69372 Lyon (France)
✆ (33) 78 09 15 40

■ AFFILIATIONS: CCPPRB Centre Léon-Bérard — CCPPRB, Lyon Cedex 08, France (*Member*).
■ TEACHING: AEU (university attestation) of Medical Philosophy and Ethics, University of Lyon-I. — Postgraduate DEA Diploma in Biomedical Ethics, René-Descartes University (Paris-V).
■ PUBLICATIONS: yes.
■ WORKING ON: ethics committees [B22], College of Physicians [B27], ethics [B48], biomedical ethics education [B78], medical ethics [C52], behavior [C73], social control [D3], imprisonment [D11], handicapped [D43], killing [D90], justifiable killing [D91], sex offenses [D92], rape [D93], torture [D94], biomedical research [F2], research subjects [F12], healthy volunteers [F13], mental health [G4], traffic accidents [G7], mental institutions [G29], public hospitals [G30], intensive care units [G33], medical records [G50], sports medicine [G64], psychiatry [G69], health care costs [H9], prenatal diagnosis [I3], organ donors [I15], organ transplantation [I24], patients' rights [J2], confidentiality [J20], jurisprudence — accountability [K8], HIV seropositivity [N36], death and resuscitation [O1], behavioral and mental disorders [P2], outpatient commitment [P18].

■ INTERESTED IN: *idem, plus:* ethics committees [B22], College of Physicians [B27], ethics [B48], biomedical ethics education [B78], medical ethics [C52], behavior [C73], social control [D3], imprisonment [D11], handicapped [D43], killing [D90], abortion [M47], justifiable killing [D91], sex offenses [D92], rape [D93], torture [D94], biomedical research [F2], research subjects [F12], healthy volunteers [F13], mental health [G4], traffic accidents [G7], mental institutions [G29], public hospitals [G30], intensive care units [G33], medical records [G50], sports medicine [G64], psychiatry [G69], health care costs [H9], prenatal diagnosis [I3], organ donors [I15], organ transplantation [I24], patients' rights [J2], jurisprudence — accountability [K8], HIV seropositivity [N36], death and resuscitation [O1], behavioral and mental disorders [P2], outpatient commitment [P18]

F 198 **VERSPIEREN** Patrick

Head of the Department of Biomedical Ethics, Centre Sèvres.

Centre Sèvres
12, rue d'Assas
75006 Paris (France)
✆ (33) 1 - 45 44 18 99 Fax: (33) 1 - 40 49 01 92 (Assas-Editions)

■ AFFILIATIONS: Association européenne des centres d'éthique médicale — AECEM, Leuven, Belgium (*Administrator*).
■ TEACHING: Biomedical ethics at the Centre Sèvres, Institut catholique de Paris , and various schools of medicine in France and Lebanon (on request), health care and nursing schools. — Training on request for physicians and nurses.
■ PUBLICATIONS: yes.

■ WORKING ON: dignity [B5], ethics [B48], biomedical ethics education [B78], Bioethics bill, 1992 (France) [B98], Roman catholic ethics [C23], medical ethics [C52], nursing ethics [C53], force feeding [D95], nontherapeutic research [F5], therapeutic research [F6], human genome project [F8], human experimentation [F11], prenatal diagnosis [I3], sperm banks [I11], organ procurement [I26], fetal tissue transplantation [I27], extraordinary treatment [I39], patient information [J4], consent to treatment [J8], treatment refusal [J15], medical secrecy [J22], reproductive technologies [M26], therapeutic abortion [M50], pain [N12], persistent vegetative state [N42], brain death [O6], terminal care [O7], euthanasia [O15], dementia [P11], genome mapping [Q23], DNA linkage [Q33], genetic counseling [Q36], human genome [Q47].
■ INTERESTED IN: data bases [A3], bibliography [A4], dignity [B5], Council of Europe [B25], biomedical ethics education [B78], codes of biomedical ethics [B90], European Convention of Bioethics [B97], Islamic ethics [C28], Jewish ethics [C29], eastern orthodox ethics [C26], history before the 20th century [C58], EC — European Communities [D62], ECC — European Community Commission [D63], force feeding [D95], human genome project [F8], resource allocation [H4], mass screening [I6], mandatory screening [I7], organ procurement [I26], disclosure [J7], advance directives [J18], living wills [J19], reproductive technologies [M26], therapeutic abortion [M50], aborted fetuses [M51], anencephaly [N25], persistent vegetative state [N42], terminal care [O7], euthanasia [O15], genome mapping [Q23], DNA linkage [Q33], genetic screening [Q37], gene therapy [Q42], human genome [Q47].

F 199 **VIAL** Michèle

M.D. Clinician, Assistant to the Head of the Neonatal Paediatrics Department, Antoine Béclère Hospital of Clamart. — Head of the Paediatrics unit of the Maternity ward. — Co-director of the Foetal Medicine Unit of the Maternity-Ward.

Hôpital Antoine-Béclère
Service de pédiatrie néonatale
157, rue de la Porte-de-Trivaux
92141 Clamart (France)
✆ (33) 1 - 45 37 46 41

■ PUBLICATIONS: yes.
■ WORKING ON: human development [L12], childhood difficulties, diseases, protection [L21], pregnancy and childbirth [M39], abortion [M47], fetal development [M54], fetal therapy [M57], autopsies [O3], death [O5], euthanasia [O15], mentally handicapped [P7], genetic defects and hereditary diseases [Q2], biotechnology — genetic engineering [Q20], medical genetics [Q35].
■ INTERESTED IN: fundamental rights of the individuals [B2], ethical rules and principles [B47], biomedical ethics education [B78], medical education [B80], bioethics [B89], medical ethics [C52], parents [D32], biology [E5], epidemiology [E9], diagnostic imaging [E32], decision making [E54], decision analysis [E55], [E56], obstetrics and gynecology [G66], pediatrics [G68], transplantation [I22], parental notification [J6], disclosure [J7], third party consent [J11], parental consent [J13], physicians and researchers accountability [J23], legal personality [K2], conflict of interest [K9], child's interest [K24], therapeutic risk [K32], wrongful life [K33], constitutional law [K48], European law [K49], international law [K50], medical law [K51], health legislation [K52], law [K56], human development [L12], childhood difficulties, diseases, protection [L21], embryos [M10], procreation [M15], pregnancy and childbirth [M39], abortion [M47], fetal development [M54], fetal therapy [M57], autopsies [O3], death [O5], euthanasia [O15], mentally handicapped [P7], genetic defects and hereditary diseases [Q2], biotechnology — genetic engineering [Q20], medical genetics [Q35].

F 200 **VIDAL** Jean-Marie

Research Fellow at the CNRS (URA 373).

CNRS — URA 373
Campus scientifique de Beaulieu
Bât. 25
35042 Rennes Cedex (France)
✆ (33) 99 28 67 71 Fax: (33) 99 28 69 27

■ PUBLICATIONS: yes.
■ WORKING ON: ethics [B48], interdisciplinary communication [B70], evolution [C47], psychology [C68], science [E2], research design [E19], computers [E33], cognitive research [F7], psychiatry [G69], patient participation [J16], investigator subject relationship [J29], mind [L20], case studies [N4], follow-up studies [N13], psychotic disorders [P10].
■ INTERESTED IN: review [A2], publications [A9], self determination [B4], human person [B12], ethics [B48], interdisciplinary communication [B70], evolution [C47], psychology [C68], science [E2], research design [E19], computers [E33], therapeutic research [F6], cognitive research [F7], psychiatry [G69], patient participation [J16], professional deontology [J25], investigator subject

relationship [J29], mind [L20], case studies [N4], follow-up studies [N13], psychotic disorders [P10], psychotherapy [P30].

F 201 **VIDAL** Raoul

Rheumatologist (retired). — Counsellor at the Ordre départemental des médecins des Pyrénées-Atlantiques.

Ordre départemental des médecins des Pyrénées-Atlantiques
Complexe de la République
Rue Carnot
64000 Pau (France)
✆ (33) 59 27 85 65 Fax: (33) 59 83 79 88

■ AFFILIATIONS: Groupe d'études d'éthique biomédicale des Pyrénées-Atlantiques, Pau, France (*Co-founder, Executive*). — Conseil départemental de l'Ordre des médecins des Pyrénées Atlantiques, France (*Executive*).
■ TEACHING: Lectures. — Training of medical students. — Courses in Nursing Schools.
■ PUBLICATIONS: yes.
■ WORKING ON: human person [B12], human body commercialization [B13], traffic of organs [B15], ethics committees [B22], CCNE — Comité Consultatif National d'Ethique (France) [B23], professional deontology organs [B26], College of Physicians [B27], ethics [B48], deontology [B59], education [B76], biomedical ethics education [B78], nursing education [B79], codes of biomedical ethics [B90], history of biomedical ethics [B94], bioethical issues [B96], philosophical ethics [C2], religious ethics [C21], Christian ethics [C22], Protestant ethics [C27], Islamic ethics [C28], Jewish ethics [C29], scriptural interpretation [C40], future generations [C44], medical ethics [C52], genetics [E6], computerized file [E35], biological containment [E39], biomedical research [F2], health occupations [G61], medicine [G63], medical secrecy [J22], physicians and researchers accountability [J23], reproductive organs and embryonic structures [M9], procreation [M15], wish of children [M16], reproduction [M17], reproductive technologies [M26], death [O5], brain death [O6], palliative care [O8], life-sustaining treatment [O9], withholding treatment [O10], allowing to die [O11], terminally ill [O12], prolongation of life [O13], quality adjusted life years [O14], euthanasia [O15], suicide [O19], attitudes to death [O20].
■ INTERESTED IN: review [A2], fundamental rights of the individuals [B2], CNIL — Commission Nationale Informatique et Libertés (France) [B19], International Charter of Human Rights [B34], European Convention on Human Rights [B36], humanity heritage [B46], press conference [B73], CCNE advice (France) [B91], ethical review [B93], European Convention of Bioethics [B97], Christian science [C37], Jehovah's witnesses [C38], biology and human future [C43], philosophy [C61], public policy [D77], justifiable killing [D91], sex offenses [D92], rape [D93], torture [D94], force feeding [D95], history of science [E3], review committees [E42], biomedical research [F2], experimentation [F9], methodology of research and experimentation [F15], medicine [G63], diagnosis [I2], biological specimens procurement, blood transfusion, organ transplantation [I8], patients' rights [J2], medical secrecy [J22], physicians and researchers accountability [J23], aged — problems related with aging [L45], reproductive technologies [M26], death [O5], brain death [O6], palliative care [O8], life-sustaining treatment [O9], withholding treatment [O10], allowing to die [O11], terminally ill [O12], prolongation of life [O13], quality adjusted life years [O14], euthanasia [O15], suicide [O19], attitudes to death [O20], biotechnology — genetic engineering [Q20], recombinant DNA research [Q31], medical genetics [Q35], population genetics [Q43].

F 202 **VIGNAT** Jean-Pierre

Head of the Psychiatry Department of Saint-Jean-de-Dieu Hospital (Lyon), Coordinator of the Psychiatric Ethics committee of the Hospital. — Executive Secretary of the Groupe français d'epidémiologie psychiatrique.

Hôpital Saint-Jean-de-Dieu
290, route de Vienne
69373 Lyon Cedex 08 (France)
✆ (33) 78 09 78 15 Fax: (33) 78 77 96 67

■ AFFILIATIONS: Conseil d'éthique en psychiatrie de l'hôpital Saint-Jean-de-Dieu, Lyon Cedex 08, France (*Coordinator*).
■ WORKING ON: human rights [B3], freedom [B10], reification [B14], patient advocacy [B42], ethics [B48], mental institutions [G29], psychiatry [G69], consent to treatment [J8], personhood [K3], old person abuse [L51], psychiatric technics [P24].
■ INTERESTED IN: epidemiology [E9], computerized file [E35], medical informatics [F21], records [E34], organizational policies [G17], mental health [G4], resource allocation [H4], medical secrecy [J22], therapeutic injunction [K64], medical evaluation [G51].

F 203 **WEIL** Eva

Clinical Psychologist in the Parisian Hospitals.

Service de médecine de la reproduction
Hôpital Necker
149, rue de Sèvres
75743 Paris Cedex (France)
✆ (33) 1 - 42 73 87 23 et 87 11 Fax: (33) 1 - 42 73 81 80

■ AFFILIATIONS: Assiciation des psychocliniciens des PMA — APPMA, Paris, France (*President*).
■ TEACHING: Postgraduate DEA diploma in Ethics, CHU (University hospital) of Necker.
■ PUBLICATIONS: yes.
■ WORKING ON: interdisciplinary communication [B70], motivation [C76], research team [E16], mental health [G4], disclosure [J7], mind [L20], transsexualism [M24], semen donors [M28], AID [M32], embryo donation [M38], suffering [N11].
■ INTERESTED IN: documentation [A8], publications [A9], ethics committees [B22], Council of Europe [B25], biomedical ethics education [B78], European Convention of Bioethics [B97], health services research [F4], WHO — World Health Organization [G36], physician patient relationship [J32], professional patient relationship [J33].

F 204 **ZITTOUN** Robert

Head of the Haematology Department, Hôtel-Dieu Hospital (Paris).

Service d'hématologie
Hôtel-Dieu
1, place du Parvis Notre-Dame
75181 Paris Cedex 04 (France)
✆ (33) 1 - 42 34 84 13

■ TEACHING: palliative care.
■ PUBLICATIONS: yes.
■ WORKING ON: terminal care [O7], euthanasia [O15], clinical trials [F3], withholding treatment [O10].
■ INTERESTED IN: terminal care [O7], clinical trials [F3], withholding treatment [O10].

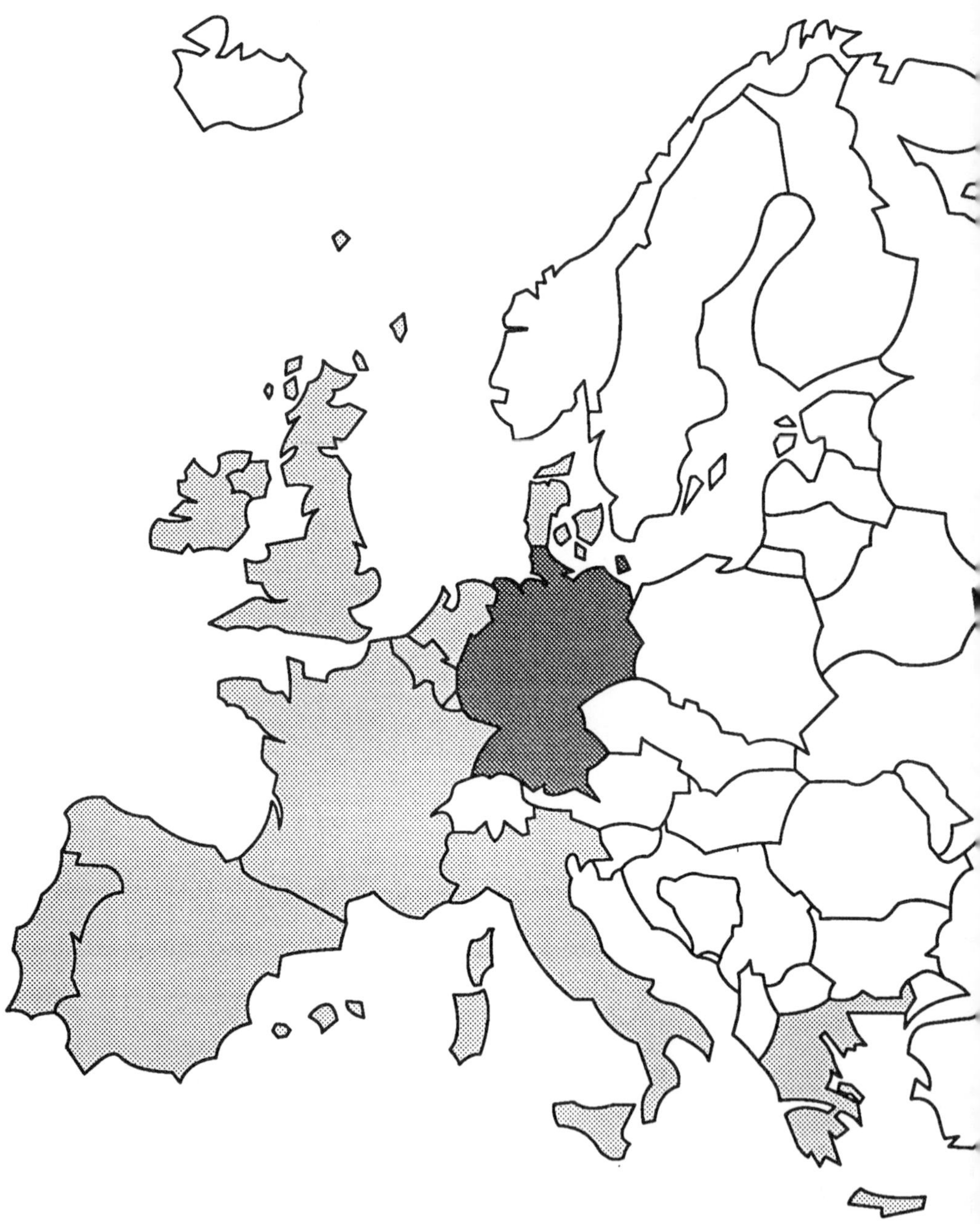

Germany

Area: 356 910 km²
Population: 78.5 millions
Gross national product: DM 2.72trn; US$ 1.6trn
Gross national product per head: 25 500 $
Gross national product growth: 1991: 2.6%
1992: 2.9%

Organisations

O D 1 **AKADEMIE DER WISSENSCHAFTEN UND DER LITERATUR**
Geschwister-Stroll-Strasse 2
6500 Mainz (Germany)

O D 2 **AKADEMIE FÜR ETHIK IN DER MEDIZIN E.V.** (AEM)
Humboldtallee 36
3400 Göttingen (Germany)
✆ (49) 551 - 39 96 80 Fax: (49) 551 - 39 39 96

■ AFFILIATION: office attached to the Institut für Geschichte der Medizin, Universität Göttingen.
■ DATE OF FOUNDATION: 05/12/1986.
■ EXECUTIVE BOARD: Prof., Dr. Eduard Seidler (*Chairperson*), Dr. Stella Reiter-Theil (*Head of Office*).
■ TEACHING: yes.
■ PUBLICATIONS: magazine Ethik in der Medizin.
■ DOCUMENTATION CENTRE: open to the public.
■ COLLOQUIUMS, SYMPOSIUMS: Ethics and legal medicine, November 1992, Salzburg, Austria. — Teaching and Training in medical ethics, October 1992, Bad Segeberg, Germany, together with the EACME.
■ DECISIONS, ADVICE: "Ethik in der ärztlichen Ausbildung" advice on the development of ethics courses witin medical studies. — "Embryonen-Forschung- zulassen oder verbieten ?" (Should we allow or forbid research on embryos?).
■ WORKING ON: data bases [A3], bibliography [A4], information centers [A5], documentation [A8], interdisciplinary communication [B70], information dissemination [B71], education [B76], health education [B77], biomedical ethics education [B78], nursing education [B79], medical education [B80], internship and residency [B81], students [B82], curriculum [B83], universities [B84], health economics, population characteristics [H1], patients' rights [J2], HIV seropositivity [N36], genetics and applications — biotechnology [Q1].
■ INTERESTED IN: applied psychology [C86], counseling [C87], children's rights [L25], child and family [L31], sexuality and procreation — unborn child [M1], sexuality and procreation [M2], reproductive technologies [M26], abortion [M47], psychiatric technics [P24].

O D 3 **ARBEITSGEMEINSCHAFT-KATHOLISCH-SOZIALER BILDUNGSWERKE**
Sigwartstrasse 17
7400 Tübingen (Germany)
✆ (49) - 7071-29 28 60

O D 4 **ARBEITSKREIS MEDIZINISCHER ETHIK-KOMMISSIONEN IN DER BUNDESREPUBLIK DEUTSCHLAND**
Herbert-Lewin-Strasse 5
5000 Köln 41 (Germany)
✆ (49) 221 - 4004 333 Fax: (49) 221 - 4004 296

O D 5 **BIOETHIK-KOMMISSION DES LANDES RHEINLAND-PFALZ**
Justizministerium
Ernst-Ludwig-Strasse 3
6500 Mainz (Germany)

O D 6 **EBERHARD-KARLS-UNIVERSITÄT TÜBINGEN ZENTRUM FÜR ETHIK IN DEN WISSENSCHAFTEN (TÜBINGEN)**
Oesterbergstrasse 9
7400 Tübingen 1 (Germany)
✆ (49) 7071-29 52 51

O D 7 **EVANG. THEOL. SEMINAR/PRAKT. -THEOL. ABT.**
Hölderlinstrasse 16
7400 Tübingen (Germany)
✆ (49) 7071 - 29 42 08

O D 8 **FORSCHUNGSSTÄTTE DER EVANGELISCHEN STUDIENGEMEINSCHAFT**
Schmeilweg 5
6900 Heidelberg 1 (Germany)
✆ (49) 62 - 211 40 61

O D 9 **FREIE UNIVERSITÄT BERLIN**
Spandauer Damm 130
1000 Berlin 19 (Germany)
✆ (49) 30 - 30 03 7 00 Fax: (49) 30 - 30 03 7 26

O D 10 **INSTITUT FÜR AUSLANDISCHES UND INTERNATIONALES STRAFRECHT**
Stefan-Meier-Strasse 26
7800 Freiburg I. Br (Germany)
✆ (49) -761-20 33 000

O D 11 **INSTITUT FÜR GESCHICHTE DER MEDIZIN — INSTITUT D'HISTOIRE DE LA MÉDECINE –**
Albert-Ludwigs-Universität
Stephan-Meir-Strasse 25
7800 Freiburg (Germany)
✆ (49) 761 - 20 33 000

O D 12 **INSTITUT FÜR GESCHICHTE DER MEDIZIN**
Bismarckstrasse 6
8520 Erlangen (Germany)
✆ (49) 80 - 615 320 Fax: (49) 80 - 541 862

O D 13 **INSTITUT FÜR HUMANGENETIK UND ANTHROPOLOGIE**
Im Neuenheimer Feld 328
6900 Heidelberg (Germany)
✆ (49) 551 - 39 96 80 Fax: (49) 551 - 39 95 54

O D 14 **INSTITUT FÜR MEDIZIN UND WISSENSCHAFTSGESCHICHTE**
Universität zu Lübeck
Königstrasse 42
2400 Lübeck (Germany)
✆ (49) 451 - 7 57 32 Fax: (49) 451 - 7 46 89

O D 15 **INSTITUT FÜR MEDIZINISCHE PSYCHOLOGIE**
Domagk-Strasse 3
4400 Münster (Germany)

O D 16 **INSTITUT FÜR THEORIE UND GESCHICHTE DER MEDIZIN**
Waldeyerstrasse 27
4400 Münster (Germany)
✆ (49) 551 - 39 96 80 Fax: (49) 551 - 399 95 54

O D 17 **LEHRSTHUL FÜR PRAKTISCHE PHILOSOPHIE**
Philosophie Fakultät
Ums Imsdatwald
7700 Saarbrücken (Germany)
✆ (49) 681 - 302 23 00

O D 18 **MEDIZIN FAKULTÄT**
Eberhard-Karl Universität
Hölderlinstrasse 16
7400 Tübingen (Germany)
✆ (49) 7071 - 29 42 08

O D 19 **SEMINAR FÜR MORALTHEOLOGIE**
Johannes Gütenberg-Universität Mainz
Postfach 3980, Saarstrasse 21
6500 Mainz (Germany)
✆ (49) 61 - 31 392 547 Fax: (49) 6131 - 393 501

■ OTHER FIELDS OF ACTIVITIES: Moral theology teaching and research.
■ AFFILIATION: Seminar, department of Catholic theology at the University of Mainz.
■ DATE OF FOUNDATION: 1946.
■ EXECUTIVE BOARD: Professor Johannes Reiter.
■ TEACHING: lectures and tutoral classes at the University.
■ PUBLICATIONS: diverse articles and treatises.
■ WORKING ON: review [A2], book review [A6], literature [A7], documentation [A8], publications [A9], editorial policies [A10], respect of human dignity and human rights [B1], specific approaches to ethics [C1], social and political issues [D1], science, technology, methods [E1], biomedical research and experimentation [F1], organization of health care; facilities, manpower and services; health occupations [G1], population [H19], health care, medical acts [I1], health care professionals, researchers and patient relationship [J1], law, legislation and jurisprudence [K1], sexuality and procreation — unborn child [M1], disease [N1], death and resuscitation [O1], genetics and applications — biotechnology [Q1].

O D 20 **THE SCIENTIFIC ADVISORY BOARD HUNTINGTON GESELLSCHAFT E. V.**
Oberstadtstrasse 23
7452 Haigerloch (Germany)
✆ (49) 747 - 47 181

O D 21 **WISSENSCHAFTSZENTRUM BERLIN FÜR SOZIALFORSCHUNG**
Reichpietschufer 50
1000 Berlin 30 (Germany)
✆ (49) 30 - 254 910

O D 22 **ZENTRUM FÜR ETHIK IN DEN WISSENSCHAFTEN**
Liebermeisterstrasse 16
7400 Tübingen (Germany)
✆ (49) 7071 - 29 52 51 Fax: (49) 7071 - 29 52 55

O D 23 **ZENTRUM FÜR MEDIZINISCHE ETHIK DER RUHR-UNIVERSITÄT**
Gebaude 3/53
Postfach 10 21 48
4630 Bochum 1 (Germany)
✆ (49) 234 - 7 00 27 50/49 Fax: (49) 234 - 7 09 4201

Individuals

D 1 **AUER** Alfons
Professor emeritus of religious ethics, Katholisch-theologische Fakultät der Universität Tübingen.

Universität Tübingen, Kath.-theol. Fakultät
Paul-Lechler-Str. 8
7400 Tübingen (Allemagne)
✆ (49) 7071 65478

■ TEACHING: Education for adults.
■ PUBLICATIONS: yes.
■ WORKING ON: aged — problems related with aging [L45].

D 2 **AUTIERO** Antonio
Director of the Seminar for Methodology, University of Munich

Universität Münster
Johannisstr. 8-10
4400 Münster (Allemagne)
✆ (49) 251 83 26 17 Fax : (49) 251 83 83 57

■ AFFILIATIONS: Società italiana de Bioetica, Trento, Italy (*Member*).
■ TEACHING: Seminars at the Fondazione Lanza (Padova).
■ PUBLICATIONS: yes.
■ WORKING ON: ethical rules and principles [B47], conscience [C13], christian ethics [C22], organ transplantation [I24], sexuality and procreation [M2], reproductive technologies [M26], terminal care [O7].
■ INTERESTED IN: reproductive technologies [M26], fetal therapy [M57], euthanasia [O15].

D 3 **BAYERTZ** Kurt

Head of Institut für System und Technologie Analysen.

Institut für System und Technologieanalysen
Wielandstr. 28a
4970 Bad Oeynhausen (Allemagne)
✆ (49) 5731- 792 152 Fax : (49) 5731- 792 162

■ AFFILIATIONS: Akademie für Ethik in der Medizin, Göttingen, Germany (*Member*).
■ TEACHING: Public Health (University of Bielefeld).
■ PUBLICATIONS: yes.
■ WORKING ON: self determination [B4], dignity [B5], justice [B9], freedom [B10], human person [B12], human body commercialization [B13], ethics [B48], moral obligations [B50], moral policy [B55], altruism [B56], beneficence [B57], authoritarianism [B58], morality [B61], virtues [B64], solidarity [B65], specific approaches to ethics [C1], philisophical ethics [C2], ethical analysis [C3], normative ethics [C5], deontological ethics [C6], teleological ethics [C7], utilitarianism [C8], ethical relativism [C10], metaethics [C11], morals [C12], moral development [C14], values [C15], philosophy of biology [C42], biology and human future [C43], future generations [C44], biological life [C49], philosophy [C61], determinism [C62], sociobiology [C63], existentialism [C64], hedonism [C65], marxism [C67], biology [E5], genetics [E6], molecular biology [E8], progress [E26], biomedical technologies [E28], human genome project [F8], human experimentation [F11], prenatal diagnosis [I3], amniocentesis [I4], chorionic villus sampling [I5], mass screening [I6], mandatory screening [I7], sperm banks [I11], reproductive technologies [M26], in vitro fertilization [M29], artificial insemination [M31], host mothers [M35], embryo transfer [M36], GIFT (Gamete Intrafallopian Transfer) [M37], embryo donation [M38], brain death [O6], genetic defects [Q3], single gene defects [Q7], recessive genetic conditions [Q10], hereditary diseases [Q18], sex linked defects [Q19], genetic intervention [Q22], genome mapping [Q23], cloning [Q24], clones [Q25], recombinant DNA research [Q31], DNA fingerprinting [Q32], DNA linkage [Q33], genetic screening [Q37], eugenics [Q39], gene therapy [Q42].
■ INTERESTED IN: *idem.*

D 4 **BIRNBACHER** Dieter

Universität Essen
Postfach 103 764
4300 Essen 1 (Allemagne)
✆ (49) 211 55 21 17

■ AFFILIATIONS: Akademie für Ethik in der Medizin, Göttingen, Germany
■ TEACHING: Philosophy, applied ethics.
■ PUBLICATIONS: yes.
■ WORKING ON: bioethical issues [B96], teleological ethics [C7], value of life [C18], containment [E38], reproductive technologies [M26], abortion [M47], euthanasia [O15], suicide [O19].

D 5 **BOCKENHEIMER—LUCIUS** Gisela

M.D. — Editing and writing for the newspaper «Ethik in der Medizin».

Finkenweg 19
6370 Oberursel 4 (Germany)
✆ (49) 617 - 23 79 83

■ AFFILIATIONS : Akademie für Ethik in der Medizin — AEM, Allemagne (*Membre*).
■ WORKING ON: information sources, data bases [A1], procreation [M15], reproductive technologies [M26], abortion [M47], euthanasia [O15].
■ INTERESTED IN: review [A2], World Medical Assembly [B20], ethics committees [B22], CCNE — Comité Consultatif National d'Ethique (France) [B23], CCPPRB — French committees for the protection of human subjects of biomedical research (France) [B32], public dabates [B75], biomedical teaching education [B78], WHO — World Health Organization [G36], childhood difficulties, diseases, protection [L21], abortion [M47], euthanasia [O15].

D 6 **CLASSEN** Claus Dieter
Assistant teacher.

Juristische Fakultät der Universität
Wilhelm str. 7
7400 Tübingen (Allemagne)
✆ (49) 707 12 92 558

■ PUBLICATIONS: yes.
■ WORKING ON: fundamental rights of the individuals [B2], human rights [B3], self determination [B4], dignity [B5], constitutional law [K48], European law [K49], international law [K50], freedom [B10], privacy [B11], ethics committees [B22], International Charter of Human Rights [B34], European Convention on Human Rights [B36].
■ INTERESTED IN: fundamental rights of the individuals [B2], human rights [B3], self determination [B4], dignity [B5], freedom [B10], privacy [B11], ethics committees [B22], International Charter of Human Rights [B34], European Convention on Human Rights [B36], constitutional law [K48], European law [K49], international law [K50].

D 7 **DOPPELFELD** Elmar
Editor in chief of the German magazine for physicians (Deutsches Ärtzteblatt). — Professor at the University of Bonn (nuclear medicine).

Deutsches Ärtzteblatt
Herbert-Lewin-Srasse 5
5000 Köln 41 (Allemagne)
✆ (49) 221 4004 333 Fax : (49) 221 40 04 296

■ AFFILIATIONS: Arbeitskreis Medizinischer Ethik-Kommissionen in der Bundesrepublik Deutschland, Köln 41, Germany (*Member of the Board*).
■ PUBLICATIONS: yes.
■ WORKING ON: editorial policies [A10], bioethics [B89], ethicists [B92], ethical review [B93], European Convention of Bioethics [B97], medical ethics [C52], nursing ethics [C53], epidemiology [E9], statistics [E10], research policy [E18], biomedical research [F2], nontherapeutic research [F5], human experimentation [F11], medicine [G63].
■ INTERESTED IN: *idem.*

D 8 **EIBACH** Ulrich
Protestant pastor at the University Hospital, Bonn. — University lecturer for Systematic Theology and Ethics at the Protestant Faculty of Theology, University of Bonn.

Evang. Seelsorger Universitätsklinik
Sigmund-Freud-Strasse 25
5300 Bonn (Allemagne)
✆ (49) 228 28 21 28

■ TEACHING: Lectures at the University of Bonn, Faculty of Protestant Theology. — Teaching Medical Ethics for nurses at the University Hospital, Bonn.
■ PUBLICATIONS: yes.
■ WORKING ON: fundamental rights of the individuals [B2], dignity [B5], human person [B12], ethical rules and principles [B47], ethics [B48], bioethics [B89], philisophical ethics [C2], utilitarianism [C8], quality of life [C16], value of life [C18], religion and religious ethics [C19], protestantism [C35], philosophy of biology [C42], evolution [C47], professional ethics [C51], sociobiology [C63], counseling [C87], pastoral care [C88], human genome project [F8], health [G2], mental health [G4], family planning [G8], artificial organs [G24], donors [I14], bone marrow transplantation [I29], beginning of life [L24], abortion [M47], mifepristone [M53], suffering [N11], pain [N12], death [O5], resuscitation [O21], medical genetics [Q35], human genome [Q47].
■ INTERESTED IN: human rights [B3], dignity [B5], human person [B12], human body commercialization [B13], protection of rights — involved institutions [B16], ethics [B48], bioethics [B89], codes of biomedical ethics [B90], ethical review [B93], bioethical issues [B96], European Convention of Bioethics [B97], philisophical ethics [C2], normative ethics [C5], value of life [C18], religion and religious ethics [C19], evolution [C47], professional ethics [C51], social sciences [C89], aliens [D23], females [D35], national socialism [D83], biology [E5], research [E14], progress [E26], teaching methods [E59], biomedical research [F2], medical informatics [F21], maternal health [G3], mental health [G4], family planning [G8], artificial organs [G24], hospital [G28], hospices [G34], WHO — World Health Organization [G36], medical staff [G41], hospitalisation [G46], patient compliance [G49], psychiatry [G69], health care costs [H9], risks and benefices [H11], biological specimens

procurement, blood transfusion, organ transplantation [I8], bone marrow transplantation [I29], patients' rights [J2], living wills [J19], inter-personal relationship [J27], parent child relationship [J34], legaly incompetent person [K4], wrongful life [K33], European law [K49], childhood difficulties, diseases, protection [L21], growth disorders [L44], aged — problems related with aging [L45], old person abuse [L51], sexuality and procreation [M2], fetal therapy [M57], cancer [N17], leukemia [N18], congenital defects [N23], spina bifida [N26], nervous system diseases [N40], temporal lobe epilepsy [N45], death [O5], brain death [O6], terminal care [O7], resuscitation [O21], neurosciences — psychiatry [P1], institutionalized persons [P23], electrical stimulation of the brain [P36], psychosurgery [P38], genetics and applications — biotechnology [Q1], human genome [Q47].

D 9 **ENGELHARDT** Dietrich von

Professor.

Medizinische Universität zu Lübeck
Institut für Medizin und Wissenschaftsgeschichte
Königstr. 42
2400 Lübeck (Allemagne)
✆ (49) -451-7 57 32 Fax : (49) -451- 7 46 89

■ AFFILIATIONS: Akademie für Ethik in der Medizin — AEM, Göttingen, Germany (*Member*). — Institut für Geschichte der Medizin, Münster, Germany (*Member*).
■ TEACHING: yes.
■ PUBLICATIONS: yes.
■ WORKING ON: bibliography [A4], human rights [B3], dignity [B5], ethics committees [B22], virtues [B64], education [B76], biomedical ethics education [B78], history of biomedical ethics [B94], medical ethics [C52], humanities [C55], history [C56], patients' rights [J2], psychotic disorders [P10].
■ INTERESTED IN: *idem.*

D 10 **ESER** Albin

Director of the Max-Planck-Institut für Auslandisches und internationales Strafrecht. — Professor at the University of Freiburg.

Max-Planck Institüt für Ausländisches und internationales Strafrecht
Güntherstalstr. 73
7800 Freiburg i. Br. (Allemagne)
✆ (49) 761 70 811

■ AFFILIATIONS: Akademie für Ethik in der Medizin — AEM, Göttingen, Germany (*Member*).
■ TEACHING: Professor (criminal law), University of Freiburg.
■ PUBLICATIONS: yes.
■ WORKING ON: human rights [B3], ethics committees [B22], punishment [D8], scientific misconduct [E25], social control of science [E64], biomedical research and experimentation [F1], misconduct [K23], medical law [K51], abortion [M47], euthanasia [O15].
■ INTERESTED IN: codes of ethics [B49], health care professionals, researchers and patient relationship [J1], sexuality and procreation — unborn child [M1], genetics and applications — biotechnology [Q1].
■ INTERESTED IN: *idem, plus:* codes of ethics [B49], health care professionals, researchers and patient relationship [J1], sexuality and procreation — unborn child [M1], genetics and applications — biotechnology [Q1]

D 11 **FUCHS** Christoph

General secretary of the German Medical Association.

Bundesärztekammer
Herbert-Lewin-Str. 1
5000 Köln 41 (Allemagne)
✆ (49) -221-4004 200/203 Fax : (49) -221-4004 388

■ AFFILIATIONS: Akademie für Ethik in der Medizin — AEM, Göttingen, Germany (*Member*).
■ PUBLICATIONS: yes.

D 12 **FURGER** Franz

Professor at the Faculté de Théologie (Catholic). — Director of the Institut d'éthique sociale.

Wilhelms-Universität Münster
Hüfferstrasse 27
4400 Münster (Germany)
✆ (49) 251 - 83 26 40

■ AFFILIATIONS: CAHBI, Strasbourg, France (*Swiss representative*).
■ TEACHING: Social ethics.
■ PUBLICATIONS: yes.
■ WORKING ON: ethical rules and principles [B47], ethics [B48], moral policy [B55], moral obligations [B50], deontology [B59], obligations of society [B53], obligations to society [B54].
■ INTERESTED IN: ethical rules and principles [B47].

D 13 **GAHL** Klaus

Director of Medical Clinic II (Internal Medicine), Professor of Internal Medicine/Cardiology of Medical School, Hanover. — Head of Medical Clinic II, Municipal Hospital (Städttischen Kliniken), Brawschweig, (academic teaching hospital affiliated to Hanover Medical School).

Medizinischen Klinik II, Städtisches Klinikum
Salzdahlumer Str. 90
3300 Braunschweig (Allemagne)
✆ (49) -531- 595 2235

■ AFFILIATIONS: Akademie für Ethik in der Medizin — AEM, Göttingen, Germany (*Member*).
■ PUBLICATIONS: yes.
■ WORKING ON: respect of human dignity and human rights [B1], human person [B12], ethical rules and principles [B47].
■ INTERESTED IN: philisophical ethics [C2], christian ethics [C22], health care [G45], patients' rights [J2], living wills [J19], inter-personal relationship [J27], professional patient relationship [J33], legal personality [K2], patients [N5], chronically ill [N7], heart diseases [N20], terminal care [O7], quality adjusted life years [O14], attitudes to death [O20], resuscitation [O21].

D 14 **GEDDERT-STEINACHER** Tatjana

Research assistant.

Universität Tübingen
Juristische Fakultät
Wilhelm Strasse 7
7400 Tübingen (Allemagne)
✆ (49) 7071 295267

■ PUBLICATIONS: yes.
■ WORKING ON: conflict of interest [K9], technical expertise [K16], generalization of expertise [K17], legitimacy [K21], moratory [K25], judicial action [K35], constitutional amendments [K36], legal rights [K45], constitutional law [K48], European law [K49], guidelines [K55], law [K56], law enforcement [K57].
■ INTERESTED IN: *idem*.

D 15 **GOETZE-CLAREN** Wolfgang

Professor and Head of Forum Medical Ethics (University of Munich), teaching Medical Ethics. — Professor of Gynecology and Obstetrics.

Universitätsfrauenklinik München - Forum Medizinische Ethik
Maistr. 11
8000 München 2 (Allemagne)
✆ (49) -203-74 05 74 Fax : (49) -203-74 70 6

■ AFFILIATIONS: The Royal College of Psychiatrists, Philosophy Group, Oxford, United Kingdom (*Member*).
■ TEACHING: Ethical problematics in modern gynaecology, at Forum Medical Ethics (University of Munich), with clinical case demonstrations.
■ PUBLICATIONS: yes.
■ WORKING ON: fundamental rights of the individuals [B2], self determination [B4], justice [B9], freedom [B10], human person [B12], protection of rights — involved institutions [B16], ethics committees [B22], professional deontology organs [B26], College of Physicians [B27], manifests and declarations concerning human rights [B33], European Convention on Human Rights [B36], Helsinki Declaration [B37], patient advocacy [B42], equal protection [B43], common good [B45], humanity heritage [B46], ethical rules and principles [B47], beneficence [B57], deontology [B59], morality [B61], virtues [B64], humanitarian organizations [B66], health and biology mediatisation [B85], ethicists [B92], European Convention of Bioethics [B97], specific approaches to ethics [C1], ancient history [C59], philosophy [C61], attitudes [C74], personality [C82], intelligence [C85], applied psychology [C86], socioeconomic factors [D18], employment [D19], handicapped [D43],

disadvantaged [D46], cultural pluralism [D51], sociology of medicine [D55], international aspects [D59], EEC — European Economic Community [D62], government and political systems [D64], science [E2], mortality [E13], investigators [E15], behavioral research [E21], scientific misconduct [E25], nuclear energy [E44], radiation [E48], decision making [E54], teaching methods [E59], biomedical research [F2], maternal health [G3], financial support [H18], prenatal diagnosis [I3], transplantation [I22], bone marrow transplantation [I29], alternative therapies [I33], rehabilitation [I42], patients' rights [J2], confidentiality [J20], jurisprudence — accountability [K8], stages of life — problems proper to childhood and elderly [L1], sexuality and procreation — unborn child [M1], disease [N1], death and ressucitation [O1], genetics and applications — biotechnology [Q1].

D 16 **GRESS- HEISTER** Markus

Scientific Editor of "Genetics and Rehabilitation"(Journal), Medical Journalist. —Head of Research Group "Education in Genetics in Medical Education-Germany" by Liga f. Bekämpfung frühzeitige Älterserscheinungen e. V., Junior Consultant of the Skillentenvoband. Ethik in der Medizin Freiburg e. V.

"Genitics and Rehabilitation"
Scientific Editor
Hegarstr. 7A
7800 Freiburg i. (Germany)
✆ (49) 761 - 27 58 95 Fax: (49) 761 - 27 88 16

■ AFFILIATIONS: European Society of Philosophy of Medicine and Health Care — ESPMH (*Member*).
■ TEACHING: Teaching Ethics in nursing school, school of physiotherapy. — Seminars and lectures.
■ PUBLICATIONS: yes.
■ WORKING ON: data bases [A3], patient advocacy [B42], competence [B51], education [B76], biomedical ethics education [B78], medical ethics [C52], decision making [E54], teaching methods [E59], physicians [G40], patient compliance [G49], patient information [J4], informed consent [J9], medical law [K51], aged — problems related with aging [L45], aging [L50], case studies [N4], euthanasia [O15], dementia [P11].

D 17 **GUENTHER** H.L.

Professor of criminal law, Judge at the Oberlandesgericht. — Member of the Ethikkommission der Fakultät für klinische Medizin der Universität Tübingen.

Juristische Fakultät
Universität Tübingen
7400 Tübingen (Germany)
✆ (49) 7071 - 12 91

■ AFFILIATIONS: Zentrum für Ethik in den Wissenschaften, Tübingen, Germany (*Mitgleid des Leitungsemins*).
■ PUBLICATIONS: yes.
■ WORKING ON: fundamental rights of the individuals [B2], human rights [B3], self determination [B4], dignity [B5], ethics committees [B22], patient advocacy [B42], medical ethics [C52], biomedical research and experimentation [F1], biomedical research [F2], clinical trials [F3], nontherapeutic research [F5], therapeutic research [F6], human genome project [F8], methodology of research and experimentation [F15], health care, medical acts [I1], diagnosis [I2], prenatal diagnosis [I3], amniocentesis [I4], biological specimens procurement, blood transfusion, organ transplantation [I8], patients' rights [J2], conflict of interest [K9], wrongful death [K12], malpractice [K19], therapeutic risk [K32], wrongful life [K33], criminal law [K53], criminal code [K54], reproductive technologies [M26], embryo donation [M38], abortion [M47], mifepristone [M53], behavioral genetics [Q34], gene therapy [Q42].

D 18 **HEISTER** Elisabeth

Physician. — Junior consultant of the Studentenverband Ethik in der Medizin. — Scientific editor of "Geriatrie und Rehabilitation" (Journal). — Research fellow (geriatrics education).

Rehaklinik Klausenbach
7618 Nordrach-Klausenbach (Germany)
✆ (49) 7838 - 82 215

■ AFFILIATIONS: Kennedy Institute of Bioethics, Washington DC, USA (*Member*).
■ PUBLICATIONS: yes.
■ WORKING ON: bioethical issues [B96], bioethics [B89], professional ethics [C51], teaching methods

[E59], clinical trials [F3], physicians [G40], physician nurse relationship [J31], aged — problems related with aging [L45].
■ INTERESTED IN: aged — problems related with aging [L45], teaching methods [E59].

D 19 **HEUBEL** Fridrich
Lecturer (Ethik in der Medizin), Faculty of Medicine, Philipps University. — Data protection registrar, Klinikum, Philipps University. — Chairperson of the "Kommission für Ethik in der ärztlischen Forschung" (Institutional Review Board), Faculty of Medicine, Philipps University.

Pharmakologisches Institut, Philipps Universität
Karl von Frisch-Strasse
3550 Marburg/Lahn (Germany)
✆ (49) 6421 - 28 59 66 Fax: (49) 6421 - 28 49 58

■ AFFILIATIONS: European Society of Philosophy of Medicine and Health Care — ESPMH, Netherlands (*Member*).
■ TEACHING: For 12 years, seminar called "Ethik in der Medizin", to medical students on an optional basis (Ethics is not yet compulsory in the German medical curriculum.).
■ PUBLICATIONS: yes.

D 20 **HINRICHSEN** Klaus Volquardt
Director of the Zentrum für Medizinische Ethik Bochum. — Professor of anatomy and embryology. — "Humanembryologie", Springer, Berlin, 1990. — Member of the European "Beratenden Ausschusses für die ärztliche ausbildung".

Zentrum für Medizinische Ethik
Ruhr-Universität Bochum
Universitätsstr. 150
4630 Bochum (Germany)
✆ (49) 234 - 70 04 553 Fax: (49) 234 - 70 94 190

■ TEACHING: Symposium on medical ethics with case study. — Lectures for medicine students..
■ PUBLICATIONS: yes.
■ WORKING ON: reproductive organs and embryonic structures [M9], embryos [M10], fetuses [M11], aborted fetuses [M51], pain [N12].
■ INTERESTED IN: reproductive organs and embryonic structures [M9], embryos [M10], fetuses [M11], excess embryos [M34].

D 21 **HÖLZLE** Christina
University teacher. — Psychologist, psychotherapist.

Inst. f. Medizinische Psychologie
Domagkstrasse 3
4400 Münster (Germany)
✆ (49) 251 - 83 54 88 Fax: (49) 251 - 83 54 94

■ TEACHING: Training for medical and psychosocial staff in counselling infertile couples (Profamilia: National Parenthood Planning).
■ PUBLICATIONS: yes.
■ WORKING ON: self determination [B4], Helsinki Declaration [B37], obligations to society [B54], public opinion [D7], government regulation [D17], couple [D26], marital relationship [D29], procreation [M15], wish of children [M16], reproduction [M17], fertility [M18], infertility [M19], sexuality [M20], in vitro fertilization [M29], AIH [M33], psychotherapy [P30], ethical relativism [C10], quality of life [C16], consequences [C45], medical ethics [C52], motivation [C76], emotions [C79], self concept [C83], counseling [C87], behavioral research [E21], health hazards [E49], therapeutic research [F6], random selection [F18].
■ INTERESTED IN: *idem, plus:* women's rights [B6]

D 22 **HONNEFELDER** Ludger
Director of the department of philosophy, section B, University of Bonn. — Chairperson, scientific working group on bioethics, Northrhine-Westphalia. — Coordinator, research project: "Genetic aberration and the normative concepts of disease and disability: a European comparison", CEC-Program "Analysis of the Human Genome". — Member, Rheno-Wesphalian Academy of Science.

Philosophisches Seminar B der Universität Bonn
Am Hof 1

5300 Bonn 1 (Germany)
✆ (49) 228 - 73 74 19 Fax: (49) 228 - 73 55 79

■ AFFILIATIONS: Akademy für Ethik in der Medizin — AEM, Göttingen, Germany (*Member*). — Kennedy Institute of Ethics, Washington, USA (*Associate member*).
■ TEACHING: Workshop "medical ethics" for students of the "Bischöfliche studienförderung Cusanusswerk", Bonn 1.
■ PUBLICATIONS: yes.
■ WORKING ON: respect of human dignity and human rights [B1], human rights [B3], dignity [B5], human equality [B7], human person [B12], Council of Europe [B25], ethical rules and principles [B47], ethics [B48], codes of ethics [B49], moral obligations [B50], morality [B61], interdisciplinary communication [B70], education [B76], biomedical ethics education [B78], students [B82], universities [B84], bioethics [B89], ethicists [B92], philisophical ethics [C2], ethical analysis [C3], normative ethics [C5], metaethics [C11], conscience [C13], values [C15], clergy [C20], christian ethics [C22], roman catholic ethics [C23], natural law [C24], roman catholicism [C31], humanities [C55], philosophy [C61], consent to treatment [J8], beginning of life [L24], euthanasia [O15], genetic defects and hereditary diseases [Q2], biotechnology [Q21], medical genetics [Q35], genetic counseling [Q36], human genome [Q47].
■ INTERESTED IN: future generations [C44], international aspects [D59], international organizations [D61], population growth [H20].

D 23 **HÜBNER** Jürgen

Senior research fellow at the Forschungsstätte der Evan. Studiengemeinschaft. — Professor for Systematic Theology, University of Heidelberg.

Forschungsstätte der Evangelischen Studiengemeinschaft
Institute for Interdisciplinary Research (FEST)
Schmeilweg 5
6900 Heidelberg (Germany)
✆ (49) 6221 - 184 061 Fax: (49) 6221 - 167 257

■ AFFILIATIONS: Akademie für Ethik in der Medizin — AEM, Göttingen (*Member*).
■ TEACHING: Lectures and seminars in the adult education (University).
■ PUBLICATIONS: yes.
■ WORKING ON: review [A2], bibliography [A4], respect of human dignity and human rights [B1], ethics [B48], altruism [B56], solidarity [B65], biomedical ethics education [B78], students [B82], universities [B84], ethicists [B92], bioethical issues [B96], clergy [C20], christian ethics [C22], protestant ethics [C27], theology [C41], biology and human future [C43], evolution [C47], biological life [C49], medical ethics [C52], history of science [E3], biology [E5], genetics [E6], research institutes [E17], ecology [E43], human genome project [F8], animal testing alternatives [F14], prenatal diagnosis [I3], physician nurse relationship [J31], physician patient relationship [J32], beginning of life [L24], in vitro fertilization [M29], embryo transfer [M36], allowing to die [O11], attitudes to death [O20], genetic defects and hereditary diseases [Q2], genetic intervention [Q22], medical genetics [Q35], human genome [Q47].
■ INTERESTED IN: information sources, data bases [A1], bibliography [A4], ethics [B48], bioethics [B89], christian ethics [C22], philosophy of biology [C42], professional ethics [C51], human genome project [F8], prenatal diagnosis [I3], beginning of life [L24], attitudes to death [O20], medical genetics [Q35], human genome [Q47].

D 24 **ILLARDT** Franz-Jozef

Ethics Consultant.

Zentrum für Geriatrie und Gerontologie Albert-Ludwigs-Universität Freiburg
Lehener Str. 88
7800 Freiburg (Germany)
✆ (49) 761 - 270 7092 Fax: (49) 761 - 270 3200

■ AFFILIATIONS: Akademie der Wissenschaften und der Literatur, Mainz, Germany (*Research fellow, Dept Medicine and Society*).
■ TEACHING: Seminars and lectures given to medical students. —Ethics rounds for givers of the Univesity Hospital. — Advanced education of physicians in the field of geriatrical ethics.
■ PUBLICATIONS: yes.
■ WORKING ON: ethical analysis [C3], christian ethics [C22], history before the 20th century [C58], placebos [I36], physician patient relationship [J32], aged — problems related with aging [L45], disease and aged [L46], aging [L50], old person abuse [L51], sterilization [M6], sexuality [M20], suicide [O19], attitudes to death [O20].

■ INTERESTED IN: ethical analysis [C3], ethicists [B92], history of biomedical ethics [B94], placebos [I36], physician patient relationship [J32], aged — problems related with aging [L45], attitudes to death [O20].

D 25 **ISTEL-LAUGELL** Karin

Doctoral candidate.

Zentrum für Ethik in den Wissenschaften
Wilhelmstr. 20
7400 Tübingen (Germany)
✆ (49) 7071 - 295 255 Fax: (49) 7071 - 295 255

■ AFFILIATIONS: The Scientific Advisory Board — Huntington Gesellschaft e. v., Haigerloch, Germany (*Member*).
■ TEACHING: Adult education (organised by the state and/or the church).
■ PUBLICATIONS: yes.
■ WORKING ON: teleological ethics [C7], roman catholic ethics [C23], roman catholicism [C31], consequences [C45], double effect [C46], medical ethics [C52], sociobiology [C63], counseling [C87], social discrimination [D53], stigmatization [D54], genetics [E6], molecular biology [E8], human genome project [F8], resource allocation [H4], costs and benefits [H10], diagnosis [I2], mandatory screening [I7], central nervous system diseases [N43], genetic defects and hereditary diseases [Q2], sex linked defects [Q19], genetic counseling [Q36], genetic screening [Q37], carriers [Q46], human genome [Q47].
■ INTERESTED IN: diagnosis [I2], mandatory screening [I7], genetic defects and hereditary diseases [Q2], sex linked defects [Q19].

D 26 **JÄGER** Lothar

Head of the department of clinical immunology, University Hospital. — Head of the Ethical committee, Medical Faculty.

Institut für Klinische immunologie
Friedrich-Schiller-Universität
Humboldtstrasse 3
6900 Jena (Germany)
✆ (49) 228 - 287 29 46 Fax: (49) 228 - 287 23 80

■ TEACHING: yes.
■ WORKING ON: review [A2], literature [A7], documentation [A8], publications [A9], human rights [B3], self determination [B4], patient advocacy [B42], obligations of society [B53], universities [B84], ethical review [B93], medical ethics [C52], research policy [E18], research design [E19], pollution [E47], clinical trials [F3], human experimentation [F11], healthy volonteers [F13], control groups [F17], random selection [F18], risks and benefices [H11], immunotherapy [I41], informed consent [J9], right to treatment [J17], biotechnology [Q21].
■ INTERESTED IN: review [A2], literature [A7], documentation [A8], patient advocacy [B42], obligations of society [B53], biomedical ethics education [B78], universities [B84], ethical review [B93], medical ethics [C52], research policy [E18], research design [E19], pollution [E47], clinical trials [F3], human experimentation [F11], healthy volonteers [F13], control groups [F17], random selection [F18], risks and benefices [H11], immunotherapy [I41], informed consent [J9], right to treatment [J17], biotechnology [Q21].

D 27 **JAHRMÄRKER** Hans

Department of Cardiology and Intensive Care Medicine, University of Munich.

Medizinische Fakultät der Universität München
Karl-Valentin-Str. 9
8022 Grünwald (Germany)
✆ (49) 86 - 41 24 34

■ AFFILIATIONS: Akademie für Ethik in der Medizin — AEM, Göttingen, Germany (*Member*).
■ PUBLICATIONS: yes.

D 28 **KAHLKE** Winfried

Direction of education, university of Hambourg, faculty of medecine.

Univ. Krakenhaus Eppendorf
Martinistr. 52

2000 Hamburg 20 (Germany)
✆ (49) 40-4717-3696 Fax : (49) 40-4717-3107

■ AFFILIATIONS: Akademie für Ethik in der Medizin — AEM, Allemagne (*Membre*).
■ TEACHING: Interdisciplinary seminar on ethics in medecine. — Advanced «Teachers Training Center» Project.
■ PUBLICATIONS: yes.
■ WORKING ON: information centers [A5], book review [A6], documentation [A8], publications [A9], fundamental rights of the individuals [B2], dignity [B5], justice [B9], human person [B12], human body commercialization [B13], protection of rights — involved institutions [B16], Amnesty International [B18], ethics committees [B22], due process [B40], ethical rules and principles [B47], ethics [B48], codes of ethics [B49], obligations of society [B53], solidarity [B65], humanitarian organizations [B66], communication [B69], interdisciplinary communication [B70], audiovisual aids [B74], public debates [B75], education [B76], health education [B77], biomedical ethics education [B78], nursing education [B79], medical education [B80], students [B82], curriculum [B83], universities [B84], codes of biomedical ethics [B90], normative ethics [C5], utilitarianism [C8], ethical relativism [C10], Christian ethics [C22], biology and human future [C43], medical ethics [C52], nursing ethics [C53], history [C56], historical aspects [C57], social groups [D22], minority groups [D42], handicapped [D43], social discrimination [D53], sociology of medicine [D55], political issues [D56], political activity [D57], North-South relationships [D60], government and political systems [D64], dehumanization [D86], war [D87], genetics [E6], scientific misconduct [E25], progress [E26], biomedical technologies [E28], drug industry [E37], decision making [E54], [E56], teaching methods [E59], human genome project [F8], public health [G5], family planning [G8], healthy policy [G13], institutional policies [G15], health legislation [G14], organ transplantation [I24], patients' rights [J2], embryos [M10], fetuses [M11], reproductive technologies [M26], congenital defects [N23], terminal care [O7], euthanasia [O15], psychoactive drugs [P33], genetic defects and hereditary diseases [Q2], biotechnology — genetic engineering [Q20].
■ INTERESTED IN: human genome project [F8], patient information [J4], informed consent [J9], treatment refusal [J15], patient participation [J16], physician's role [J26], inter-personal relationship [J27], child's interest [K24], economic value of life [K28], childhood difficulties, diseases, protection [L21], handicapped children [L38], infanticide [L42], artificial insemination [M31], embryo transfer [M36], GIFT (Gamete Intrafallopian Transfer) [M37], iatrogenic disease [N3], group therapy [P32], Huntington's chorea [Q9], Duchenne muscular dystrophy [Q12], genome mapping [Q23], genetic screening [Q37], gene therapy [Q42].

D 29 **KIELSTEIN** Rita
Physician in Internal Medicine (nephrology).

Medizinische Akademie
Leipzigerstrasse 44
3090 Magdeburg (Germany)
✆ (49) 391 - 67 24 10 Fax: (49) 391 - 67 24 40

■ TEACHING: Lectures in Ethics on renal replacement therapy (hemodialysis, transplantation), in human medicine study.
■ PUBLICATIONS: yes.
■ WORKING ON: medical ethics [C52], self concept [C83], advance directives [J18], living wills [J19].
■ INTERESTED IN: genetics [E6], human genome project [F8], artificial organs [G24], transplantation [I22], renal dialysis [I32], informed consent [J9].

D 30 **KOCH** Hans Georg
Head of the medical law department, Max-Planck-Institut.

Max-Planck-Institut für ausländisches und internationales Strafrecht
Gunterstalstr. 73
7800 Freiburg i.B (Germany)
✆ (49) 761 708 11

■ AFFILIATIONS: Akademie für Ethik in der Medizin — AEM, Göttingen, Germany (*Member*).
■ TEACHING: Lecturer at Freiburg University (medical law). — Occasional speeches.
■ PUBLICATIONS: yes.
■ WORKING ON: self determination [B4], ethics committees [B22], medical ethics [C52], communism [D81], legislation [D13], informed consent [J9], parental consent [J13], living wills [J19], medical secrecy [J22], minors [K6], child's interest [K24], medical law [K51], abortion [M47].
■ INTERESTED IN: *idem.*

D 31 LÖW Reinhard

Founding Director of the Research Institute for Philosophy, Hanover. — Professor of Philosophy of Nature and Ethics. — Pharmacist.

Forschungsinstitut für Philosophie
Lange Laube 14
3000 Hannover 1 (Germany)
✆ (49) 511 - 1 64 09 20 Fax: (49) 511 - 1 64 09 40

■ PUBLICATIONS: yes.
■ WORKING ON: information sources, data bases [A1], review [A2], book review [A6], publications [A9], editorial policies [A10], respect of human dignity and human rights [B1], fundamental rights of the individuals [B2], human rights [B3], dignity [B5], integrity [B8], freedom [B10], human person [B12], human body commercialization [B13], traffic of organs [B15], ethical rules and principles [B47], ethics [B48], codes of ethics [B49], moral obligations [B50], deontology [B59], code of deontology [B60], virtues [B64], bioethics [B89], codes of biomedical ethics [B90], specific approaches to ethics [C1], philisophical ethics [C2], ethical analysis [C3], normative ethics [C5], deontological ethics [C6], morals [C12], quality of life [C16], value of life [C18], religion and religious ethics [C19], christian ethics [C22], roman catholic ethics [C23], natural law [C24], jewish ethics [C29], roman catholicism [C31], judaism [C34], religion [C36], theology [C41], philosophy of biology [C42], biology and human future [C43], future generations [C44], consequences [C45], evolution [C47], biological life [C49], professional ethics [C51], medical ethics [C52], ethics and humanities [C54], humanities [C55], social and political issues [D1], science, technology, methods [E1], biomedical research and experimentation [F1], organization of health care; facilities, manpower and services; health occopations [G1], health care, medical acts [I1], health care professionals, researchers and patient relationship [J1], law, legislation and jurisprudence [K1], stages of life — problems proper to childhood and elderly [L1], sexuality and procreation — unborn child [M1], disease [N1], death and ressucitation [O1], neurosciences — psychiatry [P1], genetics and applications — biotechnology [Q1].
■ INTERESTED IN: bibliography [A4], documentation [A8], religion and religious ethics [C19], jewish ethics [C29], roman catholicism [C31], judaism [C34], philosophy of biology [C42], sociobiology [C63], science, technology, methods [E1], technology [E27], biological containment [E39], safety [E51], diagnosis [I2], organ transplantation [I24], professional deontology [J25], human characteristics [L19], beginning of life [L24], determination of death [O2], attitudes to death [O20], biotechnology — genetic engineering [Q20], biotechnology [Q21], cloning [Q24], clones [Q25], medical genetics [Q35].

D 32 LUNSHOF Jeantine

Research assistant.

Im Hasensprung 8
6380 Bad Homburg (Germany)
✆ (49) 6172 29 682

■ AFFILIATIONS: Universität Bamberg, Lehrstuhl für Philosophie-II, Bamberg, Germany (*Research assistant*).
■ WORKING ON: self determination [B4], women's rights [B6], privacy [B11], ethics committees [B22], professional deontology organs [B26], ethics [B48], codes of ethics [B49], moral obligations [B50], obligations of society [B53], obligations to society [B54], moral policy [B55], paternalism [B62], biomedical ethics education [B78], nursing education [B79], specific approaches to ethics [C1], genetics and applications — biotechnology [Q1], medical ethics [C52], nursing ethics [C53], government regulation [D17], scarcity [D21], females [D35], physically handicapped [D44], cultural pluralism [D51], international aspects [D59], ECC — European Community Commission [D63], state responsability [D75], genetics [E6], social control of science [E64], technology assessment [E65], clinical trials [F3], nontherapeutic research [F5], therapeutic research [F6], cognitive research [F7], human genome project [F8], human experimentation [F11], research subjects [F12], health policy [G13], patient care team [G38], nurses [G39], resource allocation [H4], health care costs [H9], costs and benefits [H10], risks and benefices [H11], prenatal diagnosis [I3], amniocentesis [I4], chorionic villus sampling [I5], mass screening [I6], mandatory screening [I7], human equality [B7], integrity [B8], justice [B9], freedom [B10], required request [I10], donors [I14], bone marrow transplantation [I29], disclosure [J7], consent to treatment [J8], informed consent [J9], presumed consent [J10], third party consent [J11], treatment refusal [J15], living wills [J19], medical secrecy [J22], investigator subject relationship [J29], nurse patient relationship [J30], physician nurse relationship [J31], personhood [K3], potentiality of personhood [K5], conflict of interest [K9], therapeutic risk [K32], wrongful life [K33], medical law [K51], health legislation [K52], beginning of life [L24], life extension [L47], health care services and aged [L48], selective abortion [M49], therapeutic abortion [M50], abortion on demand [M52], mifepristone [M53], viability [M56], chronically ill [N7], physically handicapped [N9], cancer [N17], congenital defects [N23], death [O5], brain death [O6], terminal care [O7], euthanasia [O15].

■ INTERESTED IN: self determination [B4], women's rights [B6], privacy [B11], ethics committees [B22], professional deontology organs [B26], ethics [B48], codes of ethics [B49], moral obligations [B50], obligations of society [B53], obligations to society [B54], moral policy [B55], paternalism [B62], biomedical ethics education [B78], nursing education [B79], specific approaches to ethics [C1], medical ethics [C52], nursing ethics [C53], government regulation [D17], scarcity [D21], females [D35], physically handicapped [D44], cultural pluralism [D51], international aspects [D59], ECC — European Community Commission [D63], state responsability [D75], genetics [E6], social control of science [E64], technology assessment [E65], clinical trials [F3], nontherapeutic research [F5], therapeutic research [F6], cognitive research [F7], human genome project [F8], human experimentation [F11], research subjects [F12], health policy [G13], patient care team [G38], nurses [G39], resource allocation [H4], health care costs [H9], costs and benefits [H10], risks and benefices [H11], prenatal diagnosis [I3], amniocentesis [I4], chorionic villus sampling [I5], mass screening [I6], mandatory screening [I7], required request [I10], donors [I14], bone marrow transplantation [I29], disclosure [J7], consent to treatment [J8], informed consent [J9], presumed consent [J10], third party consent [J11], treatment refusal [J15], living wills [J19], medical secrecy [J22], investigator subject relationship [J29], nurse patient relationship [J30], physician nurse relationship [J31], personhood [K3], potentiality of personhood [K5], conflict of interest [K9], therapeutic risk [K32], wrongful life [K33], medical law [K51], health legislation [K52], beginning of life [L24], life extension [L47], health care services and aged [L48], selective abortion [M49], therapeutic abortion [M50], abortion on demand [M52], mifepristone [M53], viability [M56], chronically ill [N7], physically handicapped [N9], cancer [N17], congenital defects [N23], death [O5], brain death [O6], terminal care [O7], euthanasia [O15], genetics and applications — biotechnology [Q1], human equality [B7], integrity [B8], justice [B9], freedom [B10].

D 33 **MAIO** Giovanni

Physician (internal medicine) and research fellow (medical ethics, history of medicine).

Leutkirchstrasse 43
7614 Gengenbach (Germany)
✆ (49) 771 - 64 108

■ AFFILIATIONS: Institut de l'histoire de la médecine, Freiburg, Germany (*Collaborator*).
■ TEACHING: Medical ethics for medicine students and nurses.
■ WORKING ON: justice [B9], CCNE — Comité Consultatif National d'Ethique (France) [B23], College of Physicians [B27], Nuremberg Code [B35], Helsinki Declaration [B37], code of deontology [B60], nursing education [B79], bioethics [B89], CCNE advice (France) [B91], medical ethics [C52], historical aspects [C57], biomedical research and experimentation [F1], biomedical research [F2], clinical trials [F3], nurses [G39], hospitalisation [G46], medicine [G63], consent to treatment [J8], informed consent [J9], presumed consent [J10], professional deontology [J25], health legislation [K52], aged — problems related with aging [L45], disease and aged [L46], cancer [N17], cardiovascular diseases [N19], diabetes [N22], kidney diseases [N39], persistent vegetative state [N42], central nervous system diseases [N43], palliative care [O8], life-sustaining treatment [O9], withholding treatment [O10], allowing to die [O11], terminally ill [O12], prolongation of life [O13], euthanasia [O15], active euthanasia [O16], involontary euthanasia [O17], attitudes to death [O20], health services research [F4], nontherapeutic research [F5], therapeutic research [F6], human experimentation [F11], research subjects [F12], healthy volonteers [F13], life extension [L47], health care services and aged [L48], leukemia [N18], heart diseases [N20].
■ INTERESTED IN: bibliography [A4], information centers [A5], literature [A7], publications [A9], Council of Europe [B25], public debates [B75], biomedical ethics education [B78], history of biomedical ethics [B94], bioethics movement [B95], European Convention of Bioethics [B97], Bioethics bill, 1992 (France) [B98], history [C56], historical aspects [C57], philosophy [C61], public opinion [D7], legislation [D13], history of science [E3], European law [K49], international law [K50], medical law [K51], law enforcement [K57], legal aspects [K58], bill [K60], attitudes to death [O20].

D 34 **MERAN** Johannes Gobertus

Editor of the Diskussions Forum Med. Ethik of Wien Med. Woschr. — MD in Hematology department, Med. Hochschule Hannover (Germany).

Diskussions Forum
Henniesruh 65
3000 Hannover 51 (Germany)
✆ (49) 511 - 647 98 15

■ AFFILIATIONS: Akademie für Ethik in der Medizin — AEM, Göttingen, Germany (*Member*).
■ TEACHING: Lectures in med. Hochschule Hannover.
■ PUBLICATIONS: yes.
■ WORKING ON: self determination [B4], dignity [B5], ethics committees [B22], patient advocacy [B42],

codes of ethics [B49], moral obligations [B50], communication [B69], quality of life [C16], value of life [C18], health care professionals, researchers and patient relationship [J1], patients' rights [J2], informed consent [J9], presumed consent [J10], treatment refusal [J15], advance directives [J18], living wills [J19], leukemia [N18], terminal care [O7].
■ INTERESTED IN: patient advocacy [B42], communication [B69], interdisciplinary communication [B70], quality of life [C16], health care professionals, researchers and patient relationship [J1], patient information [J4], consent to treatment [J8], informed consent [J9], living wills [J19], death and ressucitation [O1], terminal care [O7].

D 35 **MÜLLER** Albrecht
Laboratory assistant.

Universität Tübingen
Zentrum für Ethik in den Wissenschaften
Liebermeisterstrasse 20
7400 Tübingen (Germany)
✆ (49) 7071 - 295 255 Fax: (49) 7071 - 295 255

■ AFFILIATIONS: Graduiertenkolleg - Ethik in den Wissenschaften, Tübingen, Germany (*Member*).
■ WORKING ON: transgenic animals [Q29], biotechnology — genetic engineering [Q20], social and political issues [D1].
■ INTERESTED IN: *idem.*

D 36 **NIPPERT** Irmgard
Associate professor (medical sociologist) at the Institute für Humangenetik. — Minister medical school-Westfälische Wilhelm-Univesität.

Institut für Human Genetik
Westfälische Wilhelm Universität
Vesaliusweg 12-14
4400 Münster (Germany)
✆ (49) 251 - 83 54 08 Fax: (49) 251 - 83 69 95

■ AFFILIATIONS: Kommission für Offentlichkeitsarbeit und ethische fragen/German society of human genetics, Essen 1, Germany (*Member*).
■ WORKING ON: health care, medical acts [I1], organization of health care; facilities, manpower and services; health occopations [G1], maternal health [G3], state medicine [G9], maternal life [G11], decision making [E54], social control of science [E64], technology assessment [E65], genetics [E6], epidemiology [E9], statistics [E10], incidence [E11], morbidity [E12], mortality [E13], cystic fibrosis [Q15], medical genetics [Q35], genetic counseling [Q36], carriers [Q46].
■ INTERESTED IN: health care, medical acts [I1], patients' rights [J2], physician's role [J26], legal personality [K2], potentiality of personhood [K5], malpractice [K19], moratory [K25], legal liability [K30], patents [K47], economics [H2], cystic fibrosis [Q15], medical genetics [Q35], carriers [Q46].

D 37 **PIECHOWIAK** Helmut
Physician.

Mediz. Dieust d. Krankenversicherung
Margaretenstrasse 14a
8400 Regensburg (Germany)
✆ (49) 941 - 25071 Fax: (49) 941 - 28180

■ AFFILIATIONS: Akademie für Ethik in der Medizin — AEM, Göttingen, Germany (*Member*).
■ PUBLICATIONS: yes.
■ WORKING ON: social and political issues [D1], social issues [D2], aged — problems related with aging [L45].

D 38 **POLHMEIER** H.
Director of the Institut for Med. Psychol., Universität Göttingen. — Psychiatry, Med. Psychol.

Institut für Med. Psychologie Georg-August-Universität Göttigen
Humboldtallee 38
3400 Göttingen (Germany)
✆ (49) 551 - 39 81 92

■ AFFILIATIONS: Akademie für Ethik in der Medizin — AEM, Göttingen (*Member*).
■ TEACHING: Lectures and seminars in the course of Med. Psychol. for students of medicine.

■ PUBLICATIONS: yes.
■ WORKING ON: suicide [O19], euthanasia [O15].
■ INTERESTED IN: *idem.*

D 39 **PROPPING** Peter
Head of the Institute of Human Genetics, University of Bonn.

University of Bonn
Institute of Human Genetics
Wilhelmstr. 31
5300 Bonn (Germany)
✆ (49) 228 - 287 23 46 Fax: (49) 228 - 287 23 80

■ PUBLICATIONS: yes.
■ WORKING ON: hereditary diseases [Q18], behavioral genetics [Q34], genetic counseling [Q36].
■ INTERESTED IN: genetic counseling [Q36], behavioral genetics [Q34], hereditary diseases [Q18].

D 40 **RASPE** Hans Heinrich
Head of the Institute for Social Medicine, Medical University, Lübeck.

Medizinische Universität zu Lübeck
St. Jürgen-Ring 66
2400 Lübeck (Germany)
✆ (49) 451 - 530 01 40 or 451 - 530 01 42

■ AFFILIATIONS: Akademie für Ethik in der Medizin — AEM, Germany (*Founding member*).
■ PUBLICATIONS: yes.

D 41 **REITER** J.

Seminar für Moraltheologie und sozialethik Johannes-Gütenberg Universität
Johannes-Gütenberg Universität Mainz
Postfach 3980, Saarstrasse 21
6500 Mainz (Germany)
✆ (49) -6131 39 25 47 Fax: (49) -6131 39 35 01

■ AFFILIATIONS: Bioethik-Kommission des Landes Rheinland-Pfalz, Mainz, Germany (*Member*).
■ TEACHING: yes.
■ PUBLICATIONS: yes.
■ WORKING ON: review [A2], bibliography [A4], book review [A6], editorial policies [A10], respect of human dignity and human rights [B1], specific approaches to ethics [C1], social and political issues [D1], science, technology, methods [E1], biomedical research and experimentation [F1], organization of health care; facilities, manpower and services; health occopations [G1], population [H19], health care, medical acts [I1], health care professionals, researchers and patient relationship [J1], law, legislation and jurisprudence [K1], sexuality and procreation — unborn child [M1], disease [N1], death and ressucitation [O1], genetics and applications — biotechnology [Q1].
■ INTERESTED IN: review [A2], bibliography [A4], book review [A6], editorial policies [A10], respect of human dignity and human rights [B1], specific approaches to ethics [C1], social and political issues [D1], science, technology, methods [E1], biomedical research and experimentation [F1], organization of health care; facilities, manpower and services; health occopations [G1], population [H19], health care, medical acts [I1], health care professionals, researchers and patient relationship [J1], sexuality and procreation — unborn child [M1], disease [N1], death and ressucitation [O1], genetics and applications — biotechnology [Q1].

D 42 **REITER-THEIL** Stella
AEM (Akademie für Ethik in der Medizin), general secretary, director of the board, director of the documentation section. — Professor at the universities of Göttingen and Salzbourg.

Institut für Geschichte der Medizin
Humboldtallee 11
3400 Göttingen (Germany)
✆ (49) 551-39 96 80 Fax: (49) 551-39 39 96

■ AFFILIATIONS: National Committee for Professional Ethics (Bund Deutshen Psychologen. — BDP) Deutschland (*Member*).
■ TEACHING: medical ethic (University of Göttingen), ethic in psychology and clinic psychology (Faculty of Sciences, University of Salzbourg), ethic in Psychotherapy (Faculty of Medicine, University of

Vienne, Faculty of comportemental science, University of Tübingen). — Problems of ethics in familial therapies (Institute for the Marriage and Family Therapy, University of Vienne), training for teachers of medical ethic (AEM).

■ PUBLICATIONS: yes.

■ WORKING ON: data bases [A3], bibliography [A4], informations centers [A5], documentation [A8], respect of human dignity and human rights [B1], philosophical ethics [C2], professional ethics [C51], ethics and humanities [C54], biomedical research and experimentation [F1], health care professionals, researchers and patient relationship [J1], childhood difficulties, diseases, protection [L21], sexuality and procreation — unborn child [M1], neurosciences — psychiatry [P1].

■ INTERESTED IN: *idem.*

D 43 **ROESSLER** Dietrich

Chairperson of the Ethics committee.

Evang. theol. Seminar/Prakt.-theol. Abt.
Hölderlinstrasse 16,
7400 Tübingen (Germany)
✆ (49) 7071 - 29 42 08

■ AFFILIATIONS: Faculté de Médecine à l'Université Eberhard-Karl de Tübingen, Germany (*Chairperson of the Ethics committee*).

■ TEACHING: interdisciplinary seminars on medical ethics.

■ PUBLICATIONS: yes.

■ WORKING ON: bibliography [A4], information centers [A5], book review [A6], literature [A7], documentation [A8], publications [A9], self determination [B4], dignity [B5], integrity [B8], human person [B12], ethics committees [B22], CCNE — Comité Consultatif National d'Ethique (France) [B23], ethics [B48], codes of ethics [B49], moral obligations [B50], competence [B51], professional competence [B52], obligations of society [B53], moral policy [B55], altruism [B56], beneficence [B57], code of deontology [B60], morality [B61], solidarity [B65], humanitarian organizations [B66], interdisciplinary communication [B70], biomedical ethics education [B78], medical education [B80], students [B82], curriculum [B83], universities [B84], bioethical issues [B96], ethical analysis [C3], wedge argument [C4], situational ethics [C9], metaethics [C11], values [C15], quality of life [C16], value of life [C18], protestant ethics [C27], protestantism [C35], theology [C41], medical ethics [C52], counseling [C87], pastoral care [C88], handicapped [D43], sociology of medicine [D55], investigators [E15], research institutes [E17], research design [E19], occupational medicine [E46], health hazards [E49], decision making [E54], decision analysis [E55], goals [E57], standards [E60], biomedical research [F2], physicians [G40], patient compliance [G49], medical evaluation [G51], medicine [G63], health care costs [H9], health care, medical acts [I1], health care professionals, researchers and patient relationship [J1], stages of life — problems proper to childhood and elderly [L1], sexuality and procreation — unborn child [M1], disease [N1], death and ressucitation [O1], neurosciences — psychiatry [P1], genetics and applications — biotechnology [Q1].

■ INTERESTED IN: *idem.*

D 44 **SASS** Hans- Martin

Head of the Zentrum für Medizinische Ethik der Ruhr-Universität Bochum.

Zentrum für Medizinische Ethik der Ruhr-Universität Bochum
Gebaüde 3/53, Postfach 10 21 48
4630 Bochum 1 (Germany)
✆ (49) 234 - 7 00 27 50 Fax: (49) 234 -7 09 42 01

■ TEACHING: yes.

■ PUBLICATIONS: yes.

D 45 **SCHAEFER** Hans

Scientific consultant in an insurance company ("Berufsgewissenschaft"). — Professeur emeritus of physiology. — Research fellow in philosophy of medicine.

Physiologisches Institut der Universität Heidelberg
Im Neuenheimer Feld 326
6900 Heidelberg 1 (Germany)
✆ (49) 6221 - 56 38 48

■ AFFILIATIONS: Akademie für Ethik in der Medizin — AEM, Göttingen, Germany (*Member*).

■ TEACHING: conferences before scientific societies, academies and faculties.

■ PUBLICATIONS: yes.

■ WORKING ON: review [A2], codes of ethics [B49], competence [B51], moral policy [B55], health

education [B77], codes of biomedical ethics [B90], roman catholic ethics [C23], medical ethics [C52], family members [D24], sociology of medicine [D55], epidemiology [E9], hearts [L5], child and family [L31].
■ INTERESTED IN: review [A2], codes of biomedical ethics [B90], medical ethics [C52], child and family [L31].

D 46 **SCHLAUDRAFF** Udo
Pastor of the University Hospital of Göttingen. — Lecturer, faculty of Medicine. — In charge of the department of Medical Ethics, Ev. -luth. Landeskirche Hannover. — Member of the Ethics committee, Ärztekammer Niedersachsen.

Evang. Klinikpfarramt/ Universitätsklinikum
Robert-Koch-Strasse 40
3400 Göttingen (Germany)
✆ (49) 551 - 36 263

■ AFFILIATIONS: Akademie für Ethik in der Medizin — AEM, Göttingen, Germany (*Member of the Board*).
■ TEACHING: seminars for medical students.
■ PUBLICATIONS: yes.
■ WORKING ON: traffic of organs [B15], ethics [B48], interdisciplinary communication [B70], Christian ethics [C22], Protestant ethics [C27], psychology [C68], medical ethics [C52], emotions [C79], counseling [C87], pastoral care [C88], human experimentation [F11], autopsies [O3], terminal care [O7], euthanasia [O15].
■ INTERESTED IN: traffic of organs [B15], codes of ethics [B49], codes of biomedical ethics [B90], christian ethics [C22], protestant ethics [C27], humanities [C55], psychology [C68], transplantation [I22], fetal tissue transplantation [I27], medical etiquette [J28], investigator subject relationship [J29], legaly incompetent person [K4], legal gardians [K7], wrongful life [K33], Supreme Court decisions [K41], beginning of life [L24], newborns [L29], handicapped children [L38], life extension [L47], fetal therapy [M57], determination of death [O2], brain death [O6], terminal care [O7], euthanasia [O15], preimplantation genetic diagnosis [Q38], gene therapy [Q42].

D 47 **SCHÖLMERICH** Paul
Chairperson of Commission of Medical Research in the Academy of Sciences and Literature, Mainz. — Formerly Chief of Department of Internal Medicine in the University of Mainz.

Academy of Sciences and Literature
Geschwister Schollstr. 2
6500 Mainz (Germany)
✆ (49) 6131 - 577 - 0

■ AFFILIATIONS: Akademie für Ethik in der Medizin — AEM, Göttingen, Germany (*Member*).
■ TEACHING: Klinisch-ethisches Kolloquim Universitätsklinikum Mainz.
■ PUBLICATIONS: yes.
■ WORKING ON: quality of life [C16], resource allocation [H4], health care costs [H9], organ transplantation [I24], quality adjusted life years [O14].
■ INTERESTED IN: history of biomedical ethics [B94], quality of life [C16], human genome project [F8], resource allocation [H4], health care costs [H9], organ transplantation [I24], withholding treatment [O10], resuscitation [O21].

D 48 **SCHROEDER-KURTH** Traute
Director of Department of Cytogenetics, University of Heidelberg.

Institut für Humangenetik und Anthropologie
Im Neuenheimer Feld 328
6900 Heidelberg (Germany)
✆ (49) 6221 - 56 38 77 or 56 38 85 (laboratory) Fax: (49) 6221 - 56 38 98

■ AFFILIATIONS: Academy of Ethics in Medicine — AEM, Göttingen, Germany (*Member*).
■ TEACHING: Ethics in medicine, University of Heidelberg.
■ PUBLICATIONS: yes.

D 49 **SCHUBERT-LEHNHARDT** Viola
Senior Assistant.

Martin-Luther Universität, Abt. Ethik u. Geschichte d. Medizin
PSF 302

04090 Halle (Germany)
✆ (49) 28072

■ TEACHING: Teaching bioethics at University for medical students.
■ PUBLICATIONS: yes.
■ WORKING ON: self determination [B4], dignity [B5], justice [B9], ethics [B48], moral obligations [B50], obligations of society [B53], deontology [B59], morality [B61], virtues [B64], medical education [B80], values [C15], quality of life [C16], social worth [C17], value of life [C18], protestant ethics [C27], medical ethics [C52], philosophy [C61], hedonism [C65], marxism [C67], social control [D3], punishment [D8], government regulation [D17], cultural pluralism [D51], socialism [D84].

D 50 **SEIDLER** Eduard

Universität Freiburg
Stefan-Meier-Strasse 26
7800 Freiburg i. Br (Germany)
✆ (49) 203 - 3001 Fax: (49) 203 - 3007

D 51 **STEIGLEDER** Klaus
Scientific coordinator of the Zentrum für Ethik in den Wissenschaften, University of Tübingen.

Zentrum für Ethik in den Wissenschaften
Centre for Ethics in the Sciences and Humanities
Liebermeisterstr. 16
7400 Tübingen (Germany)
✆ (49) 7071 - 29 52 55

■ TEACHING: yes.
■ PUBLICATIONS: yes.
■ WORKING ON: human rights [B3], dignity [B5], normative ethics [C5], abortion [M47].
■ INTERESTED IN: *idem.*

D 52 **TOELLNER** Richard
Professor. — Head of the Institut für Theorie und Geschichte der Medizin der Westfälischen Wilhelms-Universität Münster.

Institut für Geschichte und Theorie der Medizin Universität Münster
Waldeyerstrasse 27
4400 Münster (Germany)
✆ (49) 251 - 83 5339

■ AFFILIATIONS: Arbeitskreis Medizinischer Ethik-Kommissionen in der Bundesrepublik Deutschland, Köln 41, Germany *(Member of the University committee)*
■ TEACHING: to medical students and physicians.
■ PUBLICATIONS: yes.
■ WORKING ON: fundamental rights of the individuals [B2], protection of rights — involved institutions [B16], ethical rules and principles [B47], codes of biomedical ethics [B90], history of biomedical ethics [B94], philisophical ethics [C2], religion and religious ethics [C19], medical ethics [C52], history [C56], history of science [E3], biomedical research [F2], patient information [J4], informed consent [J9], biotechnology — genetic engineering [Q20].
■ INTERESTED IN: information sources, data bases [A1], information — communication — média [B68], Bioethics bill, 1992 (France) [B98], public health [G5], health occupations [G61], patients' rights [J2], biotechnology — genetic engineering [Q20], medical genetics [Q35].

D 53 **TRÖHLER** Ulrich
Professor. — Head of the Institut für Geschichte der Medizin, University of Göttingen.

Institut für Geschichte der Medizin
Humboldtallee 36
3400 Göttingen (Germany)
✆ (49) 551 - 39 95 54

■ AFFILIATIONS: Akademie für Ethik in der Medizin — AEM, GÖttingen FRG, Germany (*General Secretary*).
■ TEACHING: Lectures and seminars at the Faculty of Medecine, University of Göttingen, and University of Basel (Switzerland).

■ PUBLICATIONS: yes.
■ WORKING ON: history of biomedical ethics [B94], history [C56], historical aspects [C57], history before the 20th century [C58], history of the 20th century [C60], animal experimentation [F10], human experimentation [F11], investigator subject relationship [J29], leukemia [N18].
■ INTERESTED IN: ethical review [B93], history of biomedical ethics [B94], bioethics movement [B95], European Convention of Bioethics [B97], history [C56], historical aspects [C57], history before the 20th century [C58], history of the 20th century [C60], human experimentation [F11], alternative therapies [I33], placebos [I36], investigator subject relationship [J29], physician nurse relationship [J31], professional patient relationship [J33], cancer [N17], leukemia [N18].

D 54 **VON EIFF** August Wilhelm
Director of Medical Clinic of University, Bonn. — German representative at the internal ethics committee for AIDS. — Member of the scientific committee KOS-Rivista di scienza e etica. — Consultant, Environment Ministry, for noise problems. — Member of the EIKC.

Universität Bonn
Haagerweg 18a
5300 Bonn 1 (Germany)
✆ (49) 228 - 28 28 47

■ AFFILIATIONS: Scientific Committee of KOS-Rivista di scienza e etica, Milan, Italy (*Member*).
■ TEACHING: Conferences for different organisations and universities.
■ PUBLICATIONS: yes.
■ WORKING ON: roman catholic ethics [C23], evolution [C47], emotions [C79], love [C80].
■ INTERESTED IN: ethical rules and principles [B47], ethics [B48], codes of ethics [B49], moral obligations [B50], deontology [B59], obligations to society [B54], interdisciplinary communication [B70], bioethics [B89], ethicists [B92], ethical review [B93], christian ethics [C22], roman catholic ethics [C23], evolution [C47], biological life [C49], speciesism [C48], emotions [C79], love [C80], personality [C82].

D 55 **VON LUTTEROTTI** Markus
Head of the Department of Internal Medicine, Lorette-Krankenhaus Freiburg (retired).

Lugostrasse 8
7800 Freiburg i.Br (Germany)
✆ (49) 761 - 70 02 40

■ AFFILIATIONS: Akademie für Ethik in der Medizin — AEM, Göttingen, Germany (*Member*).
■ PUBLICATIONS: yes.
■ WORKING ON: biology [E5], deontological ethics [C6], teleological ethics [C7], utilitarianism [C8], evolution [C47].
■ INTERESTED IN: *idem.*

D 56 **VON SCHUBERT** Hartwig
Correspondant, Forschungsstätte der Evangelischen Studiengemeinschaft. Director of the Department for Counseling of the Deaconic Office of the Evangelisch- Lutherische Kirche in Hamburg (Evangelical-Lutheran Church in Hamburg).

Forschungsstätte der Evangelischen Studiengemeinschaft
Schmeilweg 5
6900 Heidelberg (Germany)
✆ (49) 6221 - 18 40 61 Fax: (49) 6221 - 16 72 57

■ AFFILIATIONS: Akademie für Ethik inder Medizin — AEM, Göttingen, Germany (*Member*).
■ TEACHING: Lectures at various institutions, such as: Catholic University of Nijmegen, School of Medical Sciences. — Cooperation with various institutions, such as: Institute for Bioethics (Maastricht), Lutherian World Federation (Geneva).
■ PUBLICATIONS: yes.
■ WORKING ON: publications [A9], Council of Europe [B25], interdisciplinary communication [B70], biomedical ethics education [B78], religion and religious ethics [C19], christian ethics [C22], protestant ethics [C27], theology [C41], biological containment [E39], blood donation [I20], sexuality and procreation [M2], reproductive technologies [M26], fetal development [M54], fetal therapy [M57], congenital defects [N23], genetics and applications — biotechnology [Q1], biotechnology — genetic engineering [Q20], medical genetics [Q35].
■ INTERESTED IN: *idem.*

D 57 **WEBER** Ellen
Head of Department of Clinical Pharmacology, University of Heidelberg.

Medizinische Universitäts Klinik
Bergheimerstr. 58
6900 Heidelberg (Germany)
✆ (49) 6221 - 56 42 37

■ AFFILIATIONS: Akademie für Ethik in der Medizin e. V. — AEM, Göttingen, Germany (*Member*).
■ PUBLICATIONS: yes.
■ WORKING ON: ethics committees [B22], bioethics [B89], clinical trials [F3], therapeutic research [F6], methodology of research and experimentation [F15].
■ INTERESTED IN: ethics committees [B22], CCNE — Comité Consultatif National d'Ethique (France) [B23], bioethics [B89], codes of biomedical ethics [B90], clinical trials [F3], therapeutic research [F6], methodology of research and experimentation [F15].

D 58 **WEINSCHENCK** Günther

Universität Hohenheim-Stuttgart
Postfach 7005-62
7000 Stuttgart 70 (Germany)
✆ (49) 711 - 45 92 541 Fax: (49) 711 - 45 93 481

■ TEACHING: yes.
■ PUBLICATIONS: yes.

D 59 **WIESEMANN** Claudia
Research assistant, Institut für Geschichte der Medizin.

Institut für Geschichte der Medizin
Bismarckstrasse 6
8520 Erlangen (Germany)
✆ (49) 9131 - 85 23 08

■ AFFILIATIONS: The European Society for Philosophy of Medicine and Health Care — ESPMH, Nymegen, Germany (*Member*).
■ TEACHING: to medical students, Project "Medizinische Ethik für Studierende".
■ PUBLICATIONS: yes.
■ WORKING ON: medical education [B80], internship and residency [B81], students [B82], curriculum [B83], codes of biomedical ethics [B90], case studies [N4], prognosis [N14].
■ INTERESTED IN: *idem.*

D 60 **WIESING** Urban
Assistant at the "Institut für Theorie und Geschichte der Medizin", University of Münster.

Inst. f. Theorie und Geschichte der Medizin
Waldeyerstrasse 27
4400 Münster (Germany)
✆ (49) 251 - 83 52 91

■ AFFILIATIONS: Akademie für Ethik in der Medizin — AEM, Göttingen, Germany (*Member*).
■ TEACHING: Courses in Medical Ethics for students. — Member of the workgroup "Teachers training course", of the AEM.
■ PUBLICATIONS: yes.
■ WORKING ON: situational ethics [C9], medical ethics [C52], history [C56], historical aspects [C57], history before the 20th century [C58], history of the 20th century [C60], philosophy [C61], history of science [E3], diagnosis [I2], physicians and researchers accountability [J23], physician patient relationship [J32], professional patient relationship [J33], reproductive technologies [M26], in vitro fertilization [M29], abortion [M47], prognosis [N14].
■ INTERESTED IN: *idem, plus* cosmetic surgery [I38].

D 61 **WITTERN** Renate
Head of the Instituts für Geschichte der medizin der Universität Erlangen.

Institut für Geschichte der Medizin
Bismarckstrasse 6

8520 Erlangen (Germany)
✆ (49) 9131 - 85 23 08

■ TEACHING: History of medical ethics. — Conferences for the programme "ethik in der medizin".
■ PUBLICATIONS: yes.
■ WORKING ON: medical education [B80], history of biomedical ethics [B94], ancient history [C59], national socialism [D83].
■ INTERESTED IN: national socialism [D83], artificial organs [G24], prognosis [N14], euthanasia [O15].

D 62 **WOLFF** Gerhard
Head of the genetic counselling service.

Institute of Human Genetics
Breisacher str 33
7800 Freiburg (Germany)
✆ (49) 761 - 270 70 55 Fax: (49) 761 - 270 70 41

■ AFFILIATIONS: Akademie für Ethik in der Medizin — AEM, Göttingen, Germany (*Member*).
■ PUBLICATIONS: yes.
■ INTERESTED IN: genetic defects and hereditary diseases [Q2], XYY karyotype [Q5], Huntington's chorea [Q9], cystic fibrosis [Q15], sex preselection [Q30], genetic counseling [Q36], eugenics [Q39], negative eugenics [Q40], human genome [Q47], psychotherapy [P30], therapeutic abortion [M50].

Greece

Area: 131 957 km^2
Population: 10.2m
Gross domestic product: Dr 14.9trn; US$ 75bn
Gross domestic product per head: $ 7,300
Gross domestic product growth: 1991: 0.9%
1992: 2.3%

Organisations

O GR 1 **ASSOCIATION MÉDICALE PANHELLÉNIQUE**
Vissaronos Street 7
10672 Athens (Greece)
✆ (30) 1 - 36 14 913

O GR 2 **COMMISSION D'ETHIQUE DE L'EGLISE DE GRECE**
c/o 36, Sozopoleos str.
104 36 Athens (Greece)
✆ (30) 1 - 575 64 72

O GR 3 **CONSEIL NATIONAL DE LA SANTÉ** (KESI)
Ministère de la Santé
Aritotelous 17
104 33 Athens (Greece)
✆ (30) 1 - 523 28 21 or 1 - 646 97 93

O GR 4 **HELLENIC SOCIETY OF MEDICAL ETHICS**
Research Institute on Child "Spyros Doxiades" Leoforos Amalias 42
10558 Athens (Greece)
✆ (30) 1 - 32 36 783 or 1 - 32 38 807 Fax: (30) 1 - 32 42 384

O GR 5 **INSTITUT PASTEUR HELLÉNIQUE**
127, Avenue Vass. Sofias
115 21 Athens (Greece)
✆ (30) 1 - 64 65 905 Fax: (30) 1 - 64 23 498

O GR 6 **INTERNATIONAL HIPPOCRATIC FOUNDATION**
Island of Kos
85300 Kos (Greece)
✆ (30) 242- 22131

Individuals

GR 1 **ANAPLIOTOU**
Barrister at Law working in the field of Health Laws and Bioethics.

Association hellénique médicale
7 Vissaronos Street
106 72 Athènes (Greece)

GR 2 **ANTHONY** Comninos
Professor of gynaecology and obstetrics, Athens University. — Chairperson of the Hellenic Fertility and Sterility Society. — Chairperson of the Greek UNICEF. — Member of Executive Office of Hellenic Society of Medical Ethics.

Athens University
3, Koumbari Street
106 73 Athènes (Greece)
✆ (30) 1 - 3611 345 Fax: (30) 1 - 6840 894

■ AFFILIATIONS: Hellenic Society of Medical Ethics, Greece (*Member of the Board)*
■ TEACHING: ethics in human reproduction.

■ PUBLICATIONS: yes.
■ WORKING ON: professional ethics [C51], medical ethics [C52], nursing ethics [C53], life sciences [E4], technology [E27], biomedical research and experimentation [F1], clinical trials [F3], therapeutic research [F6], human experimentation [F11], maternal health [G3], obstetrics and gynecology [G66], population [H19], biological specimens procurement, blood transfusion, organ transplantation [I8], patients' rights [J2], confidentiality [J20], physicians and researchers accountability [J23], inter-personal relationship [J27], childhood difficulties, diseases, protection [L21], sexuality and procreation — unborn child [M1], sexuality and procreation [M2], reproductive technologies [M26], abortion [M47], fetal development [M54], cancer [N17], communicable diseases [N27], terminal care [O7], euthanasia [O15].
■ INTERESTED IN: ethical rules and principles [B47], bioethics [B89], health care, medical acts [I1], sperm banks [I11], donors [I14], intracerebral fetal tissue transplantation [I28], health care professionals, researchers and patient relationship [J1], confidentiality [J20], physicians and researchers accountability [J23], inter-personal relationship [J27], sexuality and procreation — unborn child [M1], sexuality and procreation [M2], reproductive technologies [M26], abortion [M47], fetal development [M54], cancer [N17], HIV seropositivity [N36], terminal care [O7], euthanasia [O15].

GR 3 DALLA-VORGIA Panaglota

Assistant Professor, Department of Hygiene and Epidemiology, University of Athens, Medical School.

Department of Hygiene and Epidemiology
University of Athens
Medical School
11527 Athens (Greece)
✆ (30) 1-770 68 77 Fax: (30) 1-770 42 25

■ AFFILIATIONS: Hellenic Society of Medical Ethics, Athens, Greece (*Secretary General*).
■ TEACHING: Teaching medical and dental Students in the context of course "Medical Responsibility and Ethics", also of the courses "Epidemiology" and "Preventive Medicine". — Also teaching in various post-graduate courses.
■ PUBLICATIONS: yes.
■ WORKING ON: fundamental rights of the individuals [B2], human rights [B3], self determination [B4], dignity [B5], integrity [B8], justice [B9], privacy [B11], human body commercialization [B13], protection of rights — involved institutions [B16], ethics committees [B22], Council of Europe [B25], Helsinki Declaration [B37], patient advocacy [B42], ethical rules and principles [B47], ethics [B48], codes of ethics [B49], code of deontology [B60], paternalism [B62], respect [B63], information — communication — media [B68], mass media [B72], education [B76], health education [B77], biomedical ethics education [B78], medical education [B80], students [B82], universities [B84], bioethics [B89], codes of biomedical ethics [B90], ethical review [B93], bioethical issues [B96], medical ethics [C52], nursing ethics [C53], ethics and humanities [C54], ancient history [C59], social and political issues [D1], legislation [D13], epidemiology [E9], containment [E38], quality of environment [E41], nuclear energy [E44], fluoridation [E45], occupational medicine [E46], pollution [E47], data protection [E52], biomedical research [F2], health services research [F4], animal experimentation [F10], healthy volunteers [F13], health policy [G13], health services [G25], intensive care units [G33], WHO — World Health Organization [G36], nurses [G39], physicians [G40], hospitalisation [G46], medical records [G50], selection for treatment [G52], immunization [G59], dentistry [G62], occupational medicine [G65], amniocentesis [I4], mass screening [I6], transplantation [I22], vaccination [I43], patient information [J4], disclosure [J7], confidentiality [J20], inter-personal relationship [J27], law, legislation and jurisprudence [K1], conflict of interest [K9], wrongful death [K12], malpractice [K19], negligence [K20], child's interest [K24], injuries [K26], therapeutic risk [K32], legislation and law [K34], stages of life — problems proper to childhood and elderly [L1], life extension [L47], sexuality and procreation — unborn child [M1], death and resuscitation [O1], psychological stress [P17], genetics and applications — biotechnology [Q1].
■ INTERESTED IN: fundamental rights of the individuals [B2], women's rights [B6], integrity [B8], justice [B9], privacy [B11], human body commercialization [B13], traffic of organs [B15], World Medical Assembly [B20], CCNE — Comité Consultatif National d'Ethique (France) [B23], Council of Europe [B25], United Nations [B29], International Court [B30], International Charter of Human Rights [B34], European Convention on Human Rights [B36], Helsinki Declaration [B37], Universal Declaration of Human Rights [B39], public advocacy [B41], patient advocacy [B42], beneficence [B57], authoritarianism [B58], history of biomedical ethics [B94], ethical analysis [C3], religion and religious ethics [C19], religion [C36], scriptural interpretation [C40], medical ethics [C52], nursing ethics [C53], ethics and humanities [C54], history [C56], history before the 20th century [C58], ancient history [C59], history of the 20th century [C60], counseling [C87], social sciences [C89], social and political issues [D1], legislation [D13], handicapped [D43], physically handicapped [D44], disadvantaged [D46], social problems [D52], stigmatization [D54], political issues [D56], international aspects [D59], international organizations [D61], ECC — European Community Commission [D63], epidemiology [E9], research [E14], containment [E38], quality of environment [E41], nuclear energy

[E44], occupational medicine [E46], pollution [E47], data protection [E52], [E56], teaching methods [E59], biomedical research [F2], animal experimentation [F10], group of vulnerable subjects [F20], medical informatics [F21], health policy [G13], health legislation [G14], health services [G25], intensive care units [G33], WHO — World Health Organization [G36], nurses [G39], physicians [G40], health care [G45], medical records [G50], emergency care [G54], immunization [G59], dentistry [G62], medicine [G63], occupational medicine [G65], amniocentesis [I4], mass screening [I6], sperm banks [I11], tissue donation [I13], donors [I14], transplantation [I22], vaccination [I43], health care professionals, researchers and patient relationship [J1], law, legislation and jurisprudence [K1], prohibition [K38], contraception [M3], reproductive organs and embryonic structures [M9], reproductive technologies [M26], abortion [M47], fetal therapy [M57], genetics and applications — biotechnology [Q1].

GR 4 **DONTAS** A.

Head of Pre-clinical and Clinical Research, Centre of Studies of Age-related Changes in Man (Athens). — Designing, programming, coordinating and supervising various facets of research in aging.

1, Stisihorou Street
106 74 Athens (Greece)
✆ (30) 1 - 72 28 239

■ AFFILIATIONS: Hellenic Society for Ethics and Deontology, Athens, Greece (*Member of the board*).
■ TEACHING: In round-table discussions of the Hellenic Society for Ethics and Deontology, independently or in common with other societies, e.g. Gerontological Association. — Teaching of classes of medical students or nurses.
■ PUBLICATIONS: yes.
■ WORKING ON: fundamental rights of the individuals [B2], self determination [B4], health education [B77], medical education [B80], biology and human future [C43], medical ethics [C52], epidemiology [E9], investigators [E15], research team [E16], clinical trials [F3], therapeutic research [F6], health care delivery [G48], life extension [L47], health care services and aged [L48].
■ INTERESTED IN: aged — problems related with aging [L45], terminal care [O7], euthanasia [O15], attitudes to death [O20].

GR 5 **DOURAKI** Thomais

Legal Adviser for European Affairs.

Ministère de l'économie nationale
Direction des relations entre la Grèce et les Communautés européennes
Nikis str. 5-7
106 72 Athens (Greece)
✆ (30) 1 - 32 42 711 Fax: (30) 1 - 32 24 071

GR 6 **DRAGONA-MONACHOU** Myrto

Head of the Department of Philosophy, University of Crete. — Former Chairperson of the Greek Philosophical Society. — Former Chairperson of the Greek State Scholarship Foundation. — Member of the Society of Medical Ethics and Deontology.

University of Crete, Society of Medical Ethics and Deontology
18 P. Tsaldari
Maroussi
151 22 Athens (Greece)
✆ (30) 1 - 80 24 350 Fax: (30) 1 - 80 65 310

■ AFFILIATIONS: Society of Medical Ethics and Deontology, Athens, Greece (*Member*).
■ TEACHING: Seminary in the University of Crete. — Teaching in the Medical School, Athens, and in various hospitals. — Participations in colloquia and conferences.
■ PUBLICATIONS: yes.
■ WORKING ON: respect of human dignity and human rights [B1], fundamental rights of the individuals [B2], human rights [B3], human equality [B7], justice [B9], Universal Declaration of Human Rights [B39], ethical rules and principles [B47], codes of ethics [B49], moral policy [B55], deontology [B59], code of deontology [B60], morality [B61], bioethics [B89], codes of biomedical ethics [B90], ethicists [B92], history of biomedical ethics [B94], philosophical ethics [C2], normative ethics [C5], metaethics [C11], morals [C12], values [C15], religious ethics [C21], theology [C41], future generations [C44], professional ethics [C51], normality [C78], politics [D76], war [D87], patients' rights [J2], patient information [J4], confidentiality [J20].
■ INTERESTED IN: human rights [B3], justice [B9], ethical rules and principles [B47], bioethics [B89], history of biomedical ethics [B94], theology [C41].

GR 7 **GARANIS** Tina N.
Lawyer.— Research and Teaching Fellow.

Athens School of Public Health
196 Alexandras Avenue
115 21 Athènes (Greece)
✆ (30) 1 - 64 65 982 Fax: (30) 1 - 64 44 260

■ AFFILIATIONS: Hellenic Foundation for Medical Ethics, Athens, Greece (*Member*).
■ TEACHING: Medical Law and Ethics at the Athens Schools of Public Health.
■ PUBLICATIONS: yes.
■ WORKING ON: ethics committees [B22], Nuremberg Code [B35], Helsinki Declaration [B37], ethics [B48], codes of ethics [B49], beneficence [B57], deontology [B59], code of deontology [B60], paternalism [B62], biomedical ethics education [B78], codes of biomedical ethics [B90], deontological ethics [C6], teleological ethics [C7], utilitarianism [C8], medical ethics [C52], decision making [E54], WHO — World Health Organization [G36], transplantation [I22], patient information [J4], treatment refusal [J15], patient participation [J16], right to treatment [J17], living wills [J19], confidentiality [J20], physician's role [J26], investigator subject relationship [J29], malpractice [K19], negligence [K20], legal liability [K30], civil code [K44], medical law [K51], infanticide [L42], coma [N41], persistent vegetative state [N42], death [O5], brain death [O6], life-sustaining treatment [O9], withholding treatment [O10], allowing to die [O11], prolongation of life [O13], euthanasia [O15].
■ INTERESTED IN: biomedical ethics education [B78], history of biomedical ethics [B94], Bioethics bill, 1992 (France) [B98], transplantation [I22], death [O5], brain death [O6], attitudes to death [O20], human genome [Q47], deontology [B59], biomedical ethics education [B78], medical ethics [C52], negligence [K20], euthanasia [O15].

GR 8 **KARAGOUGNIS**
Barrister-at-law working in the field of Health Laws Bioethics.

Association Hellénique Médicale
7, Vissaronos Street
106 72 Athènes (Greece)

GR 9 **KATSAS** Aristotle
Director of Surgery, Evangelismos Hospital.

Hôpital Evangelismos
7-9 Vrasida Street
115 28 Athens (Greece)
✆ (30) 1 - 72 39 912

■ AFFILIATIONS: Hellenic Society for Medical Ethics and Deontology, Athens, Greece (*Member*).
■ TEACHING: To surgical residents and nurses.
■ WORKING ON: review [A2], justice [B9], privacy [B11], human person [B12], ethics committees [B22], ethics [B48], deontology [B59], interdisciplinary communication [B70], biomedical ethics education [B78], nursing education [B79], medical education [B80], uncontrolled information [B88], codes of biomedical ethics [B90], bioethical issues [B96], quality of life [C16], medical ethics [C52], nursing ethics [C53], socioeconomic factors [D18], biological warfare [D88], nuclear warfare [D89], incidence [E11], morbidity [E12], mortality [E13], peer review [E63], social control of science [E64], traffic accidents [G7], hospital [G28], medical staff [G41], selection for treatment [G52], emergency care [G54], medicine [G63], resource allocation [H4], health care costs [H9], costs and benefits [H10], risks and benefices [H11], population control [H22], elderly [H23], transplantation [I22], surgery [I37], extraordinary treatment [I39], disclosure [J7], consent to treatment [J8], presumed consent [J10], medical secrecy [J22], professional deontology [J25], physician nurse relationship [J31], physician patient relationship [J32], wrongful death [K12], malpractice [K19], negligence [K20], injuries [N49], death [O5], brain death [O6], terminal care [O7], euthanasia [O15].
■ INTERESTED IN: data bases [A3], documentation [A8], publications [A9], human body commercialization [B13], traffic of organs [B15], ethics committees [B22], professional deontology organs [B26], manifests and declarations concerning human rights [B33], Helsinki Declaration [B37], common good [B45], humanity heritage [B46], codes of ethics [B49], professional competence [B52], obligations to society [B54], deontology [B59], biomedical ethics education [B78], curriculum [B83], advertising [B87], uncontrolled information [B88], European Convention of Bioethics [B97], Bioethics bill, 1992 (France) [B98], conscience [C13], social worth [C17], value of life [C18], biology and human future [C43], future generations [C44], medical ethics [C52], nursing ethics [C53], socioeconomic factors [D18], sociology of medicine [D55], review committees [E42], safety [E51], data protection

[E52], peer review [E63], social control of science [E64], human genome project [F8], human experimentation [F11], hospices [G34], resource allocation [H4], health care costs [H9], costs and benefits [H10], risks and benefices [H11], population control [H22], transplantation [I22], fetal tissue transplantation [I27], extraordinary treatment [I39], consent forms [J14], living wills [J19], medical secrecy [J22], wrongful death [K12], malpractice [K19], negligence [K20], economic value of life [K28], wrongful life [K33], sexuality [M20], sexual behavior [M21], reproductive technologies [M26], terminal care [O7], euthanasia [O15], genetic intervention [Q22], genome mapping [Q23], human genome [Q47].

GR 10 **MANIATIS** George M.

Professor and Director, Biology Laboratory, School of Health Sciences (Patras).

Faculty of Medicine
Department of Biology
261 10 Patras (Greece)
✆ (30) 61 - 997 621 Fax: (30) 61 - 997 689

■ AFFILIATIONS: Hellenic Society of Medical Ethics, Greece (*Founding member*). — ESLA of Human Genome Programme (EC) (*Member*).
■ WORKING ON: privacy [B11], human person [B12], human body commercialization [B13], ethics committees [B22], mass media [B72], medical education [B80], codes of biomedical ethics [B90], biology and human future [C43], biology [E5], genetics [E6], molecular biology [E8], human genome project [F8], prenatal diagnosis [I3], thalassemia [Q17], genetic intervention [Q22], genome mapping [Q23], gene therapy [Q42].
■ INTERESTED IN: privacy [B11], human person [B12], human body commercialization [B13], ethics committees [B22], code of deontology [B60], paternalism [B62], codes of biomedical ethics [B90], situational ethics [C9], eastern orthodox ethics [C26], biology and human future [C43], medical ethics [C52], biology [E5], genetics [E6], molecular biology [E8], human genome project [F8], population growth [H20], prenatal diagnosis [I3], mass screening [I6], mandatory screening [I7], blood transfusions [I19], blood donation [I20], blood substitutes [I21], bone marrow transplantation [I29], bone marrow [L8], sexual behavior [M21], thalassemia [Q17], genetic intervention [Q22], genome mapping [Q23], gene therapy [Q42].

GR 11 **MARKETOS** Spyros

Chairperson, International Hippocratic Foundation. — Vice-Chairperson, European Health Club (Paris). Professor of Medicine, Athens University, Medical School. — Chairperson, Hellenic Association of the History, Sociology and Philosophy of Medical Sciences.

Athens University, Medical School
20 Patr. Ioakeim
106 75 Athens (Greece)
✆ (30) 1-360 13 78 Fax: (30) 364 21 97

■ AFFILIATIONS: International Hippocratic Foundation, Greece (*Chairperson-Director*).
■ TEACHING: Annual postgraduate seminars organised by Athens University, Medical School, in close cooperation with International Hippocratic Foundation and with Hellenic Association of the History, Sociology and Philosophy of Medical Sciences.
■ PUBLICATIONS: yes.
■ WORKING ON: history [C56], historical aspects [C57], history before the 20th century [C58], humanities [C55], ancient history [C59], moral development [C14], value of life [C18], humanism [C66].
■ INTERESTED IN: medical ethics [C52], humanities [C55], health care [G45], family practice [G60], medicine [G63], political issues [D56], socialism [D84], health policy [G13], voluntary programs [G21], social worth [C17], formal social control [D10], socioeconomic factors [D18], scarcity [D21], kidney diseases [N39].

GR 12 **PAPAEVANGELOU** Elen

Barrister-at-law working in the field of Health Laws' Bioethics.

Association hellénique médicale
7, Vissaronos Street
106 72 Athens (Greece)
✆ (30) 1-361 41 45 Fax: (30) 1 - 361 00 85

GR 13 **PAPAEVANGELOU** George

Director of the National Centre for AIDS. — Professor of Preventive Medicine, University of Athens.

Centre national du SIDA
52, Skoufa Street
106 72 Athens (Greece)
✆ (30) 1 - 64 67 473 Fax: (30) 1 - 77 81 829

■ AFFILIATIONS: Hellenic Association of Bioethics, Athens, Greece (*Member*).
■ TEACHING: Teaching Bioethics at the University. — Lecturing on Bioethics. — Special interest in AIDS: Chairperson of the Hellenic Association for the Study and Control of AIDS.
■ PUBLICATIONS: yes.
■ WORKING ON: codes of biomedical ethics [B90], social discrimination [D53], epidemiology [E9], research policy [E18], clinical trials [F3], healthy volunteers [F13], contact tracing [G12], voluntary programs [G21], nurses [G39], immunization [G59], anonymous donation [I17], vaccination [I43], nurse patient relationship [J30], physician nurse relationship [J31], physician patient relationship [J32], sexual behavior [M21], homosexuality [M22], prostitution [M25], hepatitis [N30], acquired immunodeficiency syndrome [N35], HIV seropositivity [N36], virus diseases [N37].
■ INTERESTED IN: epidemiology [E9], clinical trials [F3], immunization [G59], vaccination [I43], hepatitis [N30], venereal diseases [N31], acquired immunodeficiency syndrome [N35], HIV seropositivity [N36], virus diseases [N37].

GR 14 **ROUSSOS** Michel

Chaplain of Christian Executives. — Chief-Editor of the Jesuit magazine of Athens. — Director of a students' hostel.

Cadres chrétiens
27, rue Smyrnis
104 39 Athens (Greece)
✆ (30) 1 - 88 35 911

■ AFFILIATIONS: Mouvement international des intellectuels catholiques, Athens, Greece.
■ TEACHING: Supervisor of students, education on human rights and human advancement.
■ PUBLICATIONS: yes.

GR 15 **STEFANIS** Constantin

Athens University Medical School
Department of Psychiatry Athens (Greece)
✆ (30) 1 - 72 17 763 Fax: (30) 1 - 72 43 905

GR 16 **VALASSI-ADAM** Eleni

Head of the Department of Social Pediatrics. — Head of the Committee of Bioethics ICH (Child and Family Health) — Education and Research in Health Services.

Institut de la Santé de l'Enfant
Department de la pédiatrie Sociale
29, Eakinthon str.
154 52 Athens (Greece)

■ TEACHING: The subject is included in the postgraduate course of paediatrics for nursing staff (nursing research).
■ PUBLICATIONS: yes.
■ WORKING ON: research design [E19], peer review [E63], health services research [F4], maternal health [G3], pediatrics [G68], adolescents [L14], child and family [L31].
■ INTERESTED IN: deontology [B59], health education [B77], counseling [C87], scientific misconduct [E25], computerized file [E35], pediatrics [G68], children and media [L36].

Ireland

Area: 70 283 km²
Population: 3.5m
Gross domestic product: IR£ 28.8b; US$ 44.7bn
Gross domestic product per head: $ 12,800
Gross domestic product: 1991: 1.8%
1992: 3.4%

Organisations

O IRL 1 **THE HEALTH RESEARCH BOARD**
73 Lower Baggat St.
Dublin 2 (Ireland)
✆ (353) -1- 76 11 76

O IRL 2 **UNIVERSITY COLLEGE CORK**
Medical Faculty
Cork (Ireland)
✆ (353) 21-27 68 71 Fax: (353) 21-27 59 48

O IRL 3 **UNIVERSITY COLLEGE DUBLIN**
Medical Faculty
Earlsfort Terrace
Dublin 2 (Ireland)
✆ (353) 1-706 74 40 Fax: (353) 1-753 655

Individuals

IRL 1 **BOOTHMAN** Rosemary
Medical Doctor. — Medical Officer, Department of Health.

Department of Health
Hawkins House
Hawkins St.
Dublin 2 (Ireland)
✆ (353) 1-714 711 Fax: (353) 1-711 947

■ WORKING ON: Council of Europe [B25], alcohol abuse [P4].
■ INTERESTED IN: human person [B12], traffic of organs [B15], alcohol abuse [P4].

IRL 2 **DOOLEY** Dolores
Philosophical Research in Medical Ethics. — University Lecturer in Philosophy and Medical Ethics (Medical Faculty).

University College
Department of Philosophy
Cork (Ireland)
✆ (353) 21-27 68 71 Fax: (353) 21-27 59 48

■ AFFILIATIONS: Hastings Center Institute for Ethics in the Professions of Health, New York, N.Y., (*Irish correspondent*).
■ TEACHING: Lectures to medical students (two hours per week for 25 weeks each year). — Lectures to psychiatric interns in their post-graduate course.
■ PUBLICATIONS: yes.
■ WORKING ON: history [C56], history of biomedical ethics [B94], caesarean section [M41], decision making [E54], decision analysis [E55], patient participation [J16], reproductive technologies [M26], Amnesty International [B18], European Convention of Bioethics [B97].
■ INTERESTED IN: legislation and law [K34], caesarean section [M41], patient participation [J16], decision making [E54], decision analysis [E55], human rights [B3], women's rights [B6], human person [B12], biomedical ethics education [B78], European Convention of Bioethics [B97].

IRL 3 **HARBINSON** John
State Forensic Pathologist: investigation of suspicious death for the Police (Garda). — Professor of

Forensic Medicine: teaching of Legal Medicine to medical undergraduates at two of the three Medical Schools in Dublin (Royal College of Surgeons, Dublin University).

Royal College of Surgeons — Trinity College
Department of Forensic Medicine
188 Pearse Street Dublin 2 (Ireland)
✆ (353) 1-719 835 Fax: (353) 1-67 905 05

■ AFFILIATIONS: Royal College of Surgeons in Ireland, Ireland (*Member*). — Dublin University, Trinity College, Ireland,
■ TEACHING: Medical undergraduate courses.

IRL 4 **IGLESIAS** Teresa
University Lecturer.

Department of Philosophy
University College Dublin
Belfield
Dublin 4 (Ireland)
✆ (353) 1-269 32 44 and 706 83 67 Fax: (353) 1-269 34 69

■ AFFILIATIONS: Hastings Center Institute for Ethics in the Professions of Health, New York, U.S.A. (*Member*). — The Linacre Centre, London, United Kingdom,
■ TEACHING: Head Lecturer in Medical Ethics (Faculty of Medicine). — Philosophy Lecturer: courses in Bioethics for Philosophy Students (Faculty of Arts).
■ PUBLICATIONS: yes.
■ WORKING ON: respect of human dignity and human rights [B1], ethical rules and principles [B47], biomedical technologies [E28], biomedical research and experimentation [F1], health care, medical acts [I1], patients' rights [J2], reproductive technologies [M26], genetics and applications — biotechnology [Q1].
■ INTERESTED IN: *idem, plus:* human body [L3].

IRL 5 **JEANROND** Werner G.
Professor of Theology.

University of Dublin
Trinity College
Theological Studies
Dublin 2 (Ireland)
✆ (353) 1-702 1101 Fax: (353) 1-772 694

■ TEACHING: Courses on Bioethics within Theology and Dental Science curriculums.
■ PUBLICATIONS: yes.
■ WORKING ON: Christian ethics [C22], religious beliefs [C32], religion [C36], scriptural interpretation [C40], theology [C41], professional ethics [C51], medical ethics [C52], humanities [C55], self concept [C83], international aspects [D59], international organizations [D61], health policy [G13], dentistry [G62].
■ INTERESTED IN: dentistry [G62], self concept [C83], professional ethics [C51], medical ethics [C52], humanities [C55], Christian ethics [C22], scriptural interpretation [C40], theology [C41].

IRL 6 **O'DWYER** Timothy V.
Deputy Chief Medical Officer (advising the Department of Health on medical matters).

Department of Health
Hawkins House
Hawkins Street
Dublin 2 (Ireland)
✆ (353) 1-71 47 11 Fax: (353) 1-71 45 08

■ WORKING ON: justice [B9], dignity [B5], privacy [B11], human person [B12], traffic of organs [B15].
■ INTERESTED IN: privacy [B11], fundamental rights of the individuals [B2].

IRL 7 **REIDY** Maurice
Lecturer and Author in Moral Theology and Medical Ethics.

Holy Cross College
Clonliffe Dublin 3 (Ireland)
✆ (353) 1-375 103 Fax: (353) 1-371 474

■ TEACHING: medical students, students of Diploma in Nursing Studies, students for priesthoods.
■ PUBLICATIONS: yes.
■ WORKING ON: fundamental rights of the individuals [B2], human rights [B3], religion and religious ethics [C19], Roman catholicism [C31], medical ethics [C52], nursing ethics [C53], sexuality and procreation — unborn child [M1], sexuality and procreation [M2], reproductive technologies [M26], abortion [M47], death and resuscitation [O1], determination of death [O2], terminal care [O7], euthanasia [O15].
■ INTERESTED IN: fundamental rights of the individuals [B2], human rights [B3], religion and religious ethics [C19], Roman catholicism [C31], sexuality and procreation — unborn child [M1], sexuality and procreation [M2], reproductive technologies [M26], abortion [M47], death and resuscitation [O1], determination of death [O2], terminal care [O7], euthanasia [O15].

IRL 8 **SKRABANEK** Peter
Senior Lecturer in Community Health.

University of Dublin
Trinity College
199, Pearse Street
Dublin 2 (Ireland)
✆ (353) 1-772 941 and direct line 702 1643 Fax: (353) 1-710 697

■ TEACHING: Lectures and seminars for medical undergraduates and postgraduates.
■ PUBLICATIONS: yes.
■ WORKING ON: mass screening [I6], mandatory screening [I7], medical ethics [C52], scientific misconduct [E25], preventive medicine [G58].
■ INTERESTED IN: information sources, data bases [A1], respect of human dignity and human rights [B1], public advocacy [B41], patient advocacy [B42], state interest [B44], history of biomedical ethics [B94], scientific misconduct [E25], organization of health care; facilities, manpower and services; health occupations [G1], mass screening [I6], mandatory screening [I7], patients' rights [J2].

IRL 9 **TIERNEY** Niall
Chief Medical Officer, Department of Health.

Department of Health
Hawkins House
Hawkins Street
Dublin 2 (Ireland)
✆ (353) 1-75 47 11 Fax: (353) 1-75 19 47

■ PUBLICATIONS: yes.

Italy

Area: 301 225 km^2
Population: 57,9 millions
Gross domestic product: L 1,510trn; US$ 1.18trn
Gross domestic product per head: $ 20,300
Gross domestic product: 1991: 1.5%
1992: 2.5%

Organisations

O11 **ASSOCIATION INTERNATIONALE DROIT, ÉTHIQUE ET SCIENCE (GROUPE DE MILAZZO)**
Isenb. Villa Eolian
Via Cappucini
98057 Milazzo (Italy)
✆ (39) 90-92 82 912 Fax: (39) 90-92 86 565

O12 **ASSOCIAZIONE CULTURALE PSICOANALISI CONTRO**
Via Arenula 21
00186 Roma (Italy)
✆ (39) 6-68 67 495 and 68 33 552 Fax: (39) 6-68 67 509

■ OTHER FIELDS OF ACTIVITIES: Psychoanalysis (teaching and research).
■ DATE OF FOUNDATION: 1978.
■ EXECUTIVE BOARD: Professor Sandro Gindro (*Chairperson*), Dr. Raffaele Bracalemti (*Vice Chairperson*), Renzo Rossi (*Executive Secretary*), Dr. Emilio Mordini (*Member of the Board*), Dr. Arturo Casomi (*Member of the Board*), Professor Pietro De Santis (*Member of the Board*), Giovammi Valeri (*Member of the Board*), Dr. Carlo Calzome (*Member of the Board*).
■ TEACHING: bioethics within the association's school. — Workshops and intensive courses for state institutions.

O13 **ASSOCIAZIONE "DON GIUSEPPE ZILLI" PER LA FAMIGLIA E LA COMUNICAZIONE SOCIALE**
Centro Internazionale Studi Famiglia
Via Duccio di Boninsegna 10
20145 Milano (Italy)
✆ (39) 2-48 01 20 40 Fax: (39) 2-48 00 99 38

■ OTHER FIELDS OF ACTIVITIES: Family pastoral.— Documentation, conferences, publications.
■ AFFILIATION: an activity centre within another organisation.
■ DATE OF FOUNDATION: 1974. — Became part of the Associazione Don Zilli in 1980.
■ EXECUTIVE BOARD: Leonardo Zega (*Chairperson*), Virgilio Melchiorre (*Director*).
■ PUBLICATIONS: bibliographic newsletter.
■ DOCUMENTATION CENTRE: open to the public.
■ WORKING ON: human rights [B3], Council of Europe [B25], ethical rules and principles [B47], education [B76], biomedical ethics education [B78], codes of biomedical ethics [B90], quality of life [C16], value of life [C18], religious ethics [C21], psychology [C68], family members [D24], home care [G55], elderly [H23], sperm banks [I11], anonymous donation [I17], patient information [J4], parental notification [J6], disclosure [J7], informed consent [J9], spousal consent [J12], parental consent [J13], living wills [J19], parent child relationship [J34], adolescents [L14], child and family [L31], adoption [L32], abandoned (child) [L33], aid children [L34], disease and aged [L46], life extension [L47], health care services and aged [L48], aged [L49], aging [L50], old person abuse [L51], contraception [M3], sterilization [M6], procreation [M15], sexuality [M20], homosexuality [M22], reproductive technologies [M26], abortion [M47], palliative care [O8], euthanasia [O15], suicide [O19], attitudes to death [O20].
■ INTERESTED IN: CCNE — Comité Consultatif National d'Ethique (France) [B23], biomedical ethics education [B78], CCNE advice (France) [B91], Bioethics bill, 1992 (France) [B98], family members [D24], home care [G55], elderly [H23], patient information [J4], parental notification [J6], disclosure [J7], informed consent [J9], spousal consent [J12], parental consent [J13], living wills [J19], parent child relationship [J34].

O14 **ASSOCIAZIONE ITALIANA PER L'EDUCAZIONE DEMOGRAFICA**
Via Piave 41
00187 Roma (Italy)
✆ (39) 6-48 14 646

015 ASSOCIAZIONE MEDICI CATTOLICI ITALIANI

Via della Conciliazione 15
00193 Roma (Italy)
✆ (39) 6-68 69 182

016 ASSOCIAZIONE NAZIONALE PER LA LOTTA ALL'AIDS

Via Rovereto 10
00198 Roma (Italy)
✆ (39) 6-84 19 695

017 ASSOCIAZIONE TEOLOGICA ITALIANA PER LO STUDIO DELLA MORALE

Via Nosadella 6
40123 Bologna (Italy)
✆ (39) 51 33 03 01 Fax: (39) 51 33 13 54

■ OTHER FIELDS OF ACTIVITIES: general and specialized moral (society, economics, family).
■ OFFICIALLY REGULATED ORGANISATION: statutes.
■ DATE OF FOUNDATION: 1966.
■ EXECUTIVE BOARD: Prof Luigi Lorenzetti (*Chairperson*), Prof Francesco Compagnoni (*Vice-Chairperson*), Prof Bruno Marra (*Executive Secretary*).
■ TEACHING: yes.
■ PUBLICATIONS: "Eutanasia, Il senso del vivere e del morire umano", Edizioni Dehoniane, Bologna, 1987, 143 pp.
■ COLLOQUIUMS, SYMPOSIUMS: yes.

018 CECOS ITALIA

Via dei Soldati, 25
00186 Roma (Italy)
✆ (39) 6-6540375 et 68300253 Fax: (39) 6-65 48 8706 68308870

■ AFFILIATION: Member of the international CECOS, Chairperson Professor Lansac (Tours).
■ OFFICIALLY REGULATED ORGANISATION: statutes (modified on 21/5/91).
■ DATE OF FOUNDATION: 27/04/1984.
■ EXECUTIVE BOARD: Professor Emanuele Lauricella (*Chairperson*).
■ TEACHING: Participates in all congresses held on medically assisted fertilisation.
■ PUBLICATIONS: various scientific magazines plus collaboration on Rivista Politeia.
■ WORKING ON: public opinion [D7], punishment [D8], legislation [D13], model legislation [D14], regulation [D16], government regulation [D17], biology [E5], genetics [E6], molecular biology [E8], biomedical technologies [E28], freezing [E29], preservation [E30], biological containment [E39], health services research [F4], therapeutic research [F6], family planning [G8], obstetrics and gynecology [G66], prenatal diagnosis [I3], amniocentesis [I4], chorionic villus sampling [I5], sperm banks [I11], donors [I14], anonymous donation [I17], directed donation [I18], patient information [J4], disclosure [J7], consent to treatment [J8], informed consent [J9], spousal consent [J12], consent forms [J14], medical secrecy [J22], physicians and researchers accountability [J23], professional deontology [J25], expert evaluation [K15], legitimacy [K21], child donation [K22], child's interest [K24], anatomy, physiology, development [L2], body parts and fluids [L4], beginning of life [L24], child and family [L31], aid children [L34], contraception [M3], immunologic contraception [M5], wish of children [M16], reproduction [M17], sexuality [M20], sexual behavior [M21], homosexuality [M22], homosexuals [M23], transsexualism [M24], ovum donors [M27], semen donors [M28], in vitro fertilization [M29], FIV centre [M30], AID [M32], AIH [M33], excess embryos [M34], host mothers [M35], embryo transfer [M36], GIFT (Gamete Intrafallopian Transfer) [M37], embryo donation [M38], genetics and applications — biotechnology [Q1], genetic defects [Q3], chromosomal disorders [Q4], XYY karyotype [Q5], hereditary diseases [Q18], sex linked defects [Q19], genetic intervention [Q22], sex determination [Q26], sex preselection [Q30], gene therapy [Q42].
■ INTERESTED IN: family members [D24], single persons [D25], married persons [D28], family relationship [D31], biology [E5], genetics [E6], molecular biology [E8], investigators [E15], biomedical technologies [E28], freezing [E29], preservation [E30], cryonic suspension [E31], biological containment [E39], social control of science [E64], health services research [F4], therapeutic research [F6], human genome project [F8], human experimentation [F11], family planning [G8], maternal welfare [G10], health legislation [G14], quarantine [G20], health personnel [G37], physicians [G40], obstetrics and gynecology [G66], prenatal diagnosis [I3], amniocentesis [I4], chorionic villus sampling [I5], sperm banks [I11], donors [I14], anonymous donation [I17], directed donation [I18], patient information [J4], disclosure [J7], consent to treatment [J8], informed consent

[J9], spousal consent [J12], consent forms [J14], medical secrecy [J22], physicians and researchers accountability [J23], professional deontology [J25], expert evaluation [K15], legitimacy [K21], child donation [K22], child's interest [K24], anatomy, physiology, development [L2], body parts and fluids [L4], beginning of life [L24], child and family [L31], aid children [L34], contraception [M3], procreation [M15], wish of children [M16], reproduction [M17], sexuality [M20], sexual behavior [M21], homosexuals [M23], transsexualism [M24], ovum donors [M27], semen donors [M28], in vitro fertilization [M29], FIV centre [M30], artificial insemination [M31], AID [M32], AIH [M33], excess embryos [M34], host mothers [M35], embryo transfer [M36], GIFT (Gamete Intrafallopian Transfer) [M37], embryo donation [M38], genetics and applications — biotechnology [Q1], genetic defects [Q3], chromosomal disorders [Q4], XYY karyotype [Q5], hereditary diseases [Q18], sex linked defects [Q19], genetic intervention [Q22], sex determination [Q26], sex preselection [Q30], gene therapy [Q42].

O I 9 CENTRO DI BIOETICA

Istituto di Filosofia e Sociologia del Diritto
Via Balbi 30
16126 Genova (Italy)
✆ (39) 10-26 59 02

O I 10 CENTRO DI BIOETICA DELL'ISTITUTO GRAMSCI DI ROMA

Via del Conservatorio 55
00186 Roma (Italy)
✆ (39) 6-6833 756 and 6875 405 Fax: (39) 6-68 77 736

■ AFFILIATION: Centre of Gramsci Instute Foundation.
■ EXECUTIVE BOARD: Dr. Antonio Di Meo (*Director*), Professor Giovanni Berlinguer (*Member of the Board*), Professor Stefano Rodota (*Member of the Board*), Professor Luciano Terrenato (*Member of the Board*), Professor Eugenio Lecaldano (*Member of the Board*), Dr. Claudia Mancina (*Member of the Board*), Dr. Marina Frontali (*Member of the Board*), Professor Giorgio Bignami (*Member of the Board*).
■ PUBLICATIONS: Reviews.
■ COLLOQUIUMS, SYMPOSIUMS: Occasional colloquiums for researchers (science and philosophy), politicians, and students. — Six colloquiums have been organized since 1988.
■ WORKING ON: respect of human dignity and human rights [B1], women's rights [B6], legislation [D13], socioeconomic factors [D18], social problems [D52], politics [D76], community services [D78], biomedical research [F2], experimentation [F9], methodology of research and experimentation [F15], institutional policies [G15], legal personality [K2], jurisprudence — accountability [K8], legislation and law [K34], genetics and applications — biotechnology [Q1].
■ INTERESTED IN: technology [E27], fundamental rights of the individuals [B2], protection of rights — involved institutions [B16], legislation [D13], socioeconomic factors [D18], cultural pluralism [D51], social problems [D52], biomedical research [F2], experimentation [F9], methodology of research and experimentation [F15], legal personality [K2], jurisprudence — accountability [K8], legislation and law [K34], history of science [E3].

O I 11 CENTRO DI DEONTOLOGIA E DI ETICA MEDICA DELL'UNIVERSITÀ

Dipartimento di Sanità Pubblica dell'Università di Pisa
Via Roma 55
56100 Pisa (Italy)
✆ (39) 50 55 11 91 et 55 13 98 Fax: (39) 50 55 40 37

■ OTHER FIELDS OF ACTIVITIES: Teaching, research and assistance activities in: criminal anthropology, occupational medicine, forensic medicine, social medicine, statistics and bioethics.
■ AFFILIATION: Centre of Public Health and Biostatistics Department of the Università di Pisa.
■ DATE OF FOUNDATION: 1/4/1988.
■ EXECUTIVE BOARD: Professor Umberto Palagi (*Chairperson*), Professor Remo Rossi, Dr. Maria Antonietta Lombardi, Pr Mario Giusiani, Dr. Massimo Mariani, Professor Ugo Mario Macerata.
■ TEACHING: Deontology and medical ethics within medicine and surgery diploma curriculums, in various schools of specialized medicine and in postgraduate courses on prison medicine.

O I 12 CENTRO DI STUDI FILOSOFIA CONTEMPORANEA

CNR
Via Lomellini 8/8
16126 Genova (Italy)
✆ (39) 10-29 35 11 Fax: (39) 10-24 71 184

0114 **CENTRO PER LA RICERCA BIOETICA E CLINICA — IL ISTITUTO DI CLINICA GINECOLOGICA ED OSTETRICA**
Policlinico Umberto-I
Via le Regina Elena, 324
00161 Roma (Italy)
✆ (39) 6-4460484 et 4460507 Fax: (39) 6-4469128

■ OTHER FIELDS OF ACTIVITIES: Clinical department of obstetrics and gynaecology with teaching and research.
■ AFFILIATION: A department of the University "La Sapienza" (Rome).
■ DATE OF FOUNDATION: January 1990.
■ EXECUTIVE BOARD: Professor Ermalando V Cosmi (*Chairperson*), Professor Gian Carlo Di Renzo (*Secretary*), Dr. Marialaura Ciampoli (*Member*), Dr. Elena Mancini (*Member*).

0115 **CENTRO UNIVERSITARIO DI STORIA DELL'ETICA MEDICA**
Istituto di Patologia Medica II e Medicina del Lavoro
Policlinico S. Orsola - Via Massarenti 9
40138 Bologna (Italy)
✆ (39) 51-30 89 76 Fax: (39) 51-39 22 40

■ OTHER FIELDS OF ACTIVITIES: History of medical ethics.
■ AFFILIATION: Bologna University.
■ OFFICIALLY REGULATED ORGANISATION: Decree of the Rector of the Bologna University (1990).
■ DATE OF FOUNDATION: 23/07/1990.
■ EXECUTIVE BOARD: Professor Raffaele Alberto Bernabeo (*Director*), Professor Ettore Ambrosioni, Professor Pierluigi Bisbini, Professor Vittorio Bonomini, Professor Renzo Canestrari, Professor Giuseppe D'Antuono, Professor Antonio Maria Mancini, Professor Domenico Marrano, Professor Camillo Orlandi.
■ PUBLICATIONS: Atti del 1° Corso di bioetica ed etica professionale-Chieti, Vecchio Faggio Editore, 1990, 274pp.
■ COLLOQUIUMS, SYMPOSIUMS: first year course on bioethics and professonal ethics (Chieti, 1990).
■ WORKING ON: history of biomedical ethics [B94], history [C56], historical aspects [C57], history before the 20th century [C58], ancient history [C59], history of the 20th century [C60].
■ INTERESTED IN: bibliography [A4], codes of biomedical ethics [B90], history of biomedical ethics [B94], bioethical issues [B96], European Convention of Bioethics [B97], Christian ethics [C22], Roman catholic ethics [C23], eastern orthodox ethics [C26], Protestant ethics [C27], Islamic ethics [C28], history [C56], historical aspects [C57], history before the 20th century [C58], ancient history [C59], history of the 20th century [C60].

0116 **COMITATO DI BIOETICA DELL'ORDINE PROVINCIALE DEI MEDICI**
Corso Belone 103
96100 Siracusa (Italy)
✆ (39) 931-22 09 8

0117 **COMITATO DI ETICA NELLA RICERCA DELL'UNIVERSITÀ DI PARMA**
Università di Parma
Via Gramsci 14
43100 Parma (Italy)
✆ (39) 521-29 05 00 Fax: (39) 521-29 16 76

0118 **COMITATO ETICO DELL' ISTITUTO NAZIONALE PER LA RICERCA SUL CANCRO**
Viale Benedetto XV 10
16132 Genova (Italy)
✆ (39) 10-353 40 81 Fax: (39) 10-352 888

■ OTHER FIELDS OF ACTIVITIES: Teaching, training and public information in bioethics.
■ AFFILIATION: Istituto Nazionale for research on cancer (Genova).
■ OFFICIALLY REGULATED ORGANISATION: Decree n° 68 (24/01/84) of the Institute's Board of administration.
■ DATE OF FOUNDATION: 24/01/1984.
■ EXECUTIVE BOARD: Dr. Graziella Sinaccio (*Co-ordinator*), Professor Luisa Massimo (*Member*), Professor Leonardo Santi (*Member*), Dr. Aldo Capasso (*Member*), Professor Marcello Canale

(*Member*), Dr. Enrico Bartolini (*Member*), Dr. Nicola Perrazzelli (*Member*), Dr. Paolo Bruzzi (*Member*), Dr. Patrizia Russo (*Member*), Maria Chighine (*Member*)
■ TEACHING: bioethics to physicians and nurses.

0119 **COMITATO ETICO FONDAZIONE FLORIANI** (CE/FF)
Via San Simpliciano 2
20121 Milano (Italy)
✆ (39) 2-864 60 404 and 864 63 024 Fax: (39) 2-720 22 493

■ OTHER FIELDS OF ACTIVITIES: Palliative care for terminally ill patients. — Teaching to health care professionals, public information. — Information on palliative care principles.
■ AFFILIATION: Activity centre of the Fondazione Floriani.
■ DATE OF FOUNDATION: 12/06/1991.
■ EXECUTIVE BOARD: Professor Giorgio Di Mola (*Chairperson*), Dr. Patrizia Borsellino (*Member*), Dr. Michèle Gallucci (*Member*), Dr. Alessio Gamba (*Member*), INF Costantina Regazzo (*Member*), Dr. Amedeo Santosuosso (*Member*), Dr. Marcello Tamburini (*Member*), Dr. Franco Toscani (*Member*), Dr. Gian Cristoforo Turri (*Member*).

0120 **COMITATO ETICO PER LA SPERIMENTAZIONE CLINICA DEI FORMACI E DELLE TECNICHE DI INTERVENTO DELLA REGIONE PUGLIA**
Via Caduti di tutte le guerre 15
70126 Bari (Italy)
✆ (39) 80-58 36 53 Fax: (39) 80-40 34 77

0121 **COMITATO NAZIONALE ITALIANO DI BIOETICA**
Via dei Villini 13/15
00161 Roma (Italy)
✆ (39) 6-44 04 279 Fax: (39) 6-44 04 282

0122 **COMITATO REGIONALE DI BIOETICA DELLA REGIONE SICILIA**
Via Ottavio Ziino
90145 Palermo (Italy)
✆ (39) 91-69 61 111

0123 **COMMISSIONE DI BIOETICA DEL CONSIGLIO NAZIONALE DELLE RICERCARE**
Ple A. Moro 7
00185 Roma (Italy)
✆ (39) 6-44 60 484 Fax: (39) 6-49 57 241

0124 **COMMISSIONE ETICA DELLA FACOLTÀ DI MEDICINA E CHIRURGIA**
Università di l'Aquila
Via Roma + Palazzo del Tosto
67100 L'Aquila (Italy)
✆ (39) 862-41 33 75 and 43 33 01 Fax: (39) 862-43 33 03

■ OFFICIALLY REGULATED ORGANISATION: World Medical Association declaration of Helsinki and subsequent amendments. — Italian legislation (ie. D.M. 27/04/92), according to GCP and other EEC regulations. — Deontological and ethics guidelines of Italian Medical Association.
■ DATE OF FOUNDATION: 1990.
■ EXECUTIVE BOARD: Professor Ferdinando Di Orio (*Chairperson*), Professor Paolo Marchetti (*Secretary*), Dr. Gianlorenzo Piccioli (*Member*), Dr. Giuseppe Azzarone (*Member*), Professor Giuseppe De Gennaro (*Member*), Dr. Giorgio Leone (*Member*), Professor Rosella Cardigno (*Member*), Professor Ercole Marianella (*Member*), Professor Sergio Chimenti (*Member*), Dr. Marco Valenti (*Member*)

0125 **COMMISSIONE ETICA UNIFICATA PER I DIRITTI DEL MALATO**
Università di Napoli
Via Pansini 5
80131 Napoli (Italy)
✆ (39) 81-74 61 11

0126 **CONSULTA DI BIOETICA**
Via Sirtori 33
20129 Milano (Italy)
✆ (39) 2-29 51 09 72

0127 **COORDINAMENTO MEDICI AMNESTY INTERNATIONAL - SEZIONE ITALIANA**
Viale Mazzini 146
00195 Roma (Italy)
✆ (39) 6-38 94 03 Fax: (39) 6-31 47 77

0128 **COOPERATIVA DI SOLIDARIETÀ SOCIALE "VILLA RENATA"**
Villa Renata
Via Orsera 4
30126 Lido di Venezia (Italy)
✆ (39) 41-526 88 22 Fax: (39) 41-526 78 74

■ OTHER FIELDS OF ACTIVITIES: Treatment of drug addicts. – Research on epidemiological aspects, risks for the teenage adolescent age group, destructive and self-destructive behaviors. — AIDS. — Training.
■ AFFILIATION: The research unit is affiliated to the IREFREA (Institut de recherche européenne sur le facteur de risque chez l'enfant et l'adolescence. Siège Lyon- Au Berthelot, 14, France).
■ OFFICIALLY REGULATED ORGANISATION: Law TV N° 309/90. — Regional legislation Veneto n° 55/82.
■ DATE OF FOUNDATION: 1984.
■ EXECUTIVE BOARD: Dr. Paolo Stocco (*Chairperson*), Dr. Patricia Cristofalo (*Vice-Chairperson*), Roberto Romandini.
■ TEACHING: Training of social educators regarding drug addiction, AIDS and relations between drug addicted parents and their children.

0129 **CORSO DI PERFEZIONAMENTO IN BIOETICA DELL'UNIVERSITÀ**
Clinica Medica Generale e Terapia medica I
Policlinico Careggi
Viale Morgagni 85
50134 Firenze (Italy)
✆ (39) 55-41 11 45

0130 **CORSO DI PERFEZIONAMENTO IN BIOETICA DELL'UNIVERSITÀ**
Ospedale Generale Regionale "Miulli"
Via Masselli Campagna 106
70021 Acquaviva delle Fonti (Bari) (Italy)
✆ (39) 80-76 19 03

0131 **DOTTORATO DI RICERCA IN DEONTOLOGIA E ETICA MEDICA**
Policlinico "Le Scotte"
53100 Siena (Italy)
✆ (39) 577-40 29 5 Fax: (39) 577-28 62 02

0132 **FONDAZIONE INTERNAZIONALE FATEBENEFRATELLI DIPARTIMENTO DI BIOETICA**
Lungotevere degli Anguillara 12
00153 Roma (Italy)
✆ (39) 6-58 18 895

0133 **FONDAZIONE SMITH KLINE**
Corso Magenta 59
20123 Milano (Italy)
✆ (39) 2-4817098 Fax: (39) 2-4816224

■ OTHER FIELDS OF ACTIVITIES: Medical education.
■ OFFICIALLY REGULATED ORGANISATION: Statutes of the Foundation . — Regulation of the WHO.
■ DATE OF FOUNDATION: 19/01/1980.
■ EXECUTIVE BOARD: Professor Siro Lombardini (*Chairperson*), Dr. Vittorio Ghetti (*Director*), Luisa Marchetti (*Assistant to the Director*).
■ TEACHING: yes.
■ PUBLICATIONS: Ethics committee. — Bioethics and research.
■ DOCUMENTATION CENTRE: open to the public.
■ COLLOQUIUMS, SYMPOSIUMS: two national colloquiums.
■ DECISIONS, ADVICE: advice on practical actions and research activities.
■ WORKING ON: codes of ethics [B49], competence [B51], professional competence [B52], communication [B69], education [B76], biomedical ethics education [B78], nursing education [B79],

medical education [B80], universities [B84], curriculum [B83], codes of biomedical ethics [B90], medical ethics [C52], nursing ethics [C53], genetics [E6], morbidity [E12], mortality [E13], clinical trials [F3], human experimentation [F11], public hospitals [G30], patient compliance [G49], costs and benefits [H10], patient participation [J16], physician patient relationship [J32].

■ INTERESTED IN: book review [A6], documentation [A8], publications [A9], professional deontology organs [B26], professional competence [B52], biomedical ethics education [B78], universities [B84], metaethics [C11], value of life [C18], behavioral research [E21], computers [E33], decision making [E54].

0134 **GROUPE DE TRAVAIL SUR LES PROBLEMES ÉTHIQUES POSÉS PAR LA SCIENCE**

Facolta Valdese di Teologia
Via P. Cossa 42
00193 Roma (Italy)
✆ (39) 6-321 07 89 Fax: (39) 6-320 10 40

0135 **ISTITUTO DI CULTURA POPOLARE E PROMOZIONZ SOCIALE IN CALABRIA — BRUTIUM**

Scuola Superiore di Bioetica
Via Rossini SS 19 bis
87036 Rende Consenza (Italy)
✆ (39) 984-40 13 51 Fax: (39) 984-40 13 55

■ OTHER FIELDS OF ACTIVITIES: Training for teachers.

■ AFFILIATION: Activity Centre of the Istituto di cultura popolare e promozione sociale.

■ OFFICIALLY REGULATED ORGANISATION: Statutes region Calabria (9/11/92), n° 207.

■ DATE OF FOUNDATION: 1990.

■ EXECUTIVE BOARD: Professor Giuseppe Frega (*Chairperson*), Dr. Mauro Giancaspro (*Vice Chairperson*), Professor Franco Gentile (*Secretary*), Professor Antonieta Feola (*Member*), Soc Antonio Bellusci (*Member*), Dr. Salvatore Magaro (*Member*), Soc Giuseppe Carvelli (*Member*), Dr. Pietro Buffone.

■ TEACHING: Bi-annual updating course on medical bioethics (for physicians and teachers).

■ PUBLICATIONS: yes.

■ COLLOQUIUMS, SYMPOSIUMS: yes.

■ DECISIONS, ADVICE: yes.

■ WORKING ON: fundamental rights of the individuals [B2], human person [B12], ethics [B48], moral policy [B55], deontology [B59], deontological ethics [C6], morals [C12], value of life [C18], professional ethics [C51], philosophy [C61], self concept [C83], genetics [E6], organ donors [I15], contraception [M3], procreation [M15], in vitro fertilization [M29], abortion [M47], fetal therapy [M57], euthanasia [O15], genetics and applications — biotechnology [Q1].

■ INTERESTED IN: Amnesty International [B18], International Charter of Human Rights [B34], European Convention on Human Rights [B36], health education [B77], information — communication — media [B68], codes of biomedical ethics [B90], bioethical issues [B96], applied psychology [C86], social sciences [C89], social issues [D2], political issues [D56], violence [D85], nuclear warfare [D89], life sciences [E4], research [E14], public health [G5], health policy [G13], diagnosis [I2], biological specimens procurement, blood transfusion, organ transplantation [I8], patients' rights [J2], physicians and researchers accountability [J23], inter-personal relationship [J27], jurisprudence — accountability [K8], legislation and law [K34], anatomy, physiology, development [L2], childhood difficulties, diseases, protection [L21], aged — problems related with aging [L45], sexuality and procreation [M2], reproductive technologies [M26], abortion [M47], common symptomatology [N2], cancer [N17], cardiovascular diseases [N19], diabetes [N22], congenital defects [N23], communicable diseases [N27], genetic defects and hereditary diseases [Q2], biotechnology — genetic engineering [Q20], behavioral genetics [Q34], medical genetics [Q35], population genetics [Q43].

0136 **ISTITUTO GIOVANNI PAOLO II**

Piazza S. Giovanni Laterano 4
00184 Roma (Italy)
✆ (39) 6-69 86 401

0137 **ISTITUTO INTERNAZIONALE DI TEOLOGIA PASTORALE SANITARIA CAMILLIANUM**

Largo O.Respighi 6
00135 Roma (Italy)
✆ (39) 6-32 88 608 Fax: (39) 6-32 96 352

■ OTHER FIELDS OF ACTIVITIES: teaching and research on sanitary pastoral theology.
■ AFFILIATION: Pastoral theology "Pontificia Facoltà teologica 'Teresianum'".
■ OFFICIALLY REGULATED ORGANISATION: "Costituzione Apostolica: Sapientia christiana" and pertaining to ordinance rules approved by the Holy See, 28/04/1987.
■ DATE OF FOUNDATION: 28/04/1987.
■ EXECUTIVE BOARD: Professor Giuseppe Gina (*Chairperson*), Professor Luciano Sandrin (*Vice. Chairperson*), Giovanni Terenghi (*Executive Secretary*), Giovanni Palombaro (*Director of Education*), Yokin Kamara (*Administrator*).
■ TEACHING: Life and health ethics, Ethics and pharmacology, Ethics issues.

0138 ISTITUTO DI BIOETICA, UNIVERSITÀ CATTOLICA DEL S. CUORE

L. go Francesco Vito 1
00168 Roma (Italy)
✆ (39) 6-30 15 49 60 and 30 15 42 05 Fax: (39) 6-305 11 49

■ OFFICIALLY REGULATED ORGANISATION: Statutes (published in Medicina e Morale 1985, 2: 279-280.
■ DATE OF FOUNDATION: 1985.
■ EXECUTIVE BOARD: Professor Elio Sgreccia (*Director*).
■ TEACHING: Diploma in Medicine and Surgery. — School of Specialised Medicine. — Advanced course in bioethics. — Updating course. — Nursing ethics.
■ PUBLICATIONS: Medicina e morale (official review). — Collana scienza medicina etica (various authors). — Manuale di bioetica (2 vols.) (Elio Sgreccia).
■ DOCUMENTATION CENTRE: open to the public.
■ COLLOQUIUMS, SYMPOSIUMS: Advanced course on bioethics (annual). — Updating course on bioethics (annual). —International conventions (bi-annual). — Specialized summer course.
■ DECISIONS, ADVICE: Identity and status of human embryo (1989).— General principles on prenatal diagnosis (1987).—Ethics principles on human foetal tissue removal (1989) (numerous articles published on various issues).
■ WORKING ON: bibliography [A4], documentation [A8], codes of ethics [B49], moral policy [B55], biomedical ethics education [B78], nursing education [B79], bioethical issues [B96], philosophical ethics [C2], professional ethics [C51], quality of environment [E41], methods [E53], human experimentation [F11], sexuality and procreation — unborn child [M1], death and resuscitation [O1], genetics and applications — biotechnology [Q1].
■ INTERESTED IN: data bases [A3], literature [A7], publications [A9], documentation [A8], education [B76], European Convention of Bioethics [B97], Bioethics bill, 1992 (France) [B98], advisory committees [D69], WHO — World Health Organization [G36].

0139 ISTITUTO INTERNAZIONALE DI STUDI SUI DIRITTI DELL'UOMO (INSTITUT INTERNATIONAL D'ÉTUDE DES DROITS DE L'HOMME)

Via Cesare Cantù 10
34127 Trieste (Italy)
✆ (39) 40-52 121 Fax: (39) 40-52 121

■ OTHER FIELDS OF ACTIVITIES: Conferences, research and publications in social and legal science plus the Philosophy of Human Rights.
■ DATE OF FOUNDATION: 1984.
■ EXECUTIVE BOARD: Professor Guido Gerin (*Chairperson*).
■ TEACHING: Ethical consequences of the human genome analysis. — Patient/Physician relationship. — study of European action concerning biotechnology.
■ PUBLICATIONS: yes.
■ WORKING ON: information sources, data bases [A1], fundamental rights of the individuals [B2], patient advocacy [B42], ethics [B48], codes of ethics [B49], moral obligations [B50], competence [B51], deontology [B59], bioethical issues [B96], philosophical ethics [C2], natural law [C24], ethical relativism [C10], morals [C12], values [C15], quality of life [C16], social worth [C17], value of life [C18], health care professionals, researchers and patient relationship [J1], patients' rights [J2], patient association [J3], patient information [J4], consent to treatment [J8], minors [K6], children's rights [L25], childhood defense [L26], children [L27], abortion [M47].
■ INTERESTED IN: religion and religious ethics [C19], eastern orthodox ethics [C26], Protestant ethics [C27], Islamic ethics [C28], Jewish ethics [C29], ethics and humanities [C54], historical aspects [C57], history before the 20th century [C58], ancient history [C59], social and political issues [D1], minority groups [D42], law, legislation and jurisprudence [K1], legal personality [K2], personhood [K3], legally incompetent person [K4], potentiality of personhood [K5], jurisprudence — accountability [K8], legislation and law [K34], children and media [L36], reproductive technologies [M26].

0140 **ISTITUTO SCIENTIFICO H. SAN RAFFAELE** (HSR)
Dipartimento di Medicina e Scienze Umane
Via Olgettina 60
20132 Milano (Italy)
✆ (39) 2-2643 2765 Fax: (39) 2-2643 2482

■ AFFILIATION: A department of the Scientific Institute H. San Raffaele.
■ DATE OF FOUNDATION: 1982.
■ EXECUTIVE BOARD: Professor Luigi Maria Verzé (*Chairperson*), Professor Rocco Maufia (*Chairperson's advisor for teaching activities*), Professor Paolo Cattorini (*Director of the Department*).
■ TEACHING: Basic course. — Advanced course in bioethics.
■ PUBLICATIONS: Series on AIDS. "Modelli di medicina crisi e ativalita' dell' idea di professione" (300 persons, among whom philosophers, physicians and students). — "AIDS e leggi sull' AIDS" (in October 1992).
■ DÉCISIONS, ADVICE: "La licetta' della sperimentazione sul volontari sani".
■ WORKING ON: book review [A6], documentation [A8], publications [A9], human person [B12], ethics committees [B22], ethics [B48], codes of biomedical ethics [B90], normative ethics [C5], Roman catholic ethics [C23], nontherapeutic research [F5], therapeutic research [F6], human experimentation [F11], healthy volunteers [F13], organ transplantation [I24], patient information [J4], informed consent [J9], treatment refusal [J15], living wills [J19], embryos [M10], in vitro fertilization [M29], AIH [M33], GIFT (Gamete Intrafallopian Transfer) [M37], anencephaly [N25], persistent vegetative state [N42], brain death [O6], active euthanasia [O16], gene therapy [Q42], somatotropin [L11].
■ INTERESTED IN: somatotropin [L11], persistent vegetative state [N42], living wills [J19], legislation [D13].

0141 **ISTITUTO SICILIANO DI BIOETICA**
Corso Vittorio Emanuele 463
Via SS. Crocofino 33
90134 Palermo Acireale (Italy)
✆ (39) 91-61 11 839 Fax: (39) 91-61 11 870

0142 **LEGA NAZIONALE CONTRO LA PREDAZIONE DI ORGANI E LA MORTE A CUORE BATTENTE**
Passagio Canonici Lateranansi 22
24121 Bergamo (Italy)
✆ (39) 35-21 92 55 and 24 43 37 Fax: (39) 35-21 92 55

■ AFFILIATION: Composed of the associations, cultural groups, local committees and individuals for the promotion of the "Lega Naz. contro la predazione di organi". — Promoted by the feminist association "Lega", A.E.D. (Associazione educazione demografia), of Bergamo.
■ DATE OF FOUNDATION: 1985 — Officalised on 22/06/87 (Notaio Carlo Leidi Repertorio 79713 - Raccolta N.15265).
■ EXECUTIVE BOARD: Nerina Negrello (Etlucist-*Chairperson*), Professor Dr. Massimo Bondi (*General Surgeon, Member of the Board, medical committee*) (*Member of the Board, medical committee*), Professor Dr. David Wainwright Evans (Cardiologist, *Member of the Board, medical committee*) (*Member of the Board, medical committee*), Renate Vincenti (*Member of the Board*), Lia Caprera (*Member of the Board*), Elena Barberis (*Member of the Board*), Alberto Gotti (*Member of the Board*).
■ PUBLICATIONS: Press releases, papers and articles in various medical magazines and newspapers.
■ COLLOQUIUMS, SYMPOSIUMS: Conventions. — Congress proceedings. — TV and radio programmes.
■ DÉCISIONS, ADVICE: Medical and legal advice. — Forensic expert in defence of comatose patients and families against unauthorised organ removals.

0143 **POLITEIA. — CENTRO PER LA RICERCA E LA FORMAZIONE IN POLITICA ED ETICA**
Via Casimo del Fante 13
20122 Milano (Italy)
✆ (39) 2-58 31 39 88 Fax: (39) 2-58 31 40 72

■ OTHER FIELDS OF ACTIVITIES: Rational Choice Theories. — Business ethics. — Political science and economics. — "Public" ethics (general and applied ethics).
■ DATE OF FOUNDATION: 1983.
■ EXECUTIVE BOARD: Professor Francesco Forte (*Chairperson*), Professor Paolo Martelli (*Director*),

Professor Salvatore Veca (*Member*), Professor Sebastiano Maffetone (*Member*), Dr. Elena Granaglia (*Member*), Professor Saverio Avveduto (*Member*).

■ TEACHING: Teaching classes to students and physicians and nurses.

■ PUBLICATIONS: Books and papers series. Particularly a book edited by M. Mori, "La Bioetica: questioni morali e politiche per il futuro dell'uomo".

■ DOCUMENTATION CENTRE: open to the public.

■ COLLOQUIUMS, SYMPOSIUMS: Regular seminars and symposia every two years.

■ WORKING ON: specific approaches to ethics [C1], social and political issues [D1], organization of health care; facilities, manpower and services; health occupations [G1], health economics, population characteristics [H1], sexuality and procreation — unborn child [M1], death and resuscitation [O1].

0144 PONTIFICIO CONSIGLIO DELLA PASTORALE PER GLI OPERATORI SANITARI

Via della Conciliazione 3
00193 Roma (Italy)
✆ (39) 6-68 96 793 Fax: (39) 6-68 96 841

0145 SCIENTIFIC COMMITTEE OF KOS — RIVISTA DI SCIENZA E ETICA

KOS
Europa Scienze Umane Editrice
20132 Milano (Italy)

0146 SOCIETÀ ITALIANA DI BIOETICA

Via del Proconsolo 12
50122 Firenze (Italy)
✆ (39) 55-239 80 65 Fax: (39) 55-239 80 63

■ AFFILIATION: A centre of activity.

■ OFFICIALLY REGULATED ORGANISATION: registered statutes.

■ DATE OF FOUNDATION: 1986 registered in 1987.

■ EXECUTIVE BOARD: Professor Brunetto Chiarelli (*Chairperson*), Dr. Luez Sineo (*Secretary*), Dr. Sivliana Dispinzeri (*Council member*), Dr. Siorgio Antoniacomi (*Council member*), Maria Teresa Mazzantini (*Treasurer*), Dr. Paolo Chiozzi (*Council member*).

■ TEACHING: University courses and seminars.

■ PUBLICATIONS: Global bioethics (Problemi di bioetica).

■ DOCUMENTATION CENTRE: internal.

■ COLLOQUIUMS, SYMPOSIUMS: Annually: November 10-15, in Trento (Italy).

■ DECISIONS, ADVICE: Environmental. — Biotechnological.

0147 SOCIETÀ ITALIANA PER LA BIOETICA E I COMITATI ETICI (SIBCE)

Via Olgettina, 60
20132 Milano (Italy)
✆ (39) 2-26 43 30 14 and 25 64 Fax: (39) 2-26 43 24 82

■ OFFICIALLY REGULATED ORGANISATION: La SIBCE si ispira "ai valori della tradizione culturale personalistica quali emergono dalle carte dei diritti dell' nomo dei cadici deontologici e dalla Costituzione italiana" (art. 2 dello statuto). — Lo statuto della SIBCE datato luglio 1989.

■ DATE OF FOUNDATION: 1989.

■ EXECUTIVE BOARD: Professor Paolo Cattorini, Dr. Filippo Crivelli, Professor Aldo Mazzoni, Professor Rocco Mangia, Professor Charles Vella, Professor Carlo Romano, Professor Antonino Leocata, Professor Elio Sgreccia.

■ PUBLICATIONS: No real publications yet. A bulletin is being planned.

■ DOCUMENTATION CENTRE: open to the public.

■ COLLOQUIUMS, SYMPOSIUMS: 31/01/92 I convegno SIBCE: "Valori e salute. Labioetica nella societa pluralistica", Aula Convegni, CNP, Roma.

■ WORKING ON: review [A2], bibliography [A4], documentation [A8], dignity [B5], freedom [B10], human person [B12], ethics committees [B22].

■ INTERESTED IN: ethics [B48], deontology [B59], ethical rules and principles [B47], morality [B61], paternalism [B62], respect [B63], virtues [B64], solidarity [B65], humanitarian organizations [B66], public debates [B75], education [B76], health education [B77], biomedical ethics education [B78], nursing education [B79], medical education [B80], bioethics [B89], ethicists [B92], bioethics movement [B95], bioethical issues [B96], European Convention of Bioethics [B97], philosophical ethics [C2], professional ethics [C51], medical ethics [C52], social issues [D2], cultural pluralism [D51], social problems [D52], social discrimination [D53], sociology of medicine [D55], research design [E19], behavioral research [E21], quality of environment [E41], decision making [E54], biomedical research [F2], clinical trials [F3], hospital [G28], mental institutions [G29], public hospitals [G30], private hospitals [G31], religious hospitals [G32], intensive care units [G33], health care costs [H9], costs and benefits [H10], risks and benefices [H11], biological specimens procurement, blood transfusion, organ transplantation [I8], patients' rights [J2], confidentiality [J20], inter-personal relationship [J27].

0148 **TRIBUNALE PER I DIRITTI DEL MALATO**
Via Pietro della Valle 1
00193 Roma (Italy)
✆ (39) 2-68 93 535

Individuals

I1 **AGAZZI** Evandro
Director of the Centre pour la Philosophie Contemporaine du CNR, University of Genoa (Italy). — Professor of Philosophy, universities of Genoa and Fribourg (Switzerland). — Chairperson of the Fédération internationale des Sociétés de philosophie (FISP). — Chairperson of the Académie internationale de philosophie des sciences.

Centro di Studi Filosofia Contemporanea
CNR
Via Lomellini 8-8
16126 Genova (Italy)
✆ (39) 10-29 35 11 Fax: (39) 10-24 71 184

■ AFFILIATIONS: Centre pour la philosophie contemporaine du CNR, Genova, Italy (*Director*).
■ TEACHING: University seminars, plus specific seminars organised by the Centre pour la philosophie contemporaine du CNR, department of bioethics.
■ PUBLICATIONS: yes.
■ WORKING ON: human rights [B3], dignity [B5], human person [B12], ethics [B48], moral obligations [B50], moral policy [B55], ethical review [B93], ethical analysis [C3], normative ethics [C5], values [C15], value of life [C18], consequences [C45], double effect [C46], normality [C78], science [E2], history of science [E3], technology [E27], biomedical technologies [E28], decision making [E54], decision analysis [E55], social control of science [E64], contraception [M3], procreation [M15], reproduction [M17], in vitro fertilization [M29], excess embryos [M34], embryo transfer [M36], determination of death [O2], death [O5], brain death [O6].
■ INTERESTED IN: human person [B12], human body commercialization [B13], traffic of organs [B15], codes of biomedical ethics [B90], ethical review [B93], ethical analysis [C3], normative ethics [C5], value of life [C18], consequences [C45], double effect [C46], normality [C78], decision making [E54], decision analysis [E55], contraception [M3], in vitro fertilization [M29], excess embryos [M34], embryo transfer [M36], determination of death [O2], death [O5], brain death [O6].

I2 **AMATI MEHLER** Jacqueline
Training Analyst, Italian Psychoanalytical Association (component society of International Psychoanalytical Association, IPA). — Secretary of the IPA.

Lucrezio Caro, 62
00193 Roma (Italy)
✆ (39) 6-654 58 26 Fax: (39) 6-324 4331

I3 **ARGENTIERI** Simona
Clinical activity in Psychoanalysis and Psychotherapy.

Via G.B. Martini 6
00198 Roma (Italy)
✆ (39) 6-84 16 313

■ WORKING ON: women's rights [B6], interdisciplinary communication [B70], mass media [B72], public debates [B75], deontological ethics [C6], medical ethics [C52], sex offenses [D92], patient association [J3].
■ INTERESTED IN: animal care committees [B21], volontary admission [P19], involontary commitment [P21].

14 **ARTIOLI** Rossella

Italian Member of Parliament. — Deputy. — Minister of the University, Scientific Research and Technology (Under-Secretary of State).

Ministero Università e Ricerca Scientifica e Tecnologica
Lungotevere Thaon de Revel 76
00196 Roma (Italy)
✆ (39) 6-32 34 350 and 32 34 340 Fax: (39) 6-32 34 318

■ AFFILIATIONS: VII Standing Committee, Roma, Italy (*Vice-Chairman*).
■ TEACHING: Culture, science and education. — VII Standing Committee of the Chamber of Deputies of Rome.
■ PUBLICATIONS: yes.
■ WORKING ON: respect of human dignity and human rights [B1], fundamental rights of the individuals [B2], social and political issues [D1], legislation [D13], handicapped [D43], physically handicapped [D44], disadvantaged [D46], associations [D47], social problems [D52], political issues [D56], political activity [D57], international organizations [D61], government and political systems [D64], government agencies [D74], state responsibility [D75], socialism [D84], science, technology, methods [E1], life sciences [E4], research [E14], technology [E27], biomedical technologies [E28], organization of health care; facilities, manpower and services; health occupations [G1], health policy [G13], law, legislation and jurisprudence [K1], sexuality and procreation — unborn child [M1], abortion [M47], death and resuscitation [O1], euthanasia [O15], psychiatric technics [P24], reproductive technologies [M26], artificial insemination [M31].

15 **BERLINGUER** Giovanni

Professor of Occupational Hygiene, University "La Sapienza", Rome.

Via di S. Giacomo 4
00187 Roma (Italy)
✆ (39) 6-67 80 968

■ AFFILIATIONS: Centre de Bioéthique, Istituto Gramsci, Roma, Italy (*Member*). — Consulta Bioetica, Milano, Italy (*Member*).
■ TEACHING: Seminars at the Faculty of Science, University of Rome.
■ PUBLICATIONS: yes.
■ WORKING ON: fundamental rights of individuals [B2], women's rights [B6], human equality [B7], integrity [B8], human body commercialization [B13], traffic of organs [B15], philosophy of biology [C42], biology and human future [C43], future generations [C44], history [C56], science [E2], history of science [E3], occupational medicine [E46], public health [G5], health policy [G13], occupational medicine [G65].
■ INTERESTED IN: patients' rights [J2], common symptomatology [N2], occupational diseases [N48].

16 **BERNABEO** Raffaele Alberto

Chair of Medical History, Medical School, University of Bologna.

Via Massarenti 9
Policlinico Sant'Orsola
40138 Bologna (Italy)
✆ (39) 51-30 89 76 Fax: (39) 51-39 22 40

■ AFFILIATIONS: Centre universitaire de recherche médico-éthique, Bologna, Italy (*Director*).
■ TEACHING: Chair of Medical History.
■ WORKING ON: history of biomedical ethics [B94], medical ethics [C52], history [C56].
■ INTERESTED IN: history of biomedical ethics [B94], quality of life [C16], religion [C36], history [C56], international aspects [D59], politics [D76], history of science [E3], health occupations [G61], sports medicine [G64], sports [L53].

17 **BOERI** Renato

Director of the official journal of the Société italienne de neurologie (SIN) "The Italian Society of Neurological Sciences".

Consulta di Bioetica
Via S. Andrea 5
20121 Milano (Italy)
✆ (39) 2-78 14 96 Fax: (39) 2-76 00 39 02

■ AFFILIATIONS: Consulta di Bioetica, Milano, Italy (*Chairperson*).
■ TEACHING: Lecturer at the University of Pavia.
■ PUBLICATIONS: yes.
■ WORKING ON: self determination [B4], biomedical research [F2], nervous system diseases [N40], euthanasia [O15], resuscitation [O21].
■ INTERESTED IN: *idem.*

I 8 **BOMPIANI** Adriano
Secretary of State, Social Affairs.

Comité National pour la Bioétique
Sig. ra Colomba Malerba
Via dei Villini 15
00161 Roma (Italy)
✆ (39) 6-44 04 279 and 44 04 283 Fax: (39) 6-44 04 282

■ AFFILIATIONS: Comitato Nazionale per la Bioetica — CNB, Roma, Italy (*Chairperson*).
■ TEACHING: Conferences. — forums, congresses, etc.
■ PUBLICATIONS: yes.
■ WORKING ON: social and political issues [D1], social issues [D2], political issues [D56], sexuality and procreation — unborn child [M1], sexuality and procreation [M2], reproductive technologies [M26], abortion [M47], fetal development [M54], fetal therapy [M57], death and resuscitation [O1], genetics and applications — biotechnology [Q1].
■ INTERESTED IN: information sources, data bases [A1], respect of human dignity and human rights [B1], specific approaches to ethics [C1], social and political issues [D1], science, technology, methods [E1], biomedical research and experimentation [F1], organization of health care; facilities, manpower and services; health occupations [G1], organization of health care; facilities, manpower and services; health occupations [G1], health care, medical acts [I1], health care professionals, researchers and patient relationship [J1], law, legislation and jurisprudence [K1], stages of life — problems proper to childhood and elderly [L1], sexuality and procreation — unborn child [M1], disease [N1], death and resuscitation [O1], neurosciences — psychiatry [P1], genetics and applications — biotechnology [Q1].

I 9 **CAFFARRA** Carlo
Chairperson of the Istituto Giovanni Paolo II, (marriage and family studies).

Istituto Giovanni Paolo II
Università Lateranense
Piazza S. Giovanni Laterano 4
00184 Roma (Stato Citta Vaticano) (Italy)
✆ (39) 6-69 86 401 Fax: (39) 6-69 88 61 03

■ TEACHING: Professor of Ethics.
■ PUBLICATIONS: yes.
■ WORKING ON: fundamental rights of the individuals [B2], philosophical ethics [C2], sexuality and procreation [M2], reproductive technologies [M26], abortion [M47], determination of death [O2], euthanasia [O15].
■ INTERESTED IN: legal personality [K2], potentiality of personhood [K5], conscience clause [K63].

I 10 **CASTIGNONE** Silvana
Professor of Philosophy of Law, Faculty of Law, University of Genova.

Via Cabella 28/11B
16122 Genova (Italy)
✆ (39) 10-81 53 60

■ AFFILIATIONS: Centro di Bioetica (Genova), Genova, Italy (*Chairperson*).
■ TEACHING: seminars.
■ PUBLICATIONS: yes.
■ WORKING ON: respect of human dignity and human rights [B1], human rights [B3], women's rights [B6], human person [B12], law, legislation and jurisprudence [K1].
■ INTERESTED IN: respect of human dignity and human rights [B1], law, legislation and jurisprudence [K1].

I 11 **CATTORINI** Paolo
Director, Department of Medical Humanities, Scientific Institute H. San Raffaele, Milan. — Associate Professor of Bioethics, University of Florence.

Istituto H. S. Raffaele
Via Olgettina 60
20132 Milano (Italy)
✆ (39) 2-26 43 27 65 Fax: (39) 2-26 41 34 45

■ AFFILIATIONS: Società Italiana Per la Bioetica e i Comitati Etici — SIBCE, Milan, Italy (*Founding Member of the prov. Board of governors*).
■ TEACHING: Specialisation course in Bioethics, University of Milan, Hospital San Raffaele. — Chair of Bioethics, University of Florence.
■ PUBLICATIONS: yes.
■ WORKING ON: book review [A6], publications [A9], human person [B12], ethics committees [B22], ethics [B48], codes of ethics [B49], moral obligations [B50], competence [B51], beneficence [B57], code of deontology [B60], paternalism [B62], virtues [B64], biomedical ethics education [B78], codes of biomedical ethics [B90], ethicists [B92], ethical review [B93], bioethical issues [B96], ethical analysis [C3], normative ethics [C5], morals [C12], conscience [C13], quality of life [C16], value of life [C18], Roman catholic ethics [C23], medical ethics [C52], nursing ethics [C53], legislation [D13], clinical trials [F3], nontherapeutic research [F5], therapeutic research [F6], human experimentation [F11], research subjects [F12], healthy volunteers [F13], organ transplantation [I24], organ donation [I25], informed consent [J9], treatment refusal [J15], living wills [J19], medical secrecy [J22], professional deontology [J25].
■ INTERESTED IN: ethics committees [B22], legislation [D13], living wills [J19], somatotropin [L11], persistent vegetative state [N42].

I 12 **CHIARELLI** Brunetto

Professor of Anthropology, University of Florence.

Società Italiana di Bioetica
Via del Proconsolo 12
50122 Firenze (Italy)
✆ (39) 55-23 40 49 Fax: (39) 55-239 80 65

■ AFFILIATIONS: Società italiana di bioetica, Firenze, Italy (*Chairperson*).
■ TEACHING: Anthropology courses, Faculty of Natural Sciences.
■ PUBLICATIONS: yes.
■ WORKING ON: respect of human dignity and human rights [B1], science, technology, methods [E1], biomedical research and experimentation [F1], health economics, population characteristics [H1], stages of life — problems proper to childhood and elderly [L1], sexuality and procreation — unborn child [M1], death and resuscitation [O1], neurosciences — psychiatry [P1], genetics and applications — biotechnology [Q1].
■ INTERESTED IN: *idem.*

I 13 **CITTADINI** Ettore

Head of the Department of Gynaecology and Obstetrics at the University of Palermo.

Clinica Ostetrica e Ginecologica Dell'Università
Via Villareale 54
90141 Palermo (Italy)
✆ (39) 91-589 508 Fax: (39) 91-637 33 92

■ AFFILIATIONS: Comitato Nazionale per la Bioetica — CNB, Rome, Italy (*Member*).
■ TEACHING: advanced coursed in physiopathology of procreation and the techniques of artificial procreation.
■ PUBLICATIONS: yes.
■ WORKING ON: reproductive technologies [M26].

I 14 **COMBA** Pietro

Research Fellow, Unit of Environmental Epidemiology.

Istituto Superiore di Sanita
Viale Regina Elena, 299
00161 Rome (Italy)
✆ (39) 6-49 90 Fax: (39) 6-444 00 64

■ WORKING ON: quality of environment [E41], occupational medicine [E46], goals [E57], health hazards [E49], teaching methods [E59], social control of science [E64].

I 15 **COMPAGNONI** Domenico

Largo Angelico 1
00184 Roma (Italy)
✆ (39) 6-67 021

I 16 **CORTESINI** Raffaello
Head of the Surgical Department.

Università di Roma "La Sapienza"
Viale del Policlinico 155
00161 Roma (Italy)
✆ (39) 6-44 50 741/44 50 297 Fax: (39) 6-44 63 667

■ AFFILIATIONS: Ethics Committee of Transplantation Society, Italy (*Member*).
■ TEACHING: yes.
■ PUBLICATIONS: yes.
■ WORKING ON: transplantation [I22], transplant recipients [I23], organ transplantation [I24], organ procurement [I26].

I 17 **D'AGOSTINO** Francesco
Professor of Philosophy of Law, University of Rome "Tor Versata".

Via Bormida 1
00198 Roma (Italy)
✆ (39) 6-84 16 706 Fax: (39) 6-72 36 103

■ AFFILIATIONS: Comitato Nazionale per la Bioetica — CNB, Roma, Italy (*Member*).
■ PUBLICATIONS: yes.
■ WORKING ON: aid children [L34], procreation [M15], sexuality [M20].
■ INTERESTED IN: social worth [C17], specific approaches to ethics [C1].

I 18 **DE CARLI** Luigi
University Professor of Genetics. — Director of the Postgraduate School of Applied Genetics.

Dipartimento di Genetica e Microbiologia Univ.
Via Abbiategrasso, 207
27100 Pavia (Italy)
✆ (39) 382-39 15 54 and 563 Fax: (39) 382-52 84 96

■ AFFILIATIONS: Comitato Nazionale per la Bioetica — CNB, Roma, Italy (*Member*).
■ PUBLICATIONS: yes.
■ WORKING ON: biology [E5], genetics [E6], chromosomal disorders [Q4], XYY karyotype [Q5], biotechnology — genetic engineering [Q20], biotechnology [Q21], genetic intervention [Q22], genome mapping [Q23], cloning [Q24], clones [Q25], hybrids [Q27], DNA fingerprinting [Q32].
■ INTERESTED IN: fundamental rights of the individuals [B2], human rights [B3], self determination [B4], dignity [B5], women's rights [B6], human equality [B7], integrity [B8], justice [B9], freedom [B10], privacy [B11], human person [B12], ethics committees [B22], CCNE — Comité Consultatif National d'Ethique (France) [B23], Council of Europe [B25], office of science and technology assessment [B28], manifests and declarations concerning human rights [B33], International Charter of Human Rights [B34], Helsinki Declaration [B37], Declaration on Human and Citizen Rights [B38], ethical rules and principles [B47], evolution [C47], human genome project [F8], health care, medical acts [I1], bone marrow transplantation [I29], guidelines [K55], embryos [M10], fetuses [M11], ovum [M12], placentas [M13], sperm [M14], procreation [M15], selective abortion [M49], therapeutic abortion [M50], aborted fetuses [M51], abortion on demand [M52], fetal therapy [M57], leukemia [N18], single gene defects [Q7], dominant genetic conditions [Q8], Huntington's chorea [Q9], recessive genetic conditions [Q10], sickle cell anemia [Q11], Duchenne muscular dystrophy [Q12], hemophilia [Q13], Tay Sachs disease [Q14], cystic fibrosis [Q15], phenylketonuria [Q16], thalassemia [Q17], hereditary diseases [Q18], sex linked defects [Q19], sex determination [Q26], genome mapping [Q23], genetically modified organisms [Q28], transgenic animals [Q29], sex preselection [Q30], recombinant DNA research [Q31], DNA linkage [Q33], behavioral genetics [Q34].

I 19 **DI MEO** Antonio
Deputy Chairperson of the Fondazione Istituto Gramsci, Rome.

Istituto Gramsci
Via del Conservatorio 55

00186 Roma (Italy)
✆ (39) 6-65 41 628

■ AFFILIATIONS: Centre de Bioéthique, Istituto Gramsci, Roma, Italy (*Director*).
■ PUBLICATIONS: yes.
■ WORKING ON: history of science [E3].

I 20 **DI MOLA** Giorgio
Executive Director. — Chairperson of the Floriani Foundation Ethical Committee.

Fondazione Floriani
Via San Simpliciano, 2
20121 Milano (Italy)
✆ (39) 2-86 46 30 24 and 86 46 43 05 Fax: (39) 2-72 02 24 93

■ AFFILIATIONS: Consulta di Bioetica, Milano, Italy (*Member*).
■ TEACHING: course in bioethics and HIV within the Plano Nazionale di Formazione per Operatori Sociosanitari per la Lotta alle Infezioni da HIV (Ministry of Health, Italy).
■ PUBLICATIONS: yes.
■ WORKING ON: respect of human dignity and human rights [B1], self determination [B4], dignity [B5], patient advocacy [B42], information — communication — media [B68], specific approaches to ethics [C1], professional ethics [C51], medical ethics [C52], research [E14], methods [E53], standards [E60], biomedical research and experimentation [F1], biomedical research [F2], therapeutic research [F6], methodology of research and experimentation [F15], organization of health care; facilities, manpower and services; health occupations [G1], health [G2], health facilities [G22], hospices [G34], health care [G45], home care [G55], health care professionals, researchers and patient relationship [J1], patient information [J4], parental notification [J6], disclosure [J7], consent to treatment [J8], informed consent [J9], treatment refusal [J15], right to treatment [J17], physicians and researchers accountability [J23], physician's role [J26], inter-personal relationship [J27], disease [N1], common symptomatology [N2], suffering [N11], cancer [N17], death and resuscitation [O1], terminal care [O7], palliative care [O8], life-sustaining treatment [O9], withholding treatment [O10], allowing to die [O11], terminally ill [O12], prolongation of life [O13], quality adjusted life years [O14], euthanasia [O15], attitudes to death [O20], pain [N12].
■ INTERESTED IN: respect of human dignity and human rights [B1], self determination [B4], dignity [B5], patient advocacy [B42], information — communication — media [B68], specific approaches to ethics [C1], professional ethics [C51], medical ethics [C52], research [E14], methods [E53], standards [E60], biomedical research and experimentation [F1], biomedical research [F2], therapeutic research [F6], methodology of research and experimentation [F15], organization of health care; facilities, manpower and services; health occupations [G1], health [G2], health facilities [G22], hospices [G34], health care [G45], home care [G55], health care professionals, researchers and patient relationship [J1], physicians and researchers accountability [J23], physician's role [J26], inter-personal relationship [J27], disease [N1], common symptomatology [N2], suffering [N11], pain [N12], cancer [N17], death and resuscitation [O1], terminal care [O7], palliative care [O8], life-sustaining treatment [O9], withholding treatment [O10], allowing to die [O11], terminally ill [O12], prolongation of life [O13], quality adjusted life years [O14], euthanasia [O15], attitudes to death [O20], patient information [J4], parental notification [J6], disclosure [J7], consent to treatment [J8], informed consent [J9], treatment refusal [J15], right to treatment [J17].

I 21 **DI ORIO** Ferdinando
Professor of Clinical epidemiology. — Dean of the Medical School at Aquila.

Facoltà di Medicina
Via Roma
67100 L'Aquila (Italy)
✆ (39) 862-43 33 01 Fax: (39) 862-43 33 03

■ AFFILIATIONS: Comité d'éthique de la faculté de médecine, l'Aquila, Italy (*Chairperson*).
■ PUBLICATIONS: yes.
■ WORKING ON: human experimentation [F11], research subjects [F12].

I 22 **FEGIZ** Gianfranco
Head of the Department of Surgery.

Univerità "la Sapienza"
00161 Roma (Italy)
✆ (39) 6-4913 04

■ WORKING ON: information — communication — media [B68].
■ INTERESTED IN: *idem*.

I 23 **FIORI** Angelo
Professor in Forensic Medicine. — Director of the Institut de médecine légale.

Istituto di Medicina Legale
Università Cattolica del Sacro Cuore
L.go F. Vito 1
00168 Roma (Italy)
✆ (39) 6-305 11 68

■ TEACHING: Medical deontology, seminars and courses organised by the Bioethics Institute of the University.
■ PUBLICATIONS: yes.
■ WORKING ON: bioethics [B89], methodology of research and experimentation [F15], confidentiality [J20], inter-personal relationship [J27], abortion [M47], determination of death [O2], outpatient commitment [P18], professional ethics [C51], patients' rights [J2], physicians and researchers accountability [J23], jurisprudence — accountability [K8], fetal therapy [M57], terminal care [O7], genetic defects and hereditary diseases [Q2].
■ INTERESTED IN: *idem*.

I 24 **FLAMIGNI** Carlo
Professor of Gynaecology and Obstetrics (University of Bologna). — Head of the Department of Physiopathology of Reproduction at the Hospital S. Orsola. — Chairperson of the SIFES (Società italiana Fertilità e Sterilità). — Member of the Comitato Nazionale di Bioetica. — City Councillor.

Università di Bologna
Facoltà di Medicina
Via Massarenti, 13
40138 Bologna (Italy)
✆ (39) 51-39 76 46

■ PUBLICATIONS: yes.
■ WORKING ON: ethics committees [B22], ethics [B48], codes of ethics [B49], deontology [B59], code of deontology [B60], health education [B77], biomedical ethics education [B78], medical education [B80], universities [B84], codes of biomedical ethics [B90], bioethical issues [B96], medical ethics [C52], political activity [D57], research team [E16], research institutes [E17], biomedical technologies [E28], freezing [E29], preservation [E30], cryonic suspension [E31], diagnostic imaging [E32], clinical trials [F3], health services research [F4], nontherapeutic research [F5], therapeutic research [F6], control groups [F17], random selection [F18], selection of subjects [F19], group of vulnerable subjects [F20], maternal health [G3], family planning [G8], maternal welfare [G10], maternal life [G11], health legislation [G14], health services [G25], public hospitals [G30], obstetrics and gynecology [G66], referral and consultation [J24], professional deontology [J25], physician's role [J26], contraception [M3], reproductive organs and embryonic structures [M9], embryos [M10], fetuses [M11], ovum [M12], placentas [M13], sperm [M14], procreation [M15], wish of children [M16], reproduction [M17], fertility [M18], infertility [M19], sexuality [M20], sexual behavior [M21], ovum donors [M27], semen donors [M28], in vitro fertilization [M29], FIV centre [M30], artificial insemination [M31], AID [M32], AIH [M33], excess embryos [M34], host mothers [M35], embryo transfer [M36], GIFT (Gamete Intrafallopian Transfer) [M37], embryo donation [M38], pregnancy and childbirth [M39], childbirth [M40], caesarean section [M41], pregnant women [M42], mother fetus relationship [M43], pregnancy [M44].
■ INTERESTED IN: ethics committees [B22], information dissemination [B71], bioethical issues [B96], health education [B77], biomedical ethics education [B78], philosophical ethics [C2], information — communication — media [B68], communication [B69], medical ethics [C52], social sciences [C89], social and political issues [D1], social issues [D2], social groups [D22], family members [D24], social impact [D49], social interaction [D50], sociology of medicine [D55], referral and consultation [J24], professional deontology [J25], physician's role [J26], medical etiquette [J28], potentiality of personhood [K5], sexuality and procreation — unborn child [M1], reproductive technologies [M26].

I 25 **GERACI** Domenico
Director of Research at the Italian CNR.

Istituto di Biologia dello Sviluppo
CNR
Via Archirafi 20
90123 Palermo (Italy)
✆ (39) 91-616 18 51 Fax: (39) 91-616 56 65

■ AFFILIATIONS: European Committee on Human Embryo and Research (HER), Bioethic Committee of the CNR — CNR, Roma, Italy (*Member*).
■ WORKING ON: therapeutic research [F6], human experimentation [F11], embryos [M10], sperm [M14], in vitro fertilization [M29], embryo transfer [M36], GIFT (Gamete Intrafallopian Transfer) [M37].
■ INTERESTED IN: *idem.*

I 26 **GERIN** Guido
Chairperson of the Centre d'études de bioéthique. — Chairperson of the Centre international d'études de bioéthique.

Institut international d'études des droits de l'homme
Centre International d'études de bioéthique
Rue Cantù n° 10
34127 Trieste (Italy)
✆ (39) 40-521 21 Fax: (39) 40-37 07 77

■ AFFILIATIONS: Institut international d'études des droits de l'homme — Centre International d'études de bioéthique, Trieste, Italy (*Chairperson*).
■ TEACHING: Conferences, forums in cooperation with specialists in medicine, philosophy, ethics, and law.
■ PUBLICATIONS: yes.
■ WORKING ON: information sources, data bases [A1], data bases [A3], publications [A9], fundamental rights of the individuals [B2], human rights [B3], self determination [B4], dignity [B5], women's rights [B6], human equality [B7], integrity [B8], justice [B9], freedom [B10], privacy [B11], human person [B12], human body commercialization [B13], traffic of organs [B15], patient advocacy [B42], ethics [B48], codes of ethics [B49], moral obligations [B50], competence [B51], deontology [B59], bioethical issues [B96], deontological ethics [C6], natural law [C24], consequences [C45], Marxism [C67], biological specimens procurement, blood transfusion, organ transplantation [I8], biological substances contamination [I9], required request [I10], tissue banking [I12], organ donors [I15], donor cards [I16], anonymous donation [I17], directed donation [I18], health care professionals, researchers and patient relationship [J1], patients' rights [J2], patient association [J3], patient information [J4], consent to treatment [J8], law, legislation and jurisprudence [K1], childhood defense [L26], children [L27], anatomy, physiology, development [L2], abortion [M47].
■ INTERESTED IN: religion and religious ethics [C19], eastern orthodox ethics [C26], Protestant ethics [C27], Jewish ethics [C29], Jewish ethics [C29], ethics and humanities [C54], historical aspects [C57], history before the 20th century [C58], ancient history [C59], social and political issues [D1], indigents [D48], law, legislation and jurisprudence [K1], legal personality [K2], personhood [K3], legally incompetent person [K4], potentiality of personhood [K5], jurisprudence — accountability [K8], legislation and law [K34], children and media [L36], reproductive technologies [M26].

I 27 **GINDRO** Sandro
Chairperson of the Association Psicoanalisi Contro. — Chairperson of the Institut psychanalytique pour les recherches sociales. — Director of the Observatoire sur la condition juvénile dans la ville de Monte-Carlo. — Director of the Project "Projet socio-éducatif pour l'orientation et l'intégration des adolescents à risque de déviance" (social care for teenagers), Ministry of Social Affairs.

Associazione Psicoanalisi Contro
Via Aremula 21
00186 Roma (Italy)
✆ (39) 6-68 33 552 Fax: (39) 6-68 67 509

■ AFFILIATIONS: Association européenne des Centres d'éthique médicale — AECEM (*Member*).
■ PUBLICATIONS: yes.
■ WORKING ON: behavioral and mental disorders [P2], psychiatric technics [P24], sexuality and procreation [M2], abortion [M47], philosophical ethics [C2], ethics and humanities [C54], social and political issues [D1].
■ INTERESTED IN: *idem.*

I 28 **LEOCATA** Antonio
Head of the Department Primario Divisione Pediatrica, Garibaldi Hospital, Catania.

Via P. Mascagni 24
95131 Catania (Italy)
✆ (39) 95-53 72 41 Fax: (39) 95-31 00 00

■ AFFILIATIONS: Comitato Nazionale per la Bioetica — CNB, Roma, Italy (*Presidenza del Consiglio dei Ministri*).
■ TEACHING: Medical ethics — Deontology to physicians of the Ordine dei medici della Provincia di Catania. — Nursing ethics, Scuola per Infermieri professionali dell'Ospedale Garibaldi di Catania.
■ PUBLICATIONS: yes.

I 29 **LINEHAN** Thomas

Physician (medical practice) Occupational Medicine. — Member of the Pontifical Council of Health Care Workers. — Chairperson of International Federation of Catholic Medical Association. — Governor of the Linacre Centre for Health Care Ethics.

Federation of Catholic Medical Association
Palazzo S. Calisto
Piazza S. Calisto
00120 Citta del Vaticano (Italy)
✆ (39) 6-378 98 29 and 698 7372 Fax: (39) 6-378 98 29

■ AFFILIATIONS: Guild of Catholic Doctors, London, United Kingdom (*Member of the Council*).
■ WORKING ON: human rights [B3], ethical rules and principles [B47], codes of ethics [B49], value of life [C18], Roman catholic ethics [C23], natural law [C24], married persons [D28], health services [G25], WHO — World Health Organization [G36], professional patient relationship [J33], sexual behavior [M21].
■ INTERESTED IN: religious ethics [C21], Christian ethics [C22], Roman catholic ethics [C23].

I 30 **LOMBARDI VALLAURI** Luigi

Sociology Professor of Family and Education, University of Florence, Catholic University of Milan.

Università di Firenze
Piazza Interdipendenza 9
50129 Firenze (Italy)
✆ (39) 55-496 533 Fax: (39) 55-474 756

■ AFFILIATIONS: Centro di Bioetica, Università Cattolica del Sacro Cuore, Roma, Italy (*Member*).
■ PUBLICATIONS: yes.
■ WORKING ON: reification [B14], ethics committees [B22], codes of ethics [B49], deontology [B59], ethicists [B92], Roman catholic ethics [C23], Roman catholicism [C31], mental processes [C69], self concept [C83], exclusion [D9], legislation [D13], disadvantaged [D46], legal personality [K2], potentiality of personhood [K5], rights [K42], reproductive technologies [M26], abortion [M47], euthanasia [O15], suicide [O19], biotechnology [Q21].

I 31 **LORENZETTI** Luigi

Professor of Moral Theology.

Rivista di Teologia Morale
Via Nosadella 6
40123 Bologna (Italy)
✆ (39) 51-33 03 01 Fax: (39) 51-33 13 54

■ TEACHING: Diverse conferences within theological ethics through the Faculty of Theology.
■ PUBLICATIONS: yes.
■ WORKING ON: philosophical ethics [C2], religion and religious ethics [C19], philosophy of biology [C42], professional ethics [C51], ethics and humanities [C54].
■ INTERESTED IN: *idem*.

I 32 **MANCINA** Claudia

Research Fellow (Department of Philosophy). — MP.

Centro di Bioetica
Istituto Gramsci
Via del Conservatorio 55
00186 Rome (Italy)
✆ (39) 6-654 16 28 Fax: (39) 6-687 77 36

■ AFFILIATIONS: Centro di Bioetica, Institut Gramsci, Roma, Italy (*Member*).
■ PUBLICATIONS: yes.
■ WORKING ON: fundamental rights of the individuals [B2], self determination [B4], women's rights

[B6], freedom [B10], ethics [B48], obligations of society [B53], obligations to society [B54], morals [C12], conscience [C13], moral development [C14], values [C15], quality of life [C16], social worth [C17], value of life [C18], emotions [C79], love [C80], trust [C81], personality [C82], self concept [C83], family members [D24], parents [D32], females [D35], cultural pluralism [D51], children's rights [L25], child and family [L31], adoption [L32], procreation [M15], wish of children [M16], reproduction [M17], sexuality [M20], reproductive technologies [M26], abortion [M47].
■ INTERESTED IN: rights [K42], civil laws [K43], civil code [K44], legal rights [K45], children's rights [L25], child and family [L31], adoption [L32], reproductive technologies [M26], abortion [M47].

I 33 **MANNI** Corrado
Director of the Institut d'anesthésiologie et réanimation.

Catholic University of Sacred Heart
L.go F. Vito, 1
00168 Rome (Italy)
✆ (39) 6-30154386 Fax: (39) 6-3051343

■ AFFILIATIONS: Comitato Nazionale per la Bioetica — CNB, Rome, Italy (*Member*).
■ TEACHING: yes.
■ PUBLICATIONS: yes.

I 34 **MERLI** Silvio
Professor of Forensic Medicine, University teaching. — Director of the Second Chair in Forensic Medicine, University of Rome- "La Sapienza".

Istituto di medicina legale e delle assicurazioni
Viale Regina Elena 336
Università "La Sapienza"
00161 Roma (Italy)
✆ (39) 6-44 55 335 Fax: (39) 6-44 55 335

■ AFFILIATIONS: National Bioethic Committee — CNB, Roma, Italy (*Member*).
■ TEACHING: yes.
■ PUBLICATIONS: yes.
■ WORKING ON: respect of human dignity and human rights [B1], ethics [B48], codes of ethics [B49], deontology [B59], code of deontology [B60], specific approaches to ethics [C1], deontological ethics [C6], medical ethics [C52], organ transplantation [I24], patient information [J4], informed consent [J9], medical secrecy [J22], physicians and researchers accountability [J23], professional deontology [J25], malpractice [K19], infanticide [L42], death [O5], brain death [O6], terminally ill [O12], euthanasia [O15], active euthanasia [O16], involuntary euthanasia [O17], voluntary euthanasia [O18].
■ INTERESTED IN: biomedical research and experimentation [F1], blood transfusions [I19], transplantation [I22], organ procurement [I26], living wills [J19], physicians and researchers accountability [J23], malpractice [K19], reproductive technologies [M26], death and resuscitation [O1], terminal care [O7], euthanasia [O15], biotechnology — genetic engineering [Q20].

I 35 **MORDINI** Emilio
Projects Director for the Bioethics Department of the Association Psicoanalisi Contro.

Associazione Culturale Psicoanalisi Contro
Via Arenula 21
00186 Roma (Italy)
✆ (39) -6-68 67 509

■ AFFILIATIONS: Association européenne des centres d'éthique médicale — AECEM (*Member*).
■ TEACHING: Medicine and bioethics, School of Psychoanalysis, Associazione Culturale Psicoanalisi Contro.
■ WORKING ON: deontology [B59], education [B76], biomedical ethics education [B78], health and biology mediatisation [B85], ethicists [B92], bioethics movement [B95], quality of life [C16], value of life [C18], philosophy of biology [C42], sociobiology [C63], psychology [C68], social impact [D49], social interaction [D50], cultural pluralism [D51], social problems [D52], psychiatry [G69], physician's role [J26], inter-personal relationship [J27], mind [L20], sexuality and procreation [M2], euthanasia [O15], neurosciences — psychiatry [P1].
■ INTERESTED IN: *idem*.

I 36 **MORI** Maurizio

Via Brera 18
20121 Milano (Italy)
✆ (39) 2-87 78 73

I 37 **ORSI** Luciano
Physician.

Redarto di Rianimazione
Ospedale Maggiore
26013 Cremona (Italy)
✆ (39) 373-853 268 Fax: (39) 373-853 337

■ AFFILIATIONS: Consulta di Bioetica, Milano, Italy (*Member*).
■ WORKING ON: brain death [O6], life-sustaining treatment [O9], withholding treatment [O10], allowing to die [O11], terminally ill [O12], quality adjusted life years [O14], resuscitation [O21].
■ INTERESTED IN: brain death [O6], palliative care [O8], life-sustaining treatment [O9], withholding treatment [O10], allowing to die [O11], terminally ill [O12], prolongation of life [O13], quality adjusted life years [O14], euthanasia [O15], resuscitation [O21].

I 38 **PRIVITERA** Salvatore
Director of the Istituto Siciliano di Bioetica.

Istituto Siciliano di Bioetica
Corso Vittorio Emanuele 463
90134 Palermo (Italy)
✆ (39) 91-61 11 839 Fax: (39) 91-61 11 870

■ TEACHING: Lectures on bioethics organised by the Institute, Facoltà Teologica di Sicilia.
■ PUBLICATIONS: yes.

I 39 **RODOTA** Stefano

University of Roma
Faculty of Law
00100 Roma (Italy)
✆ (39) 6-676 03 589

■ AFFILIATIONS: Centro di Bioetica, Istituto Gramsci, Roma, Italy (*Member*).
■ TEACHING: Seminars at the Faculty of Law, University of Rome.
■ PUBLICATIONS: yes.
■ WORKING ON: fundamental rights of the individuals [B2], protection of rights — involved institutions [B16], ethical rules and principles [B47], bioethics [B89], philosophical ethics [C2], ethics and humanities [C54], social issues [D2], political issues [D56], legal personality [K2], jurisprudence — accountability [K8], legislation and law [K34].
■ INTERESTED IN: information — communication — media [B68], economics [H2], sexuality and procreation [M2], reproductive technologies [M26], determination of death [O2], terminal care [O7], euthanasia [O15], genetic defects and hereditary diseases [Q2], biotechnology — genetic engineering [Q20].

I 40 **ROSSI SCIUME** Giovanna
Sociology Professor of Family and Education.

Università Cattolica del Sacro Cuore
Dipartimento di Sociologia
Lgo Gemelli 1
20123 Milano (Italy)
✆ (39) 2-72 34 23 52 and 22 75 Fax: (39) 2-72 34 25 52

■ AFFILIATIONS: Comité national pour la bioéthique — CNB, Roma, Italy (*Member*).
■ PUBLICATIONS: yes.
■ WORKING ON: social sciences [C89], social control [D3], informal social control [D4], socioeconomic factors [D18], family members [D24], married persons [D28], marital relationship [D29], family relationship [D31], parents [D32], mothers [D33], fathers [D34], females [D35], associations [D47],

social interaction [D50], social problems [D52], sociology of medicine [D55], community services [D78], maternal health [G3], home care [G55], family practice [G60], children [L27], aid children [L34], health care services and aged [L48], in vitro fertilization [M29], artificial insemination [M31], AID [M32], AIH [M33], embryo transfer [M36], GIFT (Gamete Intrafallopian Transfer) [M37], abortion on demand [M52], terminal care [O7].

■ INTERESTED IN: ethics committees [B22], CCNE — Comité Consultatif National d'Ethique (France) [B23], Council of Europe [B25], curriculum [B83], universities [B84], values [C15], quality of life [C16], social sciences [C89], social control [D3], informal social control [D4], socioeconomic factors [D18], family members [D24], married persons [D28], marital relationship [D29], family relationship [D31], parents [D32], fathers [D34], females [D35], associations [D47], social interaction [D50], social problems [D52], sociology of medicine [D55], community services [D78], maternal health [G3], patient care [G53], home care [G55], family practice [G60], children [L27], aid children [L34], health care services and aged [L48], in vitro fertilization [M29], artificial insemination [M31], AID [M32], embryo transfer [M36], GIFT (Gamete Intrafallopian Transfer) [M37], abortion on demand [M52], terminal care [O7].

I 41 **ROSTAGNO** Sergio

Professor of Theology.

Facoltà Valdese di Teologia
Via P. Cossa 42
00193 Roma (Italy)
✆ (39) 6-32 10 789 Fax: (39) 6-32 01 040

■ AFFILIATIONS: Study group on ethical problems in science, Roma, Italy (*Coordinator*).

■ WORKING ON: Protestant ethics [C27].

■ INTERESTED IN: European Convention of Bioethics [B97], Bioethics bill, 1992 (France) [B98], Protestant ethics [C27], philosophy of biology [C42], patients' rights [J2], jurisprudence — accountability [K8], reproductive technologies [M26], abortion [M47], death and resuscitation [O1].

I 42 **RUSTICHELLI** Sergio

Chief Physician of Sterility Centre, Institute of Obstetrics and Gynaecology Chair "A", University of Turin.

Università di Torino
Centro Sterilita I Clinica Ostetrica
Via Ventimiglia 3
10126 Torino (Italy)
✆ (39) 11-63 96 411 Fax: (39) 11-69 64 022

■ WORKING ON: press conference [B73], biomedical technologies [E28], control groups [F17], obstetrics and gynaecology [G66], informed consent [J9], physician patient relationship [J32], AIH [M33], infertility [M19].

■ INTERESTED IN: diagnostic imaging [E32], sperm banks [I11], physician patient relationship [J32], sexuality [M20], GIFT (Gamete Intrafallopian Transfer) [M37].

I 43 **SALVINO LEONE**

Head of Medical Humanities Department, Fatebenefratelli Hospital (Palermo).

Via D'Annunzio 9
90144 Palermo (Italy)
✆ (39) 91-625 96 80 Fax: (39) 91-47 76 25

■ AFFILIATIONS: Istituto Siciliano di Bioetica — ISB, Italy (*Member*).

■ TEACHING: Specialising in bioethics.

■ PUBLICATIONS: yes.

■ WORKING ON: review [A2], data bases [A3], publications [A9], human rights [B3], self determination [B4], human body commercialization [B13], ethics committees [B22], College of Physicians [B27], ethics [B48], codes of ethics [B49], competence [B51], code of deontology [B60], morality [B61], universities [B84], quality of life [C16], Roman catholic ethics [C23], scriptural interpretation [C40], medical ethics [C52], nursing ethics [C53], family members [D24], handicapped [D43], genetics [E6], biomedical technologies [E28], ecology [E43], family planning [G8], hospital [G28], health care delivery [G48], obstetrics and gynecology [G66], pediatrics [G68], prenatal diagnosis [I3], organ donors [I15], transplantation [I22], renal dialysis [I32], placebos [I36], consent to treatment [J8], presumed consent [J10], physician nurse relationship [J31], physician patient relationship [J32], professional patient relationship [J33], child diseases and problems [L37], handicapped children [L38], child abuse [L41], infanticide [L42], child diseases [L43], sexuality and procreation — unborn child

[M1], physically [N8], suffering [N11], pain [N12], genetic defects [Q3], biotechnology — genetic engineering [Q20], biotechnology [Q21], medical genetics [Q35].
■ INTERESTED IN: bibliography [A4], book review [A6], literature [A7], documentation [A8], publications [A9], World Medical Assembly [B20], Council of Europe [B25], virtues [B64], press conference [B73], audiovisual aids [B74], education [B76], health education [B77], biomedical ethics education [B78], medical education [B80], health and biology mediatisation [B85], famous persons [B86], advertising [B87], ethicists [B92], teleological ethics [C7], values [C15], religious ethics [C21], Christian ethics [C22], eastern orthodox ethics [C26], Protestant ethics [C27], Islamic ethics [C28], Jewish ethics [C29], Islam [C33], Judaism [C34], Protestantism [C35], Jehovah's witnesses [C38], religious sciences [C39], theology [C41], evolution [C47], history [C56], humanism [C66], capital punishment [D15], socioeconomic factors [D18], aliens [D23], couple [D26], marital relationship [D29], ethnic groups [D36], minority groups [D42], North-South relationships [D60], capitalism [D80], communism [D81], democracy [D82], national socialism [D83], socialism [D84], history of science [E3], biology [E5], private hospitals [G31], religious hospitals [G32], health personnel [G37], nurse midwives [G43], home care [G55], patient association [J3], conflict of interest [K9].

I 44 **SGRECCIA** Elio
Professor of Bioethics. — Director of the Istituto di Bioetica, Università Cattolica del Sacro Cuore.

Università Cattolica del Sacro Cuore
Istituto di Bioetica
Largo Francesco Vito 1
00168 Roma (Italy)
✆ (39) 6-30154960 Fax: (39) 6-3051149

■ TEACHING: School of Medicine, post-graduate courses. — Nursing school.
■ PUBLICATIONS: yes.
■ WORKING ON: respect of human dignity and human rights [B1], ethical rules and principles [B47], education [B76], philosophical ethics [C2], professional ethics [C51], biomedical research and experimentation [F1], resource allocation [H4], diagnosis [I2], patients' rights [J2], sexuality and procreation — unborn child [M1], anencephaly [N25], persistent vegetative state [N42], terminal care [O7], euthanasia [O15], medical genetics [Q35].
■ INTERESTED IN: literature [A7], documentation [A8], publications [A9], education [B76], ethical review [B93], history of biomedical ethics [B94], WHO — World Health Organization [G36].

I 45 **SILVESTRINI** Bruno
Full Professor of Pharmacology, School of Pharmacology, University of Rome "La Sapienza".

Istituto di Farmacologia e Farmacognosia
Università di Roma "La Sapienza"
Pl. Aldo Moro 5
00185 Roma (Italy)
✆ (39) 6-445 70 88 Fax: (39) 6-49 91 22 49

■ AFFILIATIONS: Comitato Nazionale per la Bioetica — CNB, Roma, Italy (*Member*).
■ TEACHING: Within the "Pontificio Consiglio della Pastorale per gli Operatori Sanitari" and functions as a member consultant.
■ PUBLICATIONS: yes.
■ WORKING ON: animal experimentation [F10], toxicity [E50], neurosciences — psychiatry [P1].
■ INTERESTED IN: utilitarianism [C8], drug abuse [P6].

I 46 **SINACCIO** Graziella
Assistant Scientific Director of the Institute, Coordinator of the Ethics Committee. — Member of the Oncologic Ethics Commission of the Italian league in the fight against cancer. — Consultant in Ethics for the IRCCS (Istituti di Ricovero e Cura a Carattere Scientifico).

Via Benedetto XV 10
16132 Genova (Italy)
✆ (39) 10-35 34 081 Fax: (39) 10-35 28 88

■ AFFILIATIONS: Institut National pour la Recherche sur le cancer, Genova, Italy (*Coordinator of the Ethics Committee*).
■ TEACHING: Bioethics for physicians and nurses, in hospitals and research centres. — University level courses in biotechnology and genetic therapy.
■ PUBLICATIONS: yes.
■ WORKING ON: human person [B12], ethics committees [B22], Helsinki Declaration [B37], patient advocacy [B42], ethical rules and principles [B47], information — communication — media [B68],

codes of biomedical ethics [B90], bioethical issues [B96], quality of life [C16], professional ethics [C51], history of the 20th century [C60], philosophy [C61], applied psychology [C86], biology [E5], genetics [E6], epidemiology [E9], research team [E16], biomedical technologies [E28], drug industry [E37], biomedical research and experimentation [F1], health legislation [G14], hospital [G28], hospices [G34], organization of health care [G35], hospitalisation [G46], home care [G55], placebos [I36], health care professionals, researchers and patient relationship [J1], cancer [N17], acquired immunodeficiency syndrome [N35], HIV seropositivity [N36], terminal care [O7], euthanasia [O15], biotechnology [Q21], genetic intervention [Q22], gene therapy [Q42].
■ INTERESTED IN: history of biomedical ethics [B94], religion and religious ethics [C19], ethics and humanities [C54].

I 47 **SPAGNOLO** Antonio G.
Researcher in the field of Bioethics, Università Cattolica del Sacro Cuore.

Università Cattolica del Sacro Cuore
Istituto di Bioetica
L.go Francesco Vito 1
00168 Roma (Italy)
✆ (39) 6-30 15 42 05 Fax: (39) 6-305 11 49

■ AFFILIATIONS: Istituto di Bioetica, Università Cattolica del Sacro Cuore, Roma, Italy (*Member*).
■ TEACHING: Post-graduate courses, nursing schools.
■ PUBLICATIONS: yes.
■ WORKING ON: codes of biomedical ethics [B90], ethical review [B93], professional ethics [C51], ethics committees [B22], biomedical research and experimentation [F1], acquired immunodeficiency syndrome [N35].
■ INTERESTED IN: literature [A7], documentation [A8], CCNE advice (France) [B91], ethics committees [B22], education [B76], biomedical research and experimentation [F1], acquired immunodeficiency syndrome [N35].

I 48 **SPINSANTI** Sandro
Head of the Department of Humanities, Fondazione Internazionale Fatebenefratelli.

Fondazione Internazionale Fatebenefratelli
Lungotevere degli Anguillara 11
00153 Roma (Italy)
✆ (39) 6-58 18 895

■ TEACHING: at the Fatebenefratelli Hospital.
■ PUBLICATIONS: yes.
■ WORKING ON: bioethics [B89], professional ethics [C51], ethics and humanities [C54], patients' rights [J2].

I 49 **TOSCANI** Franco
Head of the Department of Palliative Care, Cremona Hospital. — Member of the Fondazione Floriani Ethics Committee.

Sezione di Cure Palliative POC-Ussl 51
Lgo Priori 1
26100 Cremona (Italy)
✆ (39) 372-40 52 69/40 53 30 Fax: (39) 372-43 19 75/43 41 72

■ AFFILIATIONS: Comitato Etico della Fondazione Floriani, Milano, Italy (*Member*).
■ TEACHING: Courses for doctors and nurses of the Scuola Italiana di Medicina Palliativa.
■ PUBLICATIONS: yes.
■ WORKING ON: bioethics [B89], clinical trials [F3], therapeutic research [F6], human experimentation [F11], random selection [F18], group of vulnerable subjects [F20], hospitalisation [G46], patient care [G53], home care [G55], costs and benefits [H10], treatment refusal [J15], patient participation [J16], advance directives [J18], living wills [J19], disclosure [J7], palliative care [O8], life-sustaining treatment [O9], allowing to die [O11], terminally ill [O12], active euthanasia [O16], voluntary euthanasia [O18], suicide [O19], attitudes to death [O20].
■ INTERESTED IN: review [A2], data bases [A3], self determination [B4], ethics committees [B22], patient advocacy [B42], codes of biomedical ethics [B90], ethical review [B93], medical ethics [C52], nursing ethics [C53], historical aspects [C57], peer review [E63], hospices [G34], life-sustaining treatment [O9], suicide [O19], attitudes to death [O20].

I 50 **VELLA** Charles G.

General Co-ordinator for Ethics and Public Relations, Istituto Scientifico Ospedale San Raffaele and member of the Ethics Committee. — Member of the Ethics Committee, Istituto Nazionale per lo Studio e la Cura dei Tumori. — Senior Adviser to the Ministry for Social Development (Malta). — Co-ordinator for SIBCE (Società Italiana per la Bioetica e i Comitati Etici). — Chairperson of CDPS (Steering Committee for Social Policy), Council of Europe. — Member of CAHBI (Ad Hoc Committee of Experts on Bioethics), Council of Europe. — Member of National Health Ethics Committee (Malta).

Istituto Scientifico Ospedale San Raffaele
Via Olgettina 60
20132 Milano (Italy)
✆ (39) 2-26 43 24 77 Fax: (39) 2-26 43 25 76

■ AFFILIATIONS: Council of Europe, Strasbourg Cedex, France (*Member*).
■ TEACHING: Scuola Infermieri Professionali (Istituto Scientifico Ospedale San Raffaele). — Lectures in various schools, hospitals, ethics committees.
■ PUBLICATIONS: yes.
■ WORKING ON: fundamental rights of the individuals [B2], ethics committees [B22], CCNE — Comité Consultatif National d'Ethique (France) [B23], Council for International Organization of Medical Sciences [B24], Council of Europe [B25], United Nations [B29], ethical rules and principles [B47], information — communication — media [B68], religion and religious ethics [C19], professional ethics [C51], social and political issues [D1], health [G2], health policy [G13], organization of health care [G35], health care [G45], sexuality and procreation [M2], abortion [M47], terminal care [O7].
■ INTERESTED IN: information sources, data bases [A1], respect of human dignity and human rights [B1], specific approaches to ethics [C1], social and political issues [D1], science, technology, methods [E1], biomedical research and experimentation [F1], organization of health care; facilities, manpower and services; health occupations [G1], health economics, population characteristics [H1], health care, medical acts [I1], health care professionals, researchers and patient relationship [J1], law, legislation and jurisprudence [K1], stages of life — problems proper to childhood and elderly [L1], sexuality and procreation — unborn child [M1], disease [N1], death and resuscitation [O1], neurosciences — psychiatry [P1], genetics and applications — biotechnology [Q1].

I 51 **VIAFORA** Corrado

Teacher of Philosophy, — Professor of Bioethics Scuola di Specialità di Ginecologia, Ostetrica, University of Padova. — Professor of nursing ethics, Scuola Infermieri Professionali, Hospital of Padova.

Fondazione Lanza
Via Dante 55
35139 Padova (Italy)
✆ (39) 49-87 56 788

■ AFFILIATIONS: Fondazione Lanza, Padova, Italy (*Coordinator of the Etica e Medicina Project*).
■ TEACHING: Public lectures and interdisciplinary seminars in cooperation with physicians, philosophers, theologians, lawyers, journalists and paramedics (working mainly in the area of Padova).
■ PUBLICATIONS: yes.
■ WORKING ON: philosophical ethics [C2], biomedical research and experimentation [F1], health care professionals, researchers and patient relationship [J1], sexuality and procreation — unborn child [M1], death and resuscitation [O1], ethics committees [B22], biomedical ethics education [B78], nursing education [B79], medical education [B80], professional ethics [C51].
■ INTERESTED IN: philosophical ethics [C2], biomedical research and experimentation [F1], health care professionals, researchers and patient relationship [J1], sexuality and procreation — unborn child [M1], death and resuscitation [O1], ethics committees [B22], biomedical ethics education [B78], nursing education [B79], medical education [B80], professional ethics [C51], quality of life [C16], quality of environment [E41], ecology [E43].

Luxembourg

Area: 2 586 km²
Population: 378 000
Gross national product : 294,3 bn LF

Organisations

O L 1 **COMMISSION CONSULTATIVE NATIONALE D'ÉTHIQUE POUR LES SCIENCES DE LA VIE ET DE LA SANTÉ**
Ministère de la famille
2910 Luxembourg (Luxembourg)
✆ (352) 478 6534 Fax: (352) 478 6571

■ OFFICIALLY REGULATED ORGANISATION: officially created by governmental regulation (9 Sept. 1988)
■ DATE OF FOUNDATION: 9 Sept. 1988.
■ EXECUTIVE BOARD: Dr. Arsène Betz (*Chairperson*), Dr. Jean-Claude Faver (*Vice-Chairperson*), Me Ernest Arendt (*Member*), Abbé Erny Gillen (*Member*), Professor Paul Kremer (*Member*), Mill Majerus (*Coordinator*).

Individuals

L 1 **ARENDT** Ernest
Barrister-at-law, former Chairperson of the bar— Honorary Chairperson at the Conseil d'Etat, Grand Duchy of Luxembourg — Former Chairperson of the Union internationale des Avocats (UIA).

12, Boulevard Joseph II B.P. 39
1840 Luxembourg (Luxembourg)
✆ (352) 40 78 78 Fax: (352) 40 78 04

■ AFFILIATIONS: Commission consultative nationale d'éthique pour les sciences de la vie et de la santé, Luxembourg (*Member*).
■ WORKING ON: fundamental rights of the individuals [B2], code of deontology [B60], sexuality and procreation — unborn child [M1].
■ INTERESTED IN: Bioethics bill, 1992 (France) [B98], ethics and humanities [C54], reproductive technologies [M26], ovum donors [M27], semen donors [M28], in vitro fertilization [M29], FIV center [M30], artificial insemination [M31], AID [M32], AIH [M33], excess embryos [M34], host mothers [M35], embryo transfer [M36], GIFT (Gamete Intrafallopian Transfer) [M37], embryo donation [M38], abortion [M47].

L 2 **BONÉ** Edouard
Professor Emeritus, University of Louvain.

25, avenue Gaston-Diderich
1420 Luxembourg (Luxembourg)
✆ (352) 45 25 18 Fax: (352) 25 18 20

■ PUBLICATIONS: yes.
■ WORKING ON: review [A2], human person [B12], codes of biomedical ethics [B90], history of biomedical ethics [B94], quality of life [C16], Christian ethics [C22], theology [C41], human genome project [F8], contraception [M3], in vitro fertilization [M29], biotechnology [Q21].
■ INTERESTED IN: history of biomedical ethics [B94], bioethical issues [B96], Bioethics bill, 1992 (France) [B98], quality of life [C16], human genome project [F8], population control [H22], quality adjusted life years [O14], active euthanasia [O16].

L 3 **GILLEN** Erny
Director of the Luxembourg branch of Caritas. — Professor of Theological Ethics.

52, rue Jules Wilhelm
2728 Luxembourg (Luxembourg)
✆ (352) 43 60 51 24 Fax: (352) 40 21 31 21

■ AFFILIATIONS: Commission consultative nationale d'éthique pour les sciences de la vie et de la santé, Luxembourg (*Member*).
■ TEACHING: Ethics for Health Care professionals — DCV Deutscher Caritas-Verband: Updating course for chief-physicans and their assistants (managing and directing)
Weiterbildung für Chef ärzte und Mitarbeiter firmen in der Verwältung mit Führungsaufgaben.
■ PUBLICATIONS: yes.
■ WORKING ON: literature [A7], ethics committees [B22], moral policy [B55], solidarity [B65], humanitarian organizations [B66], education [B76], health education [B77], biomedical ethics education [B78], nursing education [B79], internship and residency [B81], CCNE advices (France) [B91], teleological ethics [C7], metaethics [C11], morals [C12], quality of life [C16], clergy [C20], Christian ethics [C22], Roman catholic ethics [C23], theology [C41], double effect [C46], medical ethics [C52], nursing ethics [C53], humanism [C66], attitudes [C74], intention [C75], motivation [C76], self concept [C83], social sciences [C89], public opinion [D7], scarcity [D21], family relationship [D31], disadvantaged [D46], associations [D47], social impact [D49], cultural pluralism [D51], international aspects [D59], EC — European Communities [D62], advisory committees [D69], decision making [E54], teaching methods [E59], standards [E60], institutional policies [G15], organizational policies [G17], private hospitals [G31], religious hospitals [G32], hospices [G34], patient care team [G38], nurses [G39], physicians [G40], social workers [G44].
■ INTERESTED IN: literature [A7], ethics committees [B22], patient advocacy [B42], common good [B45], moral policy [B55], altruism [B56], beneficence [B57], authoritarianism [B58], deontology [B59], solidarity [B65], interdisciplinary communication [B70], normative ethics [C5], metaethics [C11], morals [C12], quality of life [C16], Christian ethics [C22], medical ethics [C52], nursing ethics [C53], public opinion [D7], cultural pluralism [D51], international aspects [D59], EC — European Communities [D62], advisory committees [D69], government agencies [D74], dehumanization [D86], history of science [E3], epidemiology [E9], decision making [E54], teaching methods [E59], standards [E60], nontherapeutic research [F5], cognitive research [F7], human genome project [F8], methodology of research and experimentation [F15], state medicine [G9], institutional policies [G15], residential facilities [G26], nursing home [G27], private hospitals [G31], hospices [G34], home care [G55], insurance [H5], national health insurance [H7], elderly [H23], mandatory screening [I7], organ donors [I15], patient information [J4], disclosure [J7], consent to treatment [J8], informed consent [J9].

L 4 **HEMMER** Robert

Head of the Department of Infectious diseases, Centre hospitalier de Luxembourg. — Master of Seminars, University of Brussels— Chairperson of the Comité de surveillance du sida, Luxembourg.

Centre hospitalier
4, rue Barblé
1210 Luxembourg (Luxembourg)
✆ (352) 44 11 30 91 Fax: (352) 45 87 62

■ PUBLICATIONS: yes.
■ WORKING ON: internship and residency [B81], students [B82], microbiology [E7], research team [E16], research policy [E18], immunization [G59], mass screening [I6], mandatory screening [I7], hepatitis [N30], venereal diseases [N31], acquired immunodeficiency syndrome [N35], HIV seropositivity [N36].
■ INTERESTED IN: fundamental rights of the individuals [B2], human rights [B3], biomedical research [F2], clinical trials [F3], health legislation [G14], quarantine [G20], immunization [G59], patient information [J4], hepatitis [N30], venereal diseases [N31], acquired immunodeficiency syndrome [N35], HIV seropositivity [N36], withholding treatment [O10], allowing to die [O11], terminally ill [O12].

L 5 **MEYER** Alain

Professor of French litterature. — State Councillor. — Deputy Chairman of the Consistoire israélite du Luxembourg.

33, rue Jules-Fischer
1522 Luxembourg (Luxembourg)
✆ (352) 40 25 45

■ AFFILIATIONS: Commission consultative nationale d'éthique pour les sciences de la vie et de la santé, Luxembourg (*Member*).

L 6 **SCHOCKWEILER** Fernand

Judge at the European Court of Justice.

11, rue Adolphe-Fischer
1520 Luxembourg (Luxembourg)
✆ (352) 47 36 23

■ AFFILIATIONS: Commission consultative nationale d'éthique pour les sciences de la vie et de la santé, Luxembourg (*Member*).
■ WORKING ON: reproductive technologies [M26], ovum donors [M27], semen donors [M28], in vitro fertilization [M29], FIV center [M30], artificial insemination [M31], AID [M32], AIH [M33], excess embryos [M34], host mothers [M35], embryo transfer [M36], GIFT (Gamete Intrafallopian Transfer) [M37], embryo donation [M38], terminal care [O7], voluntary euthanasia [O18], biotechnology — genetic engineering [Q20], DNA linkage [Q33].
■ INTERESTED IN: ethical rules and principles [B47], solidarity [B65], religious ethics [C21], theology [C41], population genetics [Q43], human genome [Q47].

L 7 **THIBEAU** André

Chairperson of the Association des médecins et médecins-dentistes du grand-duché de Luxembourg.

B.P. 73
9001 Ettelbruck (Luxembourg)
✆ (352) 8 20 70

■ AFFILIATIONS: Commission consultative nationale d'éthique pour les sciences de la vie et de la santé, Luxembourg (*Member*).
■ WORKING ON: ethics committees [B22], World Medical Assembly [B20], professional deontology organs [B26], patient advocacy [B42], nursing education [B79], medical ethics [C52], nursing ethics [C53].
■ INTERESTED IN: bioethics [B89], hospital [G28], physicians [G40], radiology [G70], health care costs [H9], costs and benefits [H10], risks and benefices [H11], alternative therapies [I33], euthanasia [O15].

L 8 **WAGNER** Edmond

Professor of philosophy,university of Luxembourg. — Chairperson of the Department of Moral and Political Sciences at the Institut grand-ducal.

8, rue de la Libération
4210 Esch-sur-Alzette (Luxembourg)
✆ (352) 54 14 01

■ AFFILIATIONS: Commission consultative nationale d'éthique pour les sciences de la vie et de la santé, Luxembourg (*Member*).
■ TEACHING: to secondary school teachers. — Bioethics in methodology of science (university). — Public conferences on bioethics.
■ PUBLICATIONS: yes.
■ WORKING ON: respect of human dignity and human rights [B1], fundamental rights of the individuals [B2], dignity [B5], human equality [B7], human person [B12], human body commercialization [B13], reification [B14], ethical rules and principles [B47], ethics [B48], moral policy [B55], deontology [B59], morality [B61], respect [B63], bioethics [B89], bioethics movement [B95], bioethical issues [B96], European Convention of Bioethics [B97], Bioethics bill, 1992 (France) [B98], specific approaches to ethics [C1], philosophical ethics [C2], deontological ethics [C6], teleological ethics [C7], utilitarianism [C8], morals [C12], conscience [C13], religion and religious ethics [C19], Christian ethics [C22], Roman catholic ethics [C23], natural law [C24], Islamic ethics [C28], Jewish ethics [C29], theology [C41], philosophy of biology [C42], biology and human future [C43], future generations [C44], philosophy of biology [C42], primates [C50], ethics and humanities [C54], philosophy [C61], determinism [C62], sociobiology [C63], existentialism [C64], Marxism [C67], love [C80], personality [C82], science, technology, methods [E1], science [E2], biology [E5], genetics [E6], molecular biology [E8].
■ INTERESTED IN: health care professionals, researchers and patient relationship [J1], physicians and researchers accountability [J23], disease [N1].

L 9 **WERNER** Pierre

Former Prime Minister.

2, rond-point Robert-Schumann
2525 Luxembourg (Luxembourg)
✆ (352) 222 574

■ AFFILIATIONS: Commission nationale d'ethique, Luxembourg, Belgium (*Member*).
■ INTERESTED IN: political issues [D56], EC — European Communities [D62], religion and religious ethics [C19], Christian ethics [C22], Roman catholic ethics [C23].

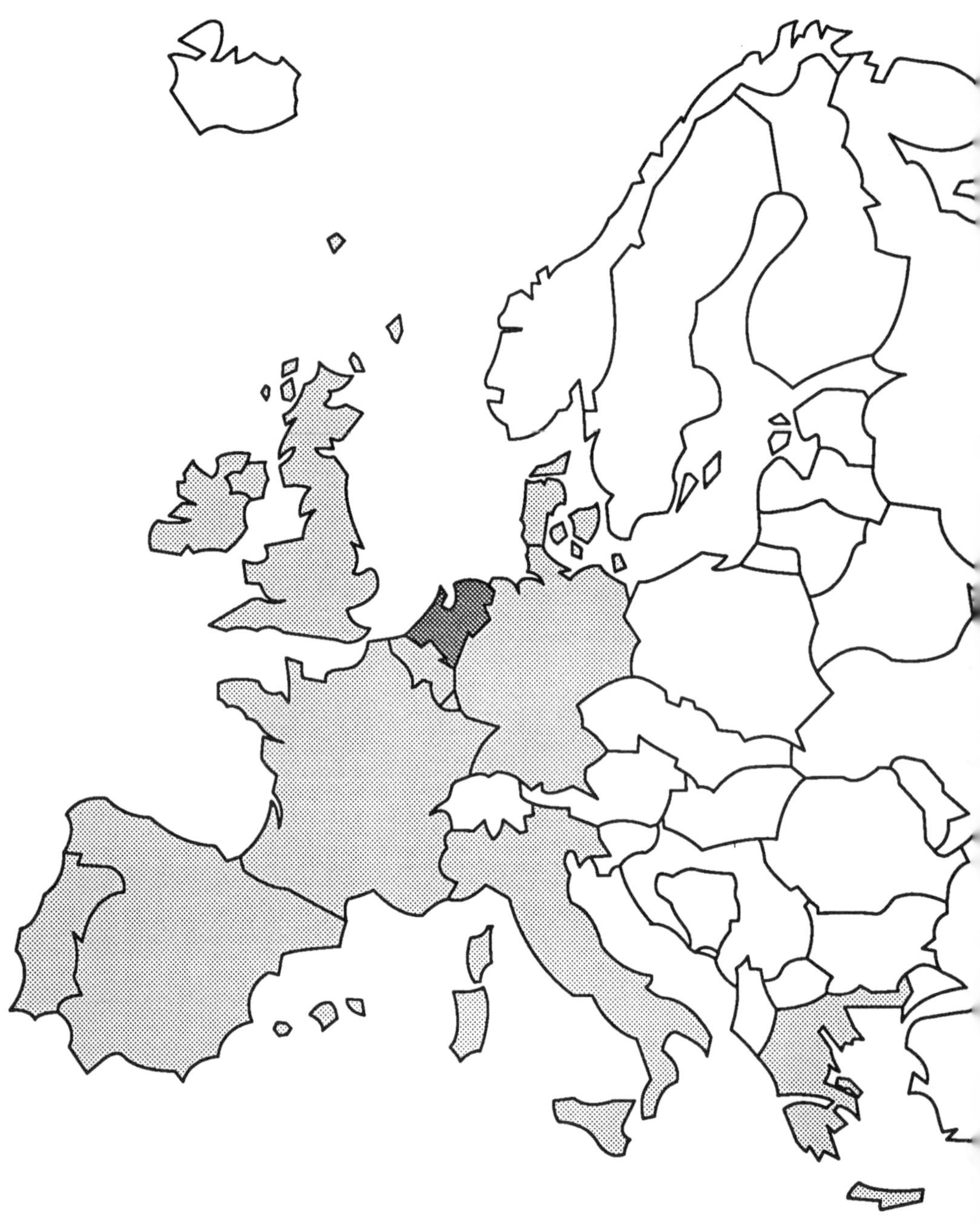

Netherlands

Area: 41 785 km²
Population: 15m
Gross domestic product: G 559bn; US$ 291bn
Gross domestic product per head: $ 19,400
Gross domestic product growth: 1991: 2.2%
1992: 2.4%

Organisations

O NL 1 **CENTER FOR APPLIED ETHICS - DEPARTMENT OF PHILOSOPHY**
Université Erasmus
P. O. Box 1738
3000 DR Rotterdam NL (Netherlands)
✆ (31) 10- 408 78 42 Fax: (31) 10- 436 77 10

O NL 2 **CENTER FOR ETHICS** (KUN)
Catholic University of Nijmegen
P.O. Box 9108
6500 HK Nijmegen (Netherlands)
✆ (31) 80 612 168

■ OTHER FIELDS OF ACTIVITIES: Environmental ethics. — Ethics in general. — Moral philosophy.
■ AFFILIATION: a centre of activity in which three departments (philosophical ethics, ethics of medicine, and theology) collaborate, devoted to ethical research.
■ DATE OF FOUNDATION: 01/09/1992 .
■ EXECUTIVE BOARD: Professor Paul Van Tongeren (*Member of board*), Professor Henk Ten Have (*Member of board*), Professor Theo Beemer (*Member of board*), Dr. Hub Zwart (*Director*).
■ TEACHING: currently investigating prospects for a postgraduate course in bioethics.
■ PUBLICATIONS: in process.
■ WORKING ON: human person [B12], ethics committees [B22], ethics [B48], ethicists [B92], history of biomedical ethics [B94], specific approaches to ethics [C1], situational ethics [C9], metaethics [C11], Roman catholic ethics [C23], natural law [C24], Protestant ethics [C27], Roman catholicism [C31], Protestantism [C35], humanism [C66], self concept [C83], social control [D3], public opinion [D7], cultural pluralism [D51], biomedical technologies [E28], quality of environment [E41], pollution [E47], decision analysis [E55], social control of science [E64], human experimentation [F11], nutrition [L16], food [L17], aged [L49], euthanasia [O15].

O NL 3 **COLLEGE D'ÉTHIQUE ET D'ASPECTS RELIGIEUX**
Oudlaan 4
B.P.9696
3506 GR Utrecht (Netherlands)

O NL 4 **DUTCH ASSOCIATION FOR HEALTH CARE AND PHILOSOPHY OF LIFE**
Thym Association /Medical Section
Secr. Gen. Dr. V. Kirkels Joannes Zwijsenlaan 121
5342 BT 055 (Netherlands)
✆ (31) (0)-4120- 21911

O NL 5 **EUROPEAN SOCIETY FOR THE PHILOSOPHY OF MEDICINE AND HEALTH CARE (UNIVERSITY OF LIMBURG)** (ESPMH)
Department of Ethics and Philosophy
P.O. Box 616
6200 Maastricht (Netherlands)
✆ (31) 152195

O NL 6 **INSTITUUT VAR GEZONHEIDSETHIEK**
P. O. B. 778
6200 Maastricht (Netherlands)
✆ (31) 43-21 75 75 Fax: 25 63 73

■ DATE DE FONDATION : 1982 .
■ DIRECTION : ***Directeur général et scientifique*** : Dr M.A.M. de Wachter (depuis 1984) ; ***comité de direction*** : Dr. L.B.J. Stuyt, Dr G. Brenninkmeijer, Mr G.J.A. Hamilton, Dr M.W.L.H. Bovy, Dr Ir.M.T. Hilhorst. Pr Dr C. van der Meer, Pr Dr E.L.Noach, Mr B. Pronk, Pr Dr Mr H.W.A. Sanders, Pr Dr H.K.A. Visser, Pr Dr W.H.G. Wolters.

■ ENSEIGNEMENT : formation en bioéthique pour 3e cycle, chercheurs, membres de comités d'éthique hospitaliers, médecins de famille, pharmaciens hospitaliers et infirmières.
■ PUBLICATIONS : séries, rapports, revues.
■ CENTRE DE DOCUMENTATION : spécialisé en bioéthique, informatisé (réseau AECEM), ouvert au public.
■ THEMES DE TRAVAUX : personne humaine [B12], comité d'éthique [B22], enseignement de l'éthique biomédicale [B78], examen éthique [B93], mouvement bioéthique [B95], CCE — Commission des Communautés Européennes [D63], comité consultatif [D69], génétique [E6], recherche internationale [E23], comité de sécurité [E42], expérimentation humaine [F11], pédiatrie [G68], population âgée [H23], don du sang [I20], formulaire de consentement [J14], individu responsable [K3], moelle osseuse [L8], enfant handicapé [L38], vieillissement [L50], lésion prénatale [N50], euthanasie [O15], euthanasie active [O16], euthanasie subie [O17], euthanasie volontaire [O18], thérapie génique [Q42].
■ CENTRES D'INTÉRET : personne humaine [B12], comité d'éthique [B22], Conseil de l'Europe [B25], enseignement de l'éthique biomédicale [B78], examen éthique [B93], mouvement bioéthique [B95], conscience morale [C13], valeur [C15], qualité de la vie [C16], éthique chrétienne [C22], éthique médicale [C52], CCE — Commission des Communautés Européennes [D63], gouvernement et système politique [D64], politique gouvernementale [D77], génétique [E6], recherche internationale [E23], industrie pharmaceutique [E37], comité de sécurité [E42], essai clinique [F3], projet génome humain [F8], expérimentation humaine [F11], unité de soins intensifs [G33], hospice [G34], pédiatrie [G68], assurance vie [H8], risque et bénéfice [H11], population âgée [H23], transfusion sanguine [I19], don du sang [I20], succédanés sanguins [I21], consentement aux soins [J8], consentement éclairé [J9], formulaire de consentement [J14], individu responsable [K3], clause de conscience [K63], moelle osseuse [L8], enfant handicapé [L38], vieillissement [L50], avortement sélectif [M49], thérapie foetale [M57], spina bifida [N26], coma [N41], état végétatif chronique [N42], lésion prénatale [N50], mort [O5], mort cérébrale [O6], soins aux mourants [O7], accompagnement des mourants [O8], acharnement thérapeutique [O9], décision médicale d'arrêt du traitement [O10], laisser-mourir [O11], phase terminale [O12], prolongation de la vie [O13], rapport qualité vie-survie [O14], euthanasie [O15], euthanasie active [O16], euthanasie subie [O17], euthanasie volontaire [O18], suicide [O19], attitude envers mort [O20], anomalie monogénique [Q7], conditions génétiques dominantes [Q8], mucoviscidose [Q15], génie génétique [Q22], carte génétique [Q23], conseil génétique [Q36], dépistage génétique [Q37], thérapie génique [Q42], génome humain [Q47].

O NL 7 **PROF. DR. G.A. LINDEBOOM INSTITUT**

Centrum voor medische ethiek
Postbus 224
6717 BE Ede (Netherlands)
✆ (31) 8380 - 30880 Fax: (31) 8380 - 24858
■ DATE OF FOUNDATION: 1987.
■ EXECUTIVE BOARD: Dr. Henk Jochensen (*Director*), Dr. Bart Cusveller (*Research Associate*), Ms. Geke Ter Velde-Van der Heide (*Secretary*).
■ TEACHING: several courses in medical ethics at Dutch universities. — Chairs in medical ethics at two Dutch universities.
■ PUBLICATIONS: series of Scientific Reports. — Lindeboom Series. — Periodical: In Perspectief.
■ DOCUMENTATION CENTRE: internal.
■ COLLOQUIUMS, SYMPOSIUMS: 1988, symposium on "Ethical and Social Aspects of Genetic Technology", attended by healthcare workers, ethicists and policy makers. — 1990, symposium on "Moral and Financial Limits of Health Care", (same public as above). — 1990, symposium on "Alternative Therapies as a symptom of crisis in health care".
■ DECISIONS, ADVICE: occasional advice to individual patients (e.g. certain reproductive technologies). — Occasional advice to patient organisations. — occasional advice to politicians and policy makers.
■ WORKING ON: information sources, data bases [A1], fundamental rights of the individuals [B2], protection of rights — involved institutions [B16], ethical rules and principles [B47], nursing education [B79], medical education [B80], bioethical issues [B96], philosophical ethics [C2], religion and religious ethics [C19], Protestant ethics [C27], religious beliefs [C32], Protestantism [C35], religion [C36], scriptural interpretation [C40], theology [C41], philosophy of biology [C42], professional ethics [C51], philosophy [C61], hedonism [C65], pastoral care [C88], scarcity [D21], marital relationship [D29], life sciences [E4], technology [E27], methods [E53], biomedical research [F2], experimentation [F9], methodology of research and experimentation [F15], health [G2], public health [G5], family planning [G8], health policy [G13], health facilities [G22], health personnel [G37], nurses [G39], health care [G45], home care [G55], immunization [G59], medicine [G63], obstetrics and gynecology [G66], economics [H2], population [H19], diagnosis [I2], prenatal diagnosis [I3], treatment [I30], patients' rights [J2], confidentiality [J20], inter-personal relationship [J27], legal personality [K2], personhood [K3], legally incompetent person [K4], child's interest [K24], accountability [K29], prohibition [K38], constitutional law [K48], patents [K47], sexuality and procreation — unborn child [M1], iatrogenic disease [N3], critically ill [N6], chronically ill [N7],

congenital defects [N23], acquired immunodeficiency syndrome [N35], HIV seropositivity [N36], nervous system diseases [N40], terminal care [O7], euthanasia [O15], genetics and applications — biotechnology [Q1].

■ INTERESTED IN: information sources, data bases [A1], fundamental rights of the individuals [B2], ethical rules and principles [B47], philosophical ethics [C2], situational ethics [C9], philosophy of biology [C42], professional ethics [C51], social issues [D2], life sciences [E4], technology [E27], containment [E38], biomedical research [F2], experimentation [F9], methodology of research and experimentation [F15], health facilities [G22], organization of health care [G35], health care [G45], health occupations [G61], economics [H2], population [H19], diagnosis [I2], biological specimens procurement, blood transfusion, organ transplantation [I8], treatment [I30], patients' rights [J2], confidentiality [J20], physicians and researchers accountability [J23], inter-personal relationship [J27], legal personality [K2], jurisprudence — accountability [K8], legislation and law [K34], childhood difficulties, diseases, protection [L21], aged — problems related with aging [L45], sexuality and procreation [M2], reproductive technologies [M26], pregnancy and childbirth [M39], abortion [M47], fetal development [M54], fetal therapy [M57], common symptomatology [N2], nervous system diseases [N40], determination of death [O2], terminal care [O7], euthanasia [O15], suicide [O19], behavioral and mental disorders [P2], outpatient commitment [P18], psychiatric technics [P24], genetic defects and hereditary diseases [Q2], biotechnology — genetic engineering [Q20], medical genetics [Q35], population genetics [Q43].

O NL 8 SECTI ETHIEK VAKGROEP FILOSOFIE, ETHIEK EN GESCHIECKUIS VAN DE GENEESKUUDE

Faculteit der Geneeskuude
Postbus 1738
3000 DR Rotterdam (Netherlands)
✆ (31) 10-408 70 62

■ AFFILIATION: Department of the Medical Faculty.

■ EXECUTIVE BOARD: Professor Inez D. de Beaufort (*Head*), Dr. Medart T. Hilhorst (*Associate*), Heleen S. Fransen-Starre (*Secretary*).

■ TEACHING: medical students. — physicians.

O NL 9 SOCIETAS ETHICA — EUROPEAN RESEARCH SOCIETY FOR ETHICS

c/o Centre for Bioethics and Health Law
University of Utrecht Heidelberg 2
3584LS Utrecht (Netherlands)
✆ (31) 30 53 43 99 Fax: (31) 30 53 32 41

O NL 10 UNIVERSITÉ D'ETAT DE LEIDEN

Department Metamedica
B.P. 2087
2301 CB Leiden (Netherlands)
✆ (31) 71 276 518

O NL 11 UNIVERSITEIT VOOR HUMANISTIEK

Van Asch van Wÿckskade 28
3512 VS Utrecht (Netherlands)
✆ (31) 31 26 74 and 36 74 37

■ OTHER FIELDS OF ACTIVITIES: teaching in research, philosophy and social sciences.

■ OFFICIALLY REGULATED ORGANISATION: yes.

■ DATE OF FOUNDATION: 01/01/1989.

■ EXECUTIVE BOARD: Professor Dr. Douwe Van Houten (*Rector*), Professor Dr. Henk Manschot (*In charge of the Department of Ethics*).

■ TEACHING: introductory courses in bioethics, intensive courses.

O NL 12 VAHGROEP GERONDHEIDSETHIEK EN WEJSBEGEERLE

Universiteissingel 50
6229 ER Maastricht (Netherlands)
✆ (31) 43-88 11 44 Fax: 67 09 32

■ OTHER FIELDS OF ACTIVITIES: Philosophy of Sciences. — Philosophy of Medicine. — History and Sociology of medicine. — Health promotion. — Women's studies.

■ AFFILIATION: a department of the Faculty of Health Sciences, State University of Limburg.

■ DATE OF FOUNDATION: 1974.

■ EXECUTIVE BOARD: Dr. Guy Widdershoven (*Chairman*).
■ TEACHING: Ethics course for the Faculty of Health Sciences. — Clinical Ethics in the Faculty of Medicine.

O NL 13 **VAKGROEP ETHIEK FILOSOFIE EN GESCHIEDENIS DER GENEESKUNDE**
Katholieke Universiteit Nijmegen
P.O. Box 9101
6500 KR Nijmegen (Netherlands)
✆ (31) 615320 Fax: (31) 540254

■ OTHER FIELDS OF ACTIVITIES: research and education in Philosophy of Medicine and History of Medicine.
■ AFFILIATION: Department of the Faculty of Medical Sciences, Catholic University of Nijmegen.
■ OFFICIALLY REGULATED ORGANISATION: University and Faculty regulations. — Educational laws.
■ DATE OF FOUNDATION: 1968.
■ EXECUTIVE BOARD: Professor Henk Ten Have (*Chairman*).
■ TEACHING: Ethics teaching for medical students, mental students, health science students. — Postgraduate courses in Medical Ethics. — Clinical Ethics conferences.
■ PUBLICATIONS: *Sempta Tironum*. — Bulletin of European Society for the Philosophy of Medicine and Health Care. — European Studies in Philosophy of Medicine. — Collections of Philosophy and Medical books.
■ DOCUMENTATION CENTRE: internal.
■ COLLOQUIUMS, SYMPOSIUMS: Nijmegen symposia in bioethics (4 times per year). — European bioethics seminars (annually).
■ WORKING ON: human person [B12], human body commercialization [B13], ethical rules and principles [B47], education [B76], history of biomedical ethics [B94], philosophical ethics [C2], religion and religious ethics [C19], research [E14], technology [E27], technology assessment [E65], biomedical research and experimentation [F1], organization of health care; facilities, manpower and services; health occupations [G1], required request [I10], renal dialysis [I32], human body [L3], aged — problems related with aging [L45], reproductive technologies [M26], kidney diseases [N39], chronically ill [N7], physically [N8], determination of death [O2], terminal care [O7], euthanasia [O15], attitudes to death [O20], resuscitation [O21], genetics and applications — biotechnology [Q1].
■ INTERESTED IN: biomedical ethics education [B78], Roman catholic ethics [C23], evaluation [E61], organization of health care; facilities, manpower and services; health occupations [G1], health policy [G13], health facilities [G22], health care [G45], resource allocation [H4], renal dialysis [I32], human body [L3], aged — problems related with aging [L45], chronically ill [N7], physically [N8], kidney diseases [N39], terminal care [O7], genetics and applications — biotechnology [Q1].

O NL 14 **VERENIGING VAN ETHICI IN NEDERLAND**
c/o J. S. Reinders De Vrise
Universiteit De Boelelaan 1105
1081 HV Amsterdam (Netherlands)

Individuals

NL 1 **ACHTERBERG W.**
Assistant Professor of Philosophy, Department of Practical Philosophy, University of Amsterdam. — Professor of Philosophy, Agricultural University of Wageningen.

Department of Practical Philosophy
University of Amsterdam
Nieuwe Doelenstraat 15
1012 CP Amsterdam (Netherlands)
✆ (31) 20-525 45 30 and 45 00 Fax: (31) 20-525 45 03

■ AFFILIATIONS: Hastings Center Institute for Ethics in the Professions of Health, New York, Briarcliff Manor, USA (*Indirect affiliation*).

■ TEACHING: Courses in Biomedical Ethics, University of Amsterdam (Faculty of Philosophy and Medical Faculty). — Courses in Environmental Ethics (including Bioethics), University of Amsterdam, Agricultural University of Wageningen.
■ WORKING ON: democracy [D82], public policy [D77], future generations [C44], speciesism [C48], biotechnology [Q21], genetically modified organisms [Q28], transgenic animals [Q29], medical ethics [C52], ethical analysis [C3], normative ethics [C5], metaethics [C11], values [C15], value of life [C18].
■ INTERESTED IN: ethical analysis [C3], normative ethics [C5], metaethics [C11], values [C15], value of life [C18], future generations [C44], speciesism [C48], biological life [C49], medical ethics [C52], humanism [C66], democracy [D82], wrongful life [K33], biotechnology [Q21], genetically modified organisms [Q28], transgenic animals [Q29], public policy [D77], democracy [D82].

NL 2 **BERGHMANS** Ron
Psychologist. — Bioethicist.

Instituut Voor Gezondheidsethiek
Postbus 778
6200 AT Maastricht (Netherlands)
✆ (31) 43 217 575 Fax: (31) 43 256 373

■ AFFILIATIONS: Institute for Bioethics, Maastricht, Netherlands (*Research associate*).
■ TEACHING: Bioethics course for members of medical ethics committees.
■ PUBLICATIONS: yes.
■ WORKING ON: ethics [B48], competence [B51], paternalism [B62], bioethical issues [B96], normative ethics [C5], medical ethics [C52], social control [D3], coercion [D5], EC — European Communities [D62], nursing home [G27], mental institutions [G29], psychiatry [G69], tissue donation [I13], blood transfusions [I19], blood donation [I20], blood substitutes [I21], fetal tissue transplantation [I27], patients' rights [J2], chronically ill [N7], euthanasia [O15], suicide [O19], outpatient commitment [P18], psychiatric technics [P24].
■ INTERESTED IN: ethics [B48], competence [B51], paternalism [B62], bioethical issues [B96], normative ethics [C5], medical ethics [C52], social control [D3], coercion [D5], EC — European Communities [D62], nursing home [G27], mental institutions [G29], psychiatry [G69], tissue donation [I13], blood transfusions [I19], blood donation [I20], blood substitutes [I21], fetal tissue transplantation [I27], patients' rights [J2], chronically ill [N7], euthanasia [O15], suicide [O19], outpatient commitment [P18], psychiatric technics [P24].

NL 3 **BORST-EILERS** E.
Vice-Chairperson of the Health Council of the Netherlands. — Professor in Evaluation of Clinical Care, University of Amsterdam and Amsterdam Academic Medical Centre.

B.P. 90517
2509 LM 'S Den Haag (Netherlands)
✆ (31) 70-34 71 441 Fax: (31) 70-38 37 109
■ AFFILIATIONS: National Committee on Ethical Aspects of Medical Research, Den Haag, Netherlands (*Chairperson*). — Institute of Bioethics (Maastricht), Maastricht, Netherlands (*Chairperson advisory board*). — Health Council of the Netherlands, Den Haag, Netherlands (*Vice-Chairperson*).
■ PUBLICATIONS: yes.
■ WORKING ON: ethics committees [B22], scarcity [D21], technology assessment [E65], health services research [F4], human experimentation [F11], medical devices [G23], artificial organs [G24], risks and benefices [H11], mass screening [I6], advance directives [J18], living wills [J19], life extension [L47], health care services and aged [L48], in vitro fertilization [M29], excess embryos [M34], GIFT (Gamete Intrafallopian Transfer) [M37], embryo donation [M38], fetal therapy [M57], active euthanasia [O16], involuntary euthanasia [O17], voluntary euthanasia [O18], genetically modified organisms [Q28], genetic screening [Q37], preimplantation genetic diagnosis [Q38], gene therapy [Q42].
■ INTERESTED IN: technology assessment [E65], risks and benefices [H11], mass screening [I6], life extension [L47], health care services and aged [L48], excess embryos [M34], genetically modified organisms [Q28], genetic screening [Q37], preimplantation genetic diagnosis [Q38], gene therapy [Q42].

NL 4 **CHRISTIENS** Marc
Lecturer, Medical Ethics.

Université catholique de Nijmegen
Département d'éthique, de philosophie et d'histoire de la médecine
Geert Grooteplein Nood 21
6525 EJ Nijmegen (Netherlands)
✆ (31) 80 61 53 20 et 613 104 Fax: (31) 80 54 02 54

■ AFFILIATIONS: Tyschrift voor Geneeskunde en Ethiek, Nymegen, Netherlands (*Editor in chief*).
■ TEACHING: yes.
■ PUBLICATIONS: yes.
■ WORKING ON: dignity [B5], human person [B12], ethics committees [B22], ethics [B48], professional competence [B52], altruism [B56], beneficence [B57], authoritarianism [B58], biomedical ethics education [B78], codes of biomedical ethics [B90], bioethical issues [B96], Christian ethics [C22], Roman Catholic ethics [C23], medical ethics [C52], nursing ethics [C53], couple [D26], marital relationship [D29], family relationship [D31], handicapped [D43], human experimentation [F11], research subjects [F12], dentistry [G62], obstetrics and gynecology [G66], pediatrics [G68], prenatal diagnosis [I3], amniocentesis [I4], chorionic villus sampling [I5], required request [I10], sperm banks [I11], donors [I14], anonymous donation [I17], patient information [J4], informed consent [J9], presumed consent [J10], third party consent [J11], conflict of interest [K9], beginning of life [L24], newborns [L29], prematurity [L30], aid children [L34], involuntary sterilization [M7], volontary sterilization [M8], embryos [M10], wish of children [M16], reproduction [M17], sexual behavior [M21], ovum donors [M27], semen donors [M28], in vitro fertilization [M29], AID [M32], AIH [M33], excess embryos [M34], host mothers [M35], embryo transfer [M36], embryo donation [M38], mother fetus relationship [M43], selective abortion [M49], therapeutic abortion [M50], abortion on demand [M52], autopsies [O3], brain death [O6].
■ INTERESTED IN: codes of ethics [B49], quality of life [C16], Christian ethics [C22], Roman Catholic ethics [C23], medical ethics [C52], nursing ethics [C53], couple [D26], decision making [E54], human experimentation [F11], WHO — World Health Organization [G36], dentistry [G62], obstetrics and gynecology [G66], pediatrics [G68], prenatal diagnosis [I3], medical secrecy [J22], conflict of interest [K9], beginning of life [L24], embryos [M10], wish of children [M16], reproduction [M17], homosexuality [M22], ovum donors [M27], semen donors [M28], artificial insemination [M31], attitudes to death [O20], hereditary diseases [Q18], genetic intervention [Q22], genome mapping [Q23], genetic counseling [Q36], eugenics [Q39].

NL 5 **DE BEAUFORT** Inez
Professor of Medical Ethics.

Medical Faculty, Department of Medical Ethics
Université Erasmus
Postbus 1738
3000 DR Rotterdam (Netherlands)
✆ (31) 10-40 88 145

■ TEACHING: Teaching medical students in different stages of their study. — Special courses for physicians.
■ PUBLICATIONS: yes.
■ WORKING ON: ethics committees [B22], moral obligations [B50], obligations of society [B53], obligations to society [B54], solidarity [B65], medical education [B80], curriculum [B83], bioethical issues [B96], wedge argument [C4], medical ethics [C52], human experimentation [F11], group of vulnerable subjects [F20], prenatal diagnosis [I3], mass screening [I6], fetal tissue transplantation [I27], investigator subject relationship [J29], potentiality of personhood [K5], child's interest [K24], preconception injuries [K27], wrongful life [K33], children's rights [L25], infanticide [L42], embryos [M10], fetuses [M11], procreation [M15], ovum donors [M27], semen donors [M28], in vitro fertilization [M29], excess embryos [M34], embryo donation [M38], pregnant women [M42], mother fetus relationship [M43], selective abortion [M49], terminal care [O7], euthanasia [O15], substance dependence [P3], alcohol abuse [P4], smoking [P5], drug abuse [P6], genetic screening [Q37], eugenics [Q39].
■ INTERESTED IN: medical education [B80], curriculum [B83], human experimentation [F11], group of vulnerable subjects [F20], prenatal diagnosis [I3], mass screening [I6], potentiality of personhood [K5], child's interest [K24], criminal law [K53], children's rights [L25], genetic screening [Q37], eugenics [Q39].

NL 6 **DE WACHTER** M.A.M
Director, Institute for Bioethics. — Chairperson, European Association of Centres of Medical Ethics (AECEM).

Institut de bioéthique
B.P. 778
6200 AT Maastricht (Netherlands)
✆ (31) 43 217 575 and 30 256 373

NL 7 **DE WERT** G.M.W.R.
Senior Investigator in Bioethics.

Institut de bioéthique
Sint Servaasklooster 39
6211 TE Maastricht (Netherlands)
✆ (31) 43-217 575 Fax: (31) 43-256 373

■ TEACHING: Teaching members of medical ethics commitees.
■ PUBLICATIONS: yes.
■ WORKING ON: normative ethics [C5], quality of life [C16], genetics [E6], pilot projects [E20], nontherapeutic research [F5], human genome project [F8], prenatal diagnosis [I3], mass screening [I6], mandatory screening [I7], fetal tissue transplantation [I27], parental consent [J13], aid children [L34], infertility [M19], semen donors [M28], in vitro fertilization [M29], AID [M32], excess embryos [M34], embryo transfer [M36], multiple pregnancy [M46], selective abortion [M49], anencephaly [N25], HIV seropositivity [N36], brain death [O6], chromosomal disorders [Q4], Huntington's chorea [Q9], sickle cell anemia [Q11], cystic fibrosis [Q15], phenylketonuria [Q16], thalassemia [Q17], sex linked defects [Q19], cloning [Q24], sex determination [Q26], sex preselection [Q30], genetic counseling [Q36], genetic screening [Q37], preimplantation genetic diagnosis [Q38], eugenics [Q39], gene therapy [Q42].
■ INTERESTED IN: involuntary sterilization [M7], mifepristone [M53], normative ethics [C5], quality of life [C16], genetics [E6], pilot projects [E20], nontherapeutic research [F5], human genome project [F8], prenatal diagnosis [I3], mass screening [I6], mandatory screening [I7], fetal tissue transplantation [I27], parental consent [J13], aid children [L34], infertility [M19], semen donors [M28], in vitro fertilization [M29], AID [M32], excess embryos [M34], embryo transfer [M36], multiple pregnancy [M46], selective abortion [M49], anencephaly [N25], HIV seropositivity [N36], brain death [O6], chromosomal disorders [Q4], Huntington's chorea [Q9], sickle cell anemia [Q11], cystic fibrosis [Q15], phenylketonuria [Q16], thalassemia [Q17], sex linked defects [Q19], cloning [Q24], sex determination [Q26], sex preselection [Q30], genetic counseling [Q36], genetic screening [Q37], preimplantation genetic diagnosis [Q38], eugenics [Q39], gene therapy [Q42].

NL 8 **DEKKERS** Wim J.M.

Member of the Department of Ethics, Phylosophy and Medical History, Catholic University of Nijmegen.

Université catholique de Nijmegen
Département d'éthique, de philosophie et d'histoire de la médecine
P.O. Box 9101
6500 EHB Nijmegen (Netherlands)
✆ (31) 80-61 53 20 Fax: (31) 80-54 02 54

■ AFFILIATIONS: European Society of Philosophy of Medicine and Health Care — ESPMH, Netherlands (*Member of the Financial Review Committee*).
■ TEACHING: Lectures in Philosophy of Medicine for students in medicine and students in health sciences.
■ PUBLICATIONS: yes.
■ WORKING ON: human person [B12], human body commercialization [B13], traffic of organs [B15], integrity [B8], medical education [B80], curriculum [B83], universities [B84], humanities [C55], philosophy [C61], existentialism [C64], biomedical technologies [E28], biological specimens procurement, blood transfusion, organ transplantation [I8], HIV seropositivity [N36], determination of death [O2], autopsies [O3], cadavers [O4], death [O5], brain death [O6].
■ INTERESTED IN: *idem*.

NL 9 **DEN HARTOGH** G.A.

Professor of Bioethics, Faculty of Medicine. — Associate Professor of Moral Philosophy, Faculty of Philosophy.

Institut de philosophie
Université d'Amsterdam
Nieuwe Doelenstraat15
1012 CP Amsterdam (Netherlands)
✆ (31) 20-525 45 34 Fax: (31) 20-525 45 03

■ AFFILIATIONS: Vereniging voor Filosofie en Geneeskunde — VFG, Netherlands (*Member*).
■ TEACHING: Lectures on bioethical aspects within the framework of the regular teaching program of the Faculty of Medicine. — Operational courses (an introduction to bioethics) for students of Medical Philosophy.
■ PUBLICATIONS: yes.

NL 10 **DIERICK** G.P.A.

Centre d'études et d'information catholique
Université catholique de Nijmegen
Erasmuslaan36
6525 GG Nijmegen (Netherlands)
✆ (31) 80-61 24 14

■ PUBLICATIONS: yes.
■ WORKING ON: bibliography [A4], information centers [A5], book review [A6], literature [A7], documentation [A8], publications [A9], self determination [B4], dignity [B5], women's rights [B6], human equality [B7], integrity [B8], justice [B9], freedom [B10], privacy [B11], human person [B12], human body commercialization [B13], reification [B14], traffic of organs [B15], protection of rights — involved institutions [B16], Amnesty International [B18], World Medical Assembly [B20], Council for International Organization of Medical Sciences [B24], Council of Europe [B25], manifests and declarations concerning human rights [B33], International Charter of Human Rights [B34], Helsinki Declaration [B37], ethical rules and principles [B47], information — communication — media [B68], bioethics [B89], codes of biomedical ethics [B90], CCNE advice (France) [B91], ethicists [B92], ethical review [B93], history of biomedical ethics [B94], bioethics movement [B95], bioethical issues [B96], deontological ethics [C6], teleological ethics [C7], utilitarianism [C8], situational ethics [C9], ethical relativism [C10], metaethics [C11], morals [C12], conscience [C13], moral development [C14], values [C15], quality of life [C16], social worth [C17], value of life [C18], religion and religious ethics [C19], clergy [C20], religious ethics [C21], Christian ethics [C22], Roman Catholic ethics [C23], natural law [C24], totality [C25], eastern orthodox ethics [C26], Protestant ethics [C27], Islamic ethics [C28], Jewish ethics [C29], religion and sects [C30], Roman catholic ism [C31], religious beliefs [C32], Islam [C33], Judaism [C34], Protestantism [C35], religion [C36], Christian science [C37], Jehovah's witnesses [C38], religious sciences [C39], scriptural interpretation [C40], theology [C41], philosophy of biology [C42], biology and human future [C43], future generations [C44], consequences [C45], sterilization [M6], involuntary sterilization [M7], volontary sterilization [M8], reproductive organs and embryonic structures [M9], embryos [M10], fetuses [M11], ovum [M12], placentas [M13], sperm [M14], procreation [M15], wish of children [M16], reproduction [M17], fertility [M18], infertility [M19], sexuality [M20], sexual behavior [M21], voluntary euthanasia [O18], suicide [O19], attitudes to death [O20], deontological ethics [C6], teleological ethics [C7], utilitarianism [C8], situational ethics [C9], ethical relativism [C10].

NL 11 **DILLMANN** R. J. M.
Staff member, Royal Dutch Medical Association (RDMA). — Lecturer, Free University, Amsterdam.

KNMG
Ethique médicale
Lomanlaan 103
3526 XD Utrecht (Netherlands)
✆ (31) 30-823 911 Fax: (31) 30-823 326

■ AFFILIATIONS: Department Metamedica, Free University Amsterdam, Amsterdam, Netherlands (*Lecturer*). — Medical ethics committee of RDMA (Royal Dutch Medical Association) — RDMA, Utrecht, Netherlands (*Staff member and secretary*).
■ TEACHING: Lectures and training sessions, Free University Amsterdam..
■ PUBLICATIONS: yes.
■ WORKING ON: College of Physicians [B27], competence [B51], biomedical ethics education [B78], codes of biomedical ethics [B90], ethical analysis [C3], philosophy of biology [C42], medical ethics [C52], science, technology, methods [E1], nontherapeutic research [F5], therapeutic research [F6], mass screening [I6], coma [N41], persistent vegetative state [N42], euthanasia [O15], dementia [P11].
■ INTERESTED IN: CCNE — Comité Consultatif National d'Ethique (France) [B23], competence [B51], biomedical ethics education [B78], codes of biomedical ethics [B90], European Convention of Bioethics [B97], situational ethics [C9], ethical relativism [C10], scarcity [D21], research policy [E18], progress [E26], technology assessment [E65], nontherapeutic research [F5], therapeutic research [F6], preventive medicine [G58], mass screening [I6], advance directives [J18], living wills [J19], coma [N41], persistent vegetative state [N42], euthanasia [O15], dementia [P11], genetic screening [Q37], eugenics [Q39].

NL 12 **DUPUIS** H.M.
Professor of Bioethics, Faculty of Medicine and Faculty of Philosophy. — Head of Department of Metamedica (which includes Ethics, History of Medicine and Public Health). — Standing member of the Medico-ethical Committee of the Faculty and the University Hospital, Leiden University.

Département Metamedica
Université d'Etat de Leiden
B.P. 2087
2301 CB Leiden (Netherlands)
✆ (31) 71-276 521 and 276 518 Fax: (31) 71-275 357

■ AFFILIATIONS: Department of Metamedica, Netherlands (*Chairman*).
■ TEACHING: Training medical students, medical doctors (postgraduate). — Courses in Bioethics for students from other disciplines (e.g. Law, Psychology, Theology).
■ PUBLICATIONS: yes.
■ WORKING ON: fundamental rights of the individuals [B2], human rights [B3], self determination [B4], integrity [B8], patient advocacy [B42], paternalism [B62], decision making [E54], decision analysis [E55], extraordinary treatment [I39], informed consent [J9], coma [N41], neural tube defects [N24], quality adjusted life years [O14], suicide [O19], withholding treatment [O10], terminal care [O7], palliative care [O8], prenatal diagnosis [I3], amniocentesis [I4], chorionic villus sampling [I5], mass screening [I6], genetics and applications — biotechnology [Q1], professional competence [B52], professional deontology [J25], physician patient relationship [J32], health economics, population characteristics [H1], economics [H2], resource allocation [H4], health care costs [H9], costs and benefits [H10], risks and benefices [H11].
■ INTERESTED IN: philosophical ethics [C2], biomedical ethics education [B78].

NL 13 **FRETZ** Led

Teaching Social Ethics and Political Philosophy at Delft University of Technology (UHD).

Delft University of Technology
Faculty of Philosophy and Technical Social Sciences
Kanaalweg 2B
2628 EB Delft (Netherlands)
✆ (31) 15-78 5143

■ PUBLICATIONS: yes.

NL 14 **HEEGER** Robert

Professor of Ethics at the State University of Utrecht. — Director of the Ethics Research Programme of the Centre for Bioethics and Health Law at the University of Utrecht.

Faculté de théologie et de philosophie
Centre de bioéthique et de législation sanitaire
Heidelberglaan2
3584 CS Utrecht (Netherlands)
✆ (31) 30-534 399 Fax: (31) 30-533 241
■ AFFILIATIONS: Societas Ethica: European Society for Research in Ethics, Utrecht, Netherlands (*Chairperson*).
■ TEACHING: Masters courses in Bioethics in the Faculties of Theology and Philosophy, Utrecht University. — Masters courses in Professional Ethics in the Faculty of Veterinary Medicine, Utrecht University. — Training of Animal Experiment Review Boards, organised by the Dutch Society of Licence-holders on Animal Experiments (NVVD).
■ PUBLICATIONS: yes.
■ WORKING ON: integrity [B8], ethics committees [B22], respect [B63], biomedical ethics education [B78], decision making [E54], animal experimentation [F10], embryo transfer [M36], transgenic animals [Q29].
■ INTERESTED IN: integrity [B8], decision making [E54], transgenic animals [Q29].

NL 15 **HILHORST** Medard

Lecturer on Medical Ethics.

Département d'éthique médicale
Université Erasmus — Faculty of Medicine and Health Sciences
B.P. 1738
3000 DR Rotterdam (Netherlands)
✆ (31) 10-436 77 10

■ TEACHING: Department of Medicine and Department of Philosophy.
■ PUBLICATIONS: yes.

NL 16 **HOUTEPEN** R.
Assistant Professor Faculty of Health Sciences, Department of Health Care Ethics and Philosophy.

Département d'éthique sanitaire et de philosophie de la médecine
Université d'Etat de Limburg
B.P. 616
6200 MD Maastricht (Netherlands)
✆ (31) 43-88 1120 and 88 1144 Fax: (31) 43-67 0932

■ TEACHING: University Ethics courses in Health Sciences and Medicine. — Clinical Ethics Education.
■ PUBLICATIONS: yes.
■ WORKING ON: common good [B45], professional competence [B52], moral policy [B55], nursing education [B79], medical education [B80], history of biomedical ethics [B94], situational ethics [C9], ethical relativism [C10], nursing ethics [C53], informal social control [D4], sociology of medicine [D55], nurses [G39], allowing to die [O11], suicide [O19], attitudes to death [O20], euthanasia [O15], terminal care [O7], ethical rules and principles [B47], philosophical ethics [C2], ethics and humanities [C54], social and political issues [D1], science, technology, methods [E1], methods [E53], biomedical research [F2], health policy [G13], health care [G45], health care professionals, researchers and patient relationship [J1], inter-personal relationship [J27].
■ INTERESTED IN: nursing education [B79], history of biomedical ethics [B94], situational ethics [C9], ethical relativism [C10], nursing ethics [C53], sociology of medicine [D55], methods [E53], health care [G45], inter-personal relationship [J27], reproductive technologies [M26], terminal care [O7], euthanasia [O15], allowing to die [O11], suicide [O19].

NL 17 **JANS** Jan
University Lecturer in Moral Theology.

Theologische Faculteit Tilburg
Postbus 90153 KUB
5000 Le Tilburg (Netherlands)
✆ (31) 13 66 25 95 Fax: (31) 13 66 31 34

■ TEACHING: University teaching. — Conferences.
■ PUBLICATIONS: yes.
■ WORKING ON: fundamental rights of the individuals [B2], moral policy [B55], ethicists [B92], conscience [C13], Roman Catholic ethics [C23], resource allocation [H4], prenatal diagnosis [I3], donors [I14], health care services and aged [L48], reproductive technologies [M26], terminal care [O7], euthanasia [O15].
■ INTERESTED IN: donors [I14].

NL 18 **JOCHEMSEN** H.
Director of the Professor-Dr. G.A. Lindeboom Institute.
Institut Prof. G.A. Lindeboom
B.P. 224
6710 BE Ede (Netherlands)
✆ (31) 8380-30 230 Fax: (31) 8380-24 858

■ AFFILIATIONS: European Association of Centres for Medical Ethics — EACME, Leuven, Belgium (*Associate member*).
■ TEACHING: Course (6x2 hrs) in a Bible Seminary. — Part of a course in Christian Medical Ethics for medical students and others interested. — Lectures at a medical faculty (on a voluntary course for the students). — Lectures for a variety of private organisations.
■ PUBLICATIONS: yes.
■ WORKING ON: *idem*.
■ INTERESTED IN: patient care [G53], fetal tissue transplantation [I27], terminal care [O7], euthanasia [O15], medical genetics [Q35].

NL 19 **MANENSCHIJN** Gerrit
Professor of Ethics.

Theological University of Kampen
Koommarkt 1
Postbox 5021
8260 GA Kampen (Netherlands)
✆ (31) 52 02 92 634

NL 20 **MUSSCHENGA** A.W.

Special Professor in Social Ethics. — Director of the Interdisciplinary Centre for the Study of Science, Society and Religion.

Institut d'éthique
Université libre
B.P. 7161
1017 MC Amsterdam (Netherlands)
✆ (31) 20-54 83 830 Fax: (31) 20-64 29 634

■ AFFILIATIONS: Institute for Ethics, Vrije Universiteit, Amsterdam, Netherlands (*Chairman of the Board*).
■ PUBLICATIONS: yes.

NL 21 **NOACH** E.L.

Professor of Pharmacology (Retired). — Chairperson of the Governmental Medicinal Drug Commission. — Chairman of the Medical Ethics Committee and of the Netherlands Organisation for Applied Research (TNO).

Van der Valk Boumanweg 234
2352 JG Leiderdorp (Netherlands)
✆ (31) 71-411 036

■ AFFILIATIONS: Instituut var Gezonheidsethiek (Institut de bioéthqiue), Maastricht,Netherlands (*Member of the Board*).
■ TEACHING: Training courses for members and future members of medical-ethics commissions and institutional review boards.
■ PUBLICATIONS: yes.
■ WORKING ON: ethics committees [B22], professional deontology organs [B26], education [B76], biomedical ethics education [B78], medical education [B80], universities [B84], medical ethics [C52], government regulation [D17], clinical trials [F3], nontherapeutic research [F5], animal experimentation [F10], human experimentation [F11], healthy volunteers [F13], consent to treatment [J8], informed consent [J9], presumed consent [J10], third party consent [J11], legislation and law [K34].
■ INTERESTED IN: animal care committees [B21], ethics committees [B22], education [B76], biomedical ethics education [B78], government regulation [D17], presumed consent [J10], third party consent [J11], legislation and law [K34].

NL 22 **ODERWALD** A. K.

Assistant Professor of Medical Philosophy and Ethics, Medical Faculty, Free University.
Département Metamedica
Université libre
Van des Boechortstraat 7
1081 BT Amsterdam (Netherlands)
✆ (31) 20-54 83 317 Fax: (31) 20-64 62 228

■ AFFILIATIONS: European Society for Philosophy of Medicine and Health Care, Nijmegen, Netherlands (*Member*).
■ TEACHING: Courses and small group teaching to medical students (part of the official curriculum).
■ PUBLICATIONS: yes.

NL 23 **PIJNENBURG** M.A.M.

Secretary General, Catholic Association of Hospitals.

Catholic Association of Hospitals
Postbus 152
5260 ADF Vught (Netherlands)
✆ (31) 73-57 90 11

■ AFFILIATIONS: Association of Ethicists (*Member*). — Association for Philosophy and Medicine (*Member*).
■ TEACHING: Courses for nurses. — Advising hospital management in ethical policy question.
■ PUBLICATIONS: yes.
■ WORKING ON: justice [B9], Roman Catholic ethics [C23], Roman catholic ism [C31], theology [C41], ethics committees [B22], solidarity [B65], human experimentation [F11], public debates [B75], death and resuscitation [O1].

■ INTERESTED IN: justice [B9], Roman Catholic ethics [C23], Roman catholic ism [C31], theology [C41], nursing education [B79], nursing ethics [C53].

NL 24 **POORTMAN Y.S.**
Executive Director.

Vereniging Sameuwerkende Ouder – en Patientenorganisaties
Vredehofstraat 31
3761 HA Svesldÿk (Netherlands)
✆ (31) 2155-28 155 Fax: (31) 2155-27 440

■ TEACHING: Production of educational materials.
■ PUBLICATIONS: yes.
■ WORKING ON: medical ethics [C52], clergy [C20], religious ethics [C21], biomedical technologies [E28], counseling [C87], pastoral care [C88], health care, medical acts [I1], congenital defects [N23], genetics and applications — biotechnology [Q1], biotechnology — genetic engineering [Q20], behavioral genetics [Q34], population genetics [Q43].

NL 25 **RAVENSCHLAG I.**
Research in Medical Ethics and Medical Law.

Institut de bioéthique
Sint Servaasklooster 39
6211 TE Maastricht (Netherlands)
✆ (31) 43-217 575 Fax: (31) 43-256 373

■ TEACHING: Training in medical ethics committees (fundamental principles of Bioethics, AIDS/HIV, the right not to know).
■ PUBLICATIONS: yes.
■ WORKING ON: fundamental rights of the individuals [B2], human rights [B3], self determination [B4], human equality [B7], justice [B9], freedom [B10], privacy [B11], ethics committees [B22], ethics [B48], respect [B63], medical ethics [C52], disclosure [J7], consent to treatment [J8], informed consent [J9], presumed consent [J10], third party consent [J11], spousal consent [J12], parental consent [J13], consent forms [J14], treatment refusal [J15], patient participation [J16], right to treatment [J17], privileged communication [J21], medical secrecy [J22], physician patient relationship [J32], professional patient relationship [J33], parent child relationship [J34], personhood [K3], legally incompetent person [K4], potentiality of personhood [K5], minors [K6], legal guardians [K7], conflict of interest [K9], battery [K10], torts [K11], wrongful death [K12], fraud [K13], deception [K14], malpractice [K19], negligence [K20], legal liability [K30], judicial action [K35], constitutional amendments [K36], contracts [K39], civil laws [K43], civil code [K44], legal rights [K45], constitutional law [K48], medical law [K51], HIV seropositivity [N36], euthanasia [O15].

NL 26 **REINDERS J. S.**
Research Fellow, Institute for Ethics. — Assistant Professor Theological Faculty. — Chairman of the Ethics Department (Free University of Amsterdam).

Institut d' éthique
Université libre
De Boelelaan 1105
1081HV Amsterdam (Netherlands)
✆ (31) 20-54 85 443

■ TEACHING: Ethics programme for graduate students. — Lectures, doctoral theses, etc. (regular academic teaching activities).
■ PUBLICATIONS: yes.
■ WORKING ON: human rights [B3], competence [B51], deontology [B59], virtues [B64], self determination [B4], human person [B12], quality of life [C16], value of life [C18], religious ethics [C21], Christian ethics [C22], Protestant ethics [C27], personality [C82], self concept [C83], pastoral care [C88], genetics [E6], microbiology [E7], biomedical technologies [E28], freezing [E29], preservation [E30], human experimentation [F11], personhood [K3], potentiality of personhood [K5], beginning of life [L24], newborns [L29], handicapped children [L38], procreation [M15], wish of children [M16], reproduction [M17], infertility [M19], in vitro fertilization [M29], excess embryos [M34], host mothers [M35], mother fetus relationship [M43], selective abortion [M49], therapeutic abortion [M50], aborted fetuses [M51], disease [N1], life-sustaining treatment [O9], withholding treatment [O10], mentally handicapped [P7], mentally retarded [P8].
■ INTERESTED IN: human rights [B3], self determination [B4], human person [B12], competence [B51], deontology [B59], virtues [B64], codes of biomedical ethics [B90], CCNE advice (France) [B91], quality of life [C16], value of life [C18], religious ethics [C21], Christian ethics [C22], Protestant ethics

[C27], Islamic ethics [C28], Jewish ethics [C29], judgement [C71], attitudes [C74], intention [C75], motivation [C76], personality [C82], self concept [C83], pastoral care [C88], genetics [E6], microbiology [E7], biomedical technologies [E28], freezing [E29], preservation [E30], human genome project [F8], human experimentation [F11], patient association [J3], patient information [J4], patient access [J5], parental consent [J13], professional deontology [J25], medical etiquette [J28], personhood [K3], potentiality of personhood [K5], children [L27], newborns [L29], handicapped children [L38], in vitro fertilization [M29], excess embryos [M34], host mothers [M35], mother fetus relationship [M43], selective abortion [M49], therapeutic abortion [M50], aborted fetuses [M51], disease [N1], life-sustaining treatment [O9], withholding treatment [O10], mentally handicapped [P7], mentally retarded [P8], genetics and applications — biotechnology [Q1], biotechnology — genetic engineering [Q20], medical genetics [Q35].

NL 27 **SCHROTEN** E.

Professor of Christian Ethics, Faculty of Theology, University of Utrecht. — Director of the Centre for Bioethics and Health Law (CBG), University of Utrecht.

Centre for Bioethics and Health Law, University of Utrecht
Heidelberglaan 2
B.P. 80 105
3508 TC Utrecht (Netherlands)
✆ (31) 30-534 399 and 532 072 Fax: (31) 30-533 241

■ TEACHING: Teaching and training of students (University of Utrecht). — Conferences.
■ PUBLICATIONS: yes.

NL 28 **STEVENS** J.A.J. (Hans)

General Practitioner. — Chairperson of the Dutch Association for Health Care and Philosophy of Life. — Member of the Board of the Foundation "Health Care and Ethics".

Thijm Association – Medical Section
Malsenlaan 3
6825 BZ Arnhem (Netherlands)
✆ (31) 85-612 361

■ AFFILIATIONS: Dutch Association for Health Care and Philosophy of Life, Netherlands, (*Chairperson*).
■ TEACHING: Dutch Association for Health Care and Philosophy of Life. — University of the Third Age, Arnhem.
■ PUBLICATIONS: yes.
■ WORKING ON: hospices [G34], organization of health care [G35], health care [G45], family practice [G60], patient association [J3], patient participation [J16], right to treatment [J17], palliative care [O8], attitudes to death [O20].
■ INTERESTED IN: medical ethics [C52], hospices [G34], organization of health care [G35], health care [G45], family practice [G60], patient association [J3], patient participation [J16], right to treatment [J17], palliative care [O8], attitudes to death [O20].

NL 29 **TEN HAVE** Henk A.M.Y.

Professor of Medical Ethics.

Département d'éthique, de philosophie et d'histoire de la Médecine
Université catholique de Nijmegen B.P. 9101
6500 HB Nijmegen (Netherlands)
✆ (31) 80-54 02 54 Fax: (31) 80-54 18 62

■ AFFILIATIONS: European Society for Philosophy of Medicine and Health Care — ESPMH, Nijmegen, Netherlands (*Secretary*).
■ TEACHING: in Faculty of Medical Sciences, Nijmegen (Medical School, School of Health Sciences, School of Dentistry, Academic Hospital Nijmegen).
■ PUBLICATIONS: yes.
■ WORKING ON: data bases [A3], bibliography [A4], publications [A9], editorial policies [A10], human person [B12], human body commercialization [B13], ethics committees [B22], Council of Europe [B25], office of science and technology assessment [B28], common good [B45], professional competence [B52], obligations of society [B53], obligations to society [B54], solidarity [B65], public debates [B75], internship and residency [B81], codes of biomedical ethics [B90], ethical review [B93], bioethical issues [B96], philosophical ethics [C2], Roman Catholic ethics [C23], philosophy of biology [C42], professional ethics [C51], ethics and humanities [C54], social issues [D2], genetics [E6], research policy [E18], pilot projects [E20], evaluation [E61], biomedical research [F2], health [G2], health policy [G13], medical evaluation [G51], preventive medicine [G58], family practice [G60],

medicine [G63], resource allocation [H4], biological specimens procurement, blood transfusion, organ transplantation [I8], treatment [I30], human body [L3].
■ INTERESTED IN: information — communication — media [B68], bioethics [B89], philosophical ethics [C2], religion and religious ethics [C19], professional ethics [C51], ethics and humanities [C54], social issues [D2], biomedical technologies [E28], evaluation [E61], health policy [G13], health facilities [G22], economics [H2], diagnosis [I2], renal dialysis [I32], surgery [I37], rehabilitation [I42], human body [L3], fetal therapy [M57], chronically ill [N7], physically [N8], kidney diseases [N39], terminal care [O7], euthanasia [O15], attitudes to death [O20], genetics and applications — biotechnology [Q1].

NL 30 **VAN ASPEREN** Geertmider M.

Professor of Ethics, Department of Philosophy (University of Amsterdam).
Université d'Amsterdam
Anna Vondelstraat 10
1052 GZ Amsterdam (Netherlands)
✆ (31) 20-52 59 111

■ TEACHING: Courses for medical students on medical ethics. — Courses for philosophy students on medical ethics.
■ PUBLICATIONS: yes.

NL 31 **VAN DEN BELD** A.

Associate Professor of Ethics, Department of Philosophy, Utrecht University.

Centre de bioéthique et de législation sanitaire
Université d'Utrecht
Heidelberglaan 2
3584 CS Utrecht (Netherlands)
✆ (31) 30-534 399 Fax: (31) 30-533 241

■ TEACHING: In the context of courses organised by the Centre for Bioethics and Health Law.
■ PUBLICATIONS: yes.
■ WORKING ON: philosophical ethics [C2], Christian ethics [C22], medical ethics [C52], philosophy [C61], determinism [C62], violence [D85], killing [D90], justifiable killing [D91], human genome project [F8], physicians and researchers accountability [J23], sexuality and procreation — unborn child [M1], death and resuscitation [O1], terminal care [O7], euthanasia [O15], genetics and applications — biotechnology [Q1], biotechnology — genetic engineering [Q20], behavioral genetics [Q34], medical genetics [Q35].
■ INTERESTED IN: *idem*.

NL 32 **VAN DEN BOER- VAN DEN BERG** J.M.A.

Scientific Researcher, Department Metamedica, State University of Leiden. — Teacher in Bioethics for nurses in several hospitals.

Université d'Etat de Leiden
Départment Metamedica
B.P. 2087
2301 CB Leiden (Netherlands)
✆ (31) 71-276 521 or 518 Fax: (31) 71-275 357

■ TEACHING: Students from the University of Leiden, physicians who specialise in hospitals, and nurses. — Training activities (in workshops) for nurses.
■ WORKING ON: genetic defects and hereditary diseases [Q2], euthanasia [O15], medical genetics [Q35], bioethics [B89], professional ethics [C51], medical ethics [C52], nursing ethics [C53], patients' rights [J2], reproductive technologies [M26], terminal care [O7].
■ INTERESTED IN: genetic defects and hereditary diseases [Q2], medical genetics [Q35].

NL 33 **VAN DER AREND** Arie

University teacher, Department of Health Care Ethics and Philosophy, Faculty of Health Sciences.

Ryksuniversiteit Limburg
Post Box 616
B.P. 616
6200 MD Maastricht (Netherlands)
✆ (31) 43-88 11 24 Fax: (31) 43-67 09 32

■ TEACHING: Educational programs in the Faculty of Health Sciences (working groups, lectures, discussion papers). — Clinical Ethics programme in the Medical Faculty (case discussions, lectures).

— Special courses and workshops (outside university).
■ PUBLICATIONS: yes.
■ WORKING ON: nursing ethics [C53], infants [L28], codes of biomedical ethics [B90], codes of ethics [B49].
■ INTERESTED IN: nursing ethics [C53], codes of biomedical ethics [B90], codes of ethics [B49].

NL 34 VAN DER BURG W.

Head of the Research Ethics and Law Centre for Bioethics and Health Law.

Centre de bioéthique et législation sanitaire
Université d'Utrecht
Heidelberglaan2
3584 CS Utrecht (Netherlands)
✆ (31) 30-534 399 Fax: (31) 30-533 241

■ AFFILIATIONS: Societas Ethica (European Research Society for Ethics), Utrecht, Netherlands (*Quaestor*).
■ TEACHING: University courses in department of Philosophy, Theology, Social Sciences, Biology, Medicine.
■ PUBLICATIONS: yes.
■ WORKING ON: human rights [B3], self determination [B4], women's rights [B6], human equality [B7], privacy [B11], traffic of organs [B15], ethics committees [B22], equal protection [B43], codes of ethics [B49], paternalism [B62], respect [B63], universities [B84], wedge argument [C4], metaethics [C11], conscience [C13], speciesism [C48], nursing ethics [C53], legislation [D13], family members [D24], physically handicapped [D44], dissent [D58], state responsibility [D75], democracy [D82], quality of environment [E41], human genome project [F8], animal experimentation [F10], mental institutions [G29], nurses [G39], obstetrics and gynecology [G66], psychiatry [G69], elderly [H23], amniocentesis [I4], chorionic villus sampling [I5], mass screening [I6], donors [I14], transplantation [I22], organ transplantation [I24], fetal tissue transplantation [I27], right to treatment [J17], nurse patient relationship [J30], physician patient relationship [J32], professional patient relationship [J33], potentiality of personhood [K5], minors [K6], child's interest [K24], legal rights [K45], medical law [K51], therapeutic injunction [K64], children's rights [L25], reproductive organs and embryonic structures [M9], procreation [M15], homosexuals [M23], ovum donors [M27], artificial insemination [M31], excess embryos [M34], GIFT (Gamete Intrafallopian Transfer) [M37], embryo donation [M38], mother fetus relationship [M43], selective abortion [M49], acquired immunodeficiency syndrome [N35], genetic defects [Q3], biotechnology [Q21], genetic intervention [Q22], transgenic animals [Q29], genetic counseling [Q36], genetic screening [Q37], preimplantation genetic diagnosis [Q38], gene therapy [Q42].

NL 35 VAN DER KLOOT MEÿBURG H.H.

Director of the N 27 Unit of Bioethics. — General Secretary of the CELAZ.

N 27 – Collège d'éthique et d'aspects religieux (CELAZ)
Oudlaan 4
B.P. 9696
3506 GR Utrecht (Netherlands)
✆ (31) 30-739 911 Fax: (31) 30-739 568

■ AFFILIATIONS: Dutch Association of Bioethics, Netherlands (*Initiator*).
■ TEACHING: Program for Moral Management for Hospital Administrators, for Hospital Ethics Committees. — Program on the Legal and Ethical Aspects of Euthanasia.
■ PUBLICATIONS: yes.

NL 36 VAN DER MEER C.

Chairman of the Committee for Ethics of Scientific Research on Human Subjects, Academic Hospital of the Vrije Universiteit (AZVU), Amsterdam.

Angsteloord 71
1391 ED Abcoude (Netherlands)
✆ (31) 2946 and 3642

■ AFFILIATIONS: Institute for Bioethics, Maastricht, Netherlands (*Member of the Board*).
■ TEACHING: Courses of Bioethics. — Lectures for students of the Faculty of Medicine of the Vrije Universiteit.
■ PUBLICATIONS: yes.
■ WORKING ON: medical ethics [C52], human experimentation [F11], euthanasia [O15], active

euthanasia [O16], involuntary euthanasia [O17], voluntary euthanasia [O18], terminal care [O7].
■ INTERESTED IN: euthanasia [O15].

NL 37 **VAN ES** Adriaan
Medical doctor. — Chairperson of the Johannes-Wier Foundation.

Fondation Johannes-Wier pour les droits humains et les soins médicaux
B.P. 1551
3800 BN Amersfoort (Netherlands)
✆ (31) 33-726 749 Fax: (31) 33-726 811

■ TEACHING: Human Rights education to medical students and postgraduate courses.
■ PUBLICATIONS: yes.

NL 38 **VAN HOUTEN** Douwinus Johannes
Rector, University for Humanist Studies.

Université des Sciences Humaines
Van Asch van Wijckskade 28
3512 VS Utrecht (Netherlands)
✆ (31) 30-36 74 37 Fax: (31) 30-34 07 38

■ TEACHING: Courses (Ethics and Management, System and Welfare).
■ PUBLICATIONS: yes.

NL 39 **VAN LEEUWEN** Evert
Professor Member of the Ethics Committee of the University Hospital.

Department of Metamedica
Faculty of Medicine
Free University
1081 BT Amsterdam (Netherlands)
✆ (31) 20-548 3340 Fax: (31) 20-646 2228

■ AFFILIATIONS: Kennedy Institute (*Member*). — IBI, Union Néerlandaise d'Ethique (*Member*).
■ TEACHING: Clinical ethics to medical students (undergraduate and postgraduate).
■ PUBLICATIONS: yes.
■ WORKING ON: quality of life [C16], value of life [C18], Protestant ethics [C27], biology and human future [C43], future generations [C44], judgement [C71], trust [C81], codes of ethics [B49], education [B76], health education [B77], medical education [B80], internship and residency [B81], ethical analysis [C3], sibling [D30], family relationship [D31], genetics [E6], epidemiology [E9], industrial research [E22], drug industry [E37], decision analysis [E55], human experimentation [F11], random selection [F18], group of vulnerable subjects [F20], population growth [H20], developing countries [H21], population control [H22], elderly [H23], prenatal diagnosis [I3], mass screening [I6], informed consent [J9], consent forms [J14], health care services and aged [L48], aging [L50], fetal therapy [M57], genetic defects [Q3], Huntington's chorea [Q9], Duchenne muscular dystrophy [Q12], cystic fibrosis [Q15], cloning [Q24], genetic counseling [Q36], genetic screening [Q37], preimplantation genetic diagnosis [Q38], positive eugenics [Q41], human genome [Q47].
■ INTERESTED IN: ethical analysis [C3], quality of life [C16], ethical relativism [C10], Protestant ethics [C27], future generations [C44], codes of ethics [B49], sibling [D30], family relationship [D31], genetics [E6], epidemiology [E9], decision analysis [E55], population growth [H20], developing countries [H21], population control [H22], elderly [H23], prenatal diagnosis [I3], mass screening [I6], fetal tissue transplantation [I27], fetal therapy [M57], genetic defects [Q3], Huntington's chorea [Q9], Duchenne muscular dystrophy [Q12], cystic fibrosis [Q15], cloning [Q24], genetic counseling [Q36], genetic screening [Q37], preimplantation genetic diagnosis [Q38], positive eugenics [Q41], human genome [Q47].

NL 40 **VAN TONGEREN** P.
Full Professor of Moral Philosophy, Head of the Department of Philosophy, Member of the Board of the Centre of Ethics of the Catholic University Nijmegen (CECUN).

Faculté de philosophie
Université catholique de Nijmegen
B.P. 9108
6500 HK Nymegen (Netherlands)
✆ (31) 80-612 168 and 612 250 Fax: (31) 80-615 564

■ AFFILIATIONS: Centre of Ethics, Catholic University Nymegen — CECUN, Nymegen, Netherlands (*Staff member, member of the board*).
■ PUBLICATIONS: yes.
■ WORKING ON: fundamental rights of the individuals [B2], ethics committees [B22], manifests and declarations concerning human rights [B33], common good [B45], ethical rules and principles [B47], bioethics [B89], philosophical ethics [C2], normative ethics [C5], teleological ethics [C7], morals [C12], values [C15], Christian ethics [C22], Roman Catholic ethics [C23], Roman catholic ism [C31], professional ethics [C51], philosophy [C61], emotions [C79], personality [C82], family members [D24], cultural pluralism [D51], violence [D85], war [D87], patients' rights [J2], physicians and researchers accountability [J23], inter-personal relationship [J27], child and family [L31], life extension [L47], wish of children [M16], attitudes to death [O20], biotechnology — genetic engineering [Q20].
■ INTERESTED IN: information sources, data bases [A1], fundamental rights of the individuals [B2], ethics committees [B22], manifests and declarations concerning human rights [B33], common good [B45], ethical rules and principles [B47], biomedical ethics education [B78], bioethics [B89], philosophical ethics [C2], normative ethics [C5], teleological ethics [C7], morals [C12], values [C15], Christian ethics [C22], Roman Catholic ethics [C23], Roman catholic ism [C31], professional ethics [C51], philosophy [C61], emotions [C79], personality [C82], family members [D24], cultural pluralism [D51], violence [D85], patients' rights [J2], physicians and researchers accountability [J23], inter-personal relationship [J27], child and family [L31], life extension [L47], sexuality and procreation [M2], reproductive technologies [M26], physically [N8], suffering [N11], pain [N12], suicide [O19], attitudes to death [O20], biotechnology — genetic engineering [Q20].

NL 41 **VEDDER** Anton Herman
Research Assistant for Bioethics.

Centre for Applied Ethics
P.O. Box 1738
3000 DR Rotterdam (Netherlands)
✆ (31) 10-408 11 59 Fax: (31) 10-452 02 04

■ TEACHING: Teaching on AIDS, Human Genetics, Personal Autonomy for Faculties of Philosophy, Theology, Medicine, Management.
■ PUBLICATIONS: yes.
■ WORKING ON: fundamental rights of the individuals [B2], bioethics [B89], philosophical ethics [C2], professional ethics [C51], social issues [D2], political issues [D56], confidentiality [J20], physicians and researchers accountability [J23], sexuality and procreation [M2], communicable diseases [N27], genetic defects and hereditary diseases [Q2].

NL 42 **VELDHUIS** Ruurd
Professor of Philosophy of Religion, Philosophical Ethics, Faculty of Urology. — Professor of Biomedical Ethics, Faculty of Medicine, Université de Groningen.

Faculté de théologie
Université d'Etat de Groningen
Nieuw Kijk in het Jatstraat104
9712 5L Groningen (Netherlands)
✆ (31) 50-635 568 Fax: (31) 50-636 200

■ TEACHING: Lectures to students in Theology (future Chaplain of Hospitals) and Medicine.
■ WORKING ON: fundamental rights of the individuals [B2], ethics committees [B22], ethical rules and principles [B47], solidarity [B65], biomedical ethics education [B78], ethical review [B93], philosophical ethics [C2], Christian ethics [C22], coercion [D5], North-South relationships [D60], political systems [D79], violence [D85], biomedical research and experimentation [F1], nontherapeutic research [F5], therapeutic research [F6], human experimentation [F11], pediatrics [G68], physicians and researchers accountability [J23], newborns [L29], prematurity [L30], handicapped children [L38], infanticide [L42], congenital defects [N23], euthanasia [O15], suicide [O19].

NL 43 **VERBRUGH** H.S.
Lecturer, Philosophy of Medicine.

Département de philosophie, éthique et histoire de la médecine
Université Erasmus
B.P. 1738
3000 DR Rotterdam (Netherlands)
✆ (31) 10-40 87 958

■ TEACHING: medical students.
■ PUBLICATIONS: yes.

NL 44 **VERKEK** M.A.
Philosophy Assistant, Senior Lecturer in Ethics.

Centre d'éthique appliquée, Département de philosophie
Université Erasmus
B.P. 1738
3000 DR Rotterdam (Netherlands)
✆ (31) 10-408 1159

■ AFFILIATIONS: Centre d'éthique appliquée, Rotterdam, Netherlands (*Chairman*).
■ TEACHING: Normal courses in the Department of Philosophy.
■ PUBLICATIONS: yes.

NL 45 **VORSTENBOSCH** J.M.G.
Researcher in Ethics and Philosophy of Technology, State University of Utrecht. — Member of the Animal Experiment Committee of the Veterinary Faculty, State University of Utrecht.

Centre for Bioethics and Health Law
Université d'Utrecht
Heidelberglaan 2
3584 CS Utrecht (Netherlands)
✆ (31) 30-53 42 80 Fax: (31) 30-53 32 41

■ AFFILIATIONS: Centre for Bioethics and Health Law, State University of Utrecht, Utrecht, Netherlands (*Leader of the Research Group on Animal Biotechnology*).
■ TEACHING: In academic context, courses on bioethics to students in several sciences.
■ PUBLICATIONS: yes.
■ WORKING ON: ethical analysis [C3], normative ethics [C5], public policy [D77], informed consent [J9], biotechnology — genetic engineering [Q20], transgenic animals [Q29].

NL 46 **WELIE** Jos V. M.
Executive Director, International Program in Bioethics Education and Research.

Université Catholique de Nijmegen
Département d'éthique, de philosophie et d'histoire de la médecine
B.P. 9101
6500 HB Nijmegen (Netherlands)
✆ (31) 80-540 254

■ AFFILIATIONS: European Society for Philosophy of Medicine and Health Care — ESPMH, Nijmegen, Netherlands (*Member*).
■ TEACHING: European Intensive Bioethics Seminar (annually), Catholic University of Nijmegen.
■ PUBLICATIONS: yes.
■ WORKING ON: data bases [A3], literature [A7], documentation [A8], publications [A9], human rights [B3], human person [B12], Council of Europe [B25], ethical rules and principles [B47], bioethical issues [B96], philosophical ethics [C2], Roman catholic ism [C31], medical ethics [C52], philosophy [C61], personality [C82], legislation [D13], decision making [E54], mental health [G4], health legislation [G14], biological substances contamination [I9], patients' rights [J2], confidentiality [J20], professional deontology [J25], physician's role [J26], medical etiquette [J28], physician patient relationship [J32], professional patient relationship [J33], legal personality [K2], conflict of interest [K9], battery [K10], wrongful death [K12], misconduct [K23], legal liability [K30], judicial action [K35], contracts [K39], medical law [K51], criminal law [K53], disease and aged [L46], life extension [L47], chronically ill [N7], death and resuscitation [O1].
■ INTERESTED IN: death and resuscitation [O1].

NL 47 **WIDDERSHOVEN** Guy
Associate Professor in Philosophy, Head of the Department of Medical Ethics.

Department of Health Ethics
Université de Limburg
B.P. 616
6200 MD Maastricht (Netherlands)
✆ (31) 43-88 11 44 Fax: (31) 43-67 09 32

■ TEACHING: Coordinator of Health Science Theory (graduate programme). — Clinical Ethics courses.
■ PUBLICATIONS: yes.
■ WORKING ON: deontology [B59], solidarity [B65], biomedical ethics education [B78], deontological

ethics [C6], teleological ethics [C7], moral development [C14], values [C15], humanities [C55], history [C56], existentialism [C64], cultural pluralism [D51], sociology of medicine [D55], decision making [E54], decision analysis [E55], mental health [G4], patient compliance [G49], diagnosis [I2], prenatal diagnosis [I3], patient information [J4], inter-personal relationship [J27], physician patient relationship [J32], case studies [N4], alcohol abuse [P4], psychotic disorders [P10], psychotherapy [P30].
■ INTERESTED IN: biomedical ethics education [B78], ethics and humanities [C54], inter-personal relationship [J27].

NL 48 WINNUBST-VANDEN BERGHE J.M.

Junior Investigator on the Ethical Aspects of Medical Genetics (with a special concern for the Interest of the fetus).

Département Metamedica
Université d'Etat de Leiden
B.P..2087
2301 CB Leiden (Netherlands)
✆ (31) 71-276 520 and 276 518 Fax: (31) 71-275 357

■ AFFILIATIONS: Vereniging van Ethici in Nederland — VVEN, Amsterdam, Netherlands (*Member*).
■ TEACHING: Lectures for medical and other students.
■ PUBLICATIONS: yes.
■ WORKING ON: medical genetics [Q35], prenatal diagnosis [I3], selective abortion [M49], wrongful life [K33].
■ INTERESTED IN: medical genetics [Q35], prenatal diagnosis [I3], selective abortion [M49].

NL 49 ZWART H.A.E.

University teacher, Research Associate, Ethics Committee Secretary.

Catholic University of Nijmegen (KUN)
School of Philosophy
P.O. Box 9108
6500 HK Nijmegen (Netherlands)
Fax: (31) 80-61 55 64

■ AFFILIATIONS: European Society for Philosophical Medicine and Health Care — ESPMH, Nijmegen, Netherlands (*Member*).
■ TEACHING: Ethics of psychological intervention and research.
■ PUBLICATIONS: yes.
■ WORKING ON: self determination [B4], human person [B12], ethics [B48], moral obligations [B50], competence [B51], obligations to society [B54], moral policy [B55], respect [B63], communication [B69], public debates [B75], biomedical ethics education [B78], ethicists [B92], ethical review [B93], history of biomedical ethics [B94], philosophical ethics [C2], natural law [C24], humanism [C66], psychology [C68], social control [D3], exclusion [D9], regulation [D16], cultural pluralism [D51], health policy [G13], health care [G45], blood donation [I20], aged — problems related with aging [L45], death and resuscitation [O1], death [O5], terminal care [O7], withholding treatment [O10], allowing to die [O11], terminally ill [O12], prolongation of life [O13], quality adjusted life years [O14], euthanasia [O15].
■ INTERESTED IN: self determination [B4], human person [B12], ethics [B48], moral obligations [B50], competence [B51], obligations to society [B54], moral policy [B55], respect [B63], communication [B69], public debates [B75], biomedical ethics education [B78], ethicists [B92], ethical review [B93], history of biomedical ethics [B94], philosophical ethics [C2], natural law [C24], Protestant ethics [C27], humanism [C66], psychology [C68], social control [D3], exclusion [D9], regulation [D16], cultural pluralism [D51], health policy [G13], health care [G45], blood donation [I20], aged — problems related with aging [L45], death and resuscitation [O1], death [O5], terminal care [O7], withholding treatment [O10], allowing to die [O11], terminally ill [O12], prolongation of life [O13], quality adjusted life years [O14], euthanasia [O15].

Portugal

Area: 92 072 km²
Population: 10,5m
Gross domestic product: Esc 11.2trn; US$ 75.8bn
Gross domestic product per head: $ 7,240
Gross domestic product growth: 1991: 2.9%
1992: 3.5%

Organisations

O P 1 **CENTRE DE DROIT BIOMÉDICAL**
Faculté de Droit
Rua Machado de Castro 173, 4° Dt
3049 Coimbra Corex (Portugal)
✆ (351) 39 22 113/4/5/6 Fax: (351) 39 23 353

O P 2 **CENTRO DE ESTUDOS DE BIOÉTICA (COIMBRA)**
Instituto Justiça e Paz
Rua Gen. Humberto Delgado 444-7° DTO
3000 Coimbra (Portugal)
✆ (351) 39 71 28 06

O P 3 **CONSELHO NACIONAL DE ETICA E DE DEONTOLOGIA DA ORDEM DOS MÉDICOS**
Av. Almirante Gago Coutinho 151
1700 Lisboa (Portugal)
✆ (351) 1-847 06 54 and 39-71 28 06 Fax: (351) 1-847 12 15

O P 4 **CONSELHO NACIONAL DE ETICA PARA AS CIENCIAS DA VIDA**
Presidencia do Conselho de Ministros
Rua Professor Gomes Teixeira
1300 Lisboa (Portugal)
✆ (351) 1-52 87 28 and 52 94 03 Fax: (351) 1-57 08 16

■ OFFICIALLY REGULATED ORGANISATION: Foundation law n° 14/90, 1990.
■ DATE OF FOUNDATION: 9/06/1990 and 30/01/1991.
■ EXECUTIVE BOARD: Dr Mario Radoso (*Chairperson*), *Members :* Pr Yosé Rueff Tavares, Pr Antonio Barbosa de Melo, Pr Michel Renaud, Pr Manuel Braga da Cruz, Pr Lucio Craveiro da Silva, Dr Paula Martinho da Silva, Dr Maria Ivone Leal, Pr Antonio Dereira Coelho, Pr José Esperança Dina, Pr Daniel Serraõ, Pr Manuel Malhado Maledo, Pr José Toscana Rico, Pr Joaõ Queiróz e Helo, Pr José Pinto da Costa, Pr Luis Fircher, Pr Joachim Pintomachado, Mme Maria de Lundes Pintassilgo, Pr Victor Pinto, Pr António Falcaô de Freitas, Pr Joachim Cerqueira Gonçalves.
■ PUBLICATIONS: yes.
■ DOCUMENTATION CENTRE: open to the public.
■ COLLOQUIUMS, SYMPOSIUMS: colloquium on informed consent and medical responsibility, march 1992.
■ DECISIONS, ADVICE: advisory notifications on organ tranplantations (for two bills). — Advisory notices on the use of cadavers in research and education. — Project: "Advisory Notice on Assisted Medical Reproduction".
■ WORKING ON: information sources, data bases [A1], review [A2], information centers [A5], documentation [A8], publications [A9], respect of human dignity and human rights [B1], fundamental rights of the individuals [B2], self determination [B4], dignity [B5], privacy [B11], human person [B12], human body commercialization [B13], protection of rights — involved institutions [B16], CCNE — Comité Consultatif National d'Ethique (France) [B23], due process [B40], public advocacy [B41], patient advocacy [B42], equal protection [B43], ethical rules and principles [B47], information — communication — media [B68], communication [B69], interdisciplinary communication [B70], information dissemination [B71], mass media [B72], press conference [B73], public debates [B75], bioethics [B89], CCNE advice (France) [B91], bioethical issues [B96], professional ethics [C51], medical ethics [C52], social and political issues [D1], legislation [D13], science, technology, methods [E1], life sciences [E4], biology [E5], research [E14], progress [E26], technology [E27], biomedical technologies [E28], biological specimens procurement, blood transfusion, organ transplantation [I8], sperm banks [I11], donors [I14], organ donors [I15], donor cards [I16], anonymous donation [I17], directed donation [I18], blood transfusions [I19], blood donation [I20], blood substitutes [I21], transplantation [I22], transplant recipients [I23], organ transplantation [I24], organ donation [I25], organ procurement [I26], fetal tissue transplantation [I27], intracerebral fetal tissue transplantation [I28], bone marrow transplantation [I29], health care professionals, researchers and patient relationship [J1], confidentiality [J20], physicians and researchers accountability [J23], inter-personal relationship [J27], procreation [M15], wish of children [M16], reproduction [M17], fertility [M18], infertility [M19], reproductive technologies [M26], ovum donors [M27], semen donors [M28], in vitro

fertilization [M29], FIV center [M30], artificial insemination [M31], AID [M32], AIH [M33], excess embryos [M34], host mothers [M35], embryo transfer [M36], GIFT (Gamete Intrafallopian Transfer) [M37], embryo donation [M38], biotechnology — genetic engineering [Q20], biotechnology [Q21], genetically modified organisms [Q28].
■ INTERESTED IN: information sources, data bases [A1], data bases [A3], bibliography [A4], documentation [A8], publications [A9], respect of human dignity and human rights [B1], fundamental rights of the individuals [B2], protection of rights — involved institutions [B16], ethical rules and principles [B47], information — communication — media [B68], bioethics [B89], social and political issues [D1], science, technology, methods [E1], life sciences [E4], research [E14], progress [E26], technology [E27], biomedical technologies [E28], biological specimens procurement, blood transfusion, organ transplantation [I8], sperm banks [I11], donors [I14], organ donors [I15], donor cards [I16], anonymous donation [I17], directed donation [I18], blood transfusions [I19], blood donation [I20], blood substitutes [I21], transplantation [I22], transplant recipients [I23], organ transplantation [I24], organ donation [I25], organ procurement [I26], fetal tissue transplantation [I27], intracerebral fetal tissue transplantation [I28], bone marrow transplantation [I29], treatment [I30], anesthesia [I31], renal dialysis [I32], alternative therapies [I33], drugs [I34], procreation [M15], wish of children [M16], reproduction [M17], fertility [M18], infertility [M19], sexuality [M20], sexual behavior [M21], homosexuality [M22], homosexuals [M23], transsexualism [M24], prostitution [M25], reproductive technologies [M26], ovum donors [M27], semen donors [M28], in vitro fertilization [M29], reproductive technologies [M26], ovum donors [M27], semen donors [M28], in vitro fertilization [M29], FIV center [M30], artificial insemination [M31], AID [M32], AIH [M33], excess embryos [M34], host mothers [M35], embryo transfer [M36], GIFT (Gamete Intrafallopian Transfer) [M37], embryo donation [M38], biotechnology — genetic engineering [Q20], biotechnology [Q21], genetically modified organisms [Q28].

Individuals

P 1 **AGUIAR** Antonio José
Head of Department.

Hospital militar de Coimbra
Rua Rodrigues de Gusmão 6
3000 Coimbra (Portugal)
✆ (351) 39 48 22 97 Fax: (351) 39 40 30 80

■ AFFILIATIONS: Centro de Estudos de Bioética, Coimbra, Portugal (*Member*).

P 2 **ARCHER** Luis Jorge Peixeto
Full Professor of Molecular Genetics, Head of the Department of Biotechnology, School of Sciences and Technology of the New University of Lisbon.

Faculdade de Ciências e Tecnologia
Quinta da Torre
2825 Monté Da Caparica (Portugal)
✆ (351) 1-295 44 64 Fax: (351) 1-295 44 61

■ AFFILIATIONS: Conselho Nacional de Etica para as Ciências da Vida, Lisboa, Portugal (*Member*).
■ TEACHING: optional course on Biosafety and Bioethics, School of Sciences and Technology. — Course on Human Person and Life (integrated in a Master Course of the Catholic University). — Lectures on Ethics of artificial reproduction in specialized courses for nurses.
■ PUBLICATIONS: yes.
■ WORKING ON: self determination [B4], freedom [B10], determinism [C62], genetics [E6], molecular biology [E8], human genome project [F8], reproductive technologies [M26], genome mapping [Q23], recombinant DNA research [Q31], gene therapy [Q42].
■ INTERESTED IN: genetics [E6], molecular biology [E8], human genome project [F8], genome mapping [Q23], recombinant DNA research [Q31], DNA fingerprinting [Q32], gene therapy [Q42].

P 3 **BARRETO** João
Director of the Department of Medical Psychology, Faculty of Medicine of Porto. — Professor, Faculty of Medicine of Porto. — Consultant in psychiatry, Hospital of S. João.

Hospital de S. João
Faculté de Médecine
Rua da Graciosa 82- 2°
4000 Porto (Portugal)
✆ (351) 2-48 39 63 Fax: (351) 2-49 81 19

■ AFFILIATIONS: Centro de Estudos de Bioética, Coimbra, Portugal (*Member*).
■ PUBLICATIONS: yes.
■ WORKING ON: aging [L50], smoking [P5], dementia [P11].
■ INTERESTED IN: epidemiology [E9], mental health [G4], psychiatry [G69], legal liability [K30], aging [L50], voluntary euthanasia [O18], smoking [P5], dementia [P11].

P 4 **BERNARDO, O.P.** Frei

Residencia Dominicana
Rua D. Afonso V
4100 Porto (Portugal)
✆ (351) 2-618 11 55

■ TEACHING: Faculty of Theology, Nursing School, Porto.
■ PUBLICATIONS: yes.
■ WORKING ON: respect of human dignity and human rights [B1], specific approaches to ethics [C1], social and political issues [D1], biomedical research and experimentation [F1], health care professionals, researchers and patient relationship [J1], stages of life — problems proper to childhood and elderly [L1], sexuality and procreation — unborn child [M1], disease [N1], death and resuscitation [O1].

P 5 **BISCAIA** Jorge
Head of the Department of Neurology, Maternity ward, Hospital Benayer Barret.— Committee of Bioethics.

Rua General Humberto Delgado 444-7 Dt,
3000 Coimbra (Portugal)

■ AFFILIATIONS: Centro de Estudos de Bioética, Coimbra, Portugal (*Chairperson*).
■ TEACHING: yes.
■ PUBLICATIONS: yes.

P 6 **BORGES** Anselmo
Catholic priest. — Teaching Philosophy at the University of Coimbra.

Seminario da Boa Nova
Apartado 10
4408 Valadares (Portugal)
✆ (351) 2-7110502

■ AFFILIATIONS: Centro de Estudos de Bioética, Coimbra, Portugal (*Member*).
■ TEACHING: yes.
■ PUBLICATIONS: yes.
■ WORKING ON: human person [B12], ethics [B48], morals [C12], religious ethics [C21], biology and human future [C43], attitudes to death [O20].
■ INTERESTED IN: bibliography [A4], death [O5], attitudes to death [O20], resuscitation [O21].

P 7 **CABRAL P.** Roque
Professor of philosophy.

Faculdade de Filosofia da UCP
Rua de S. Barnabé 42
4710 Braga Codex (Portugal)
✆ (351) 53 61 62 00 Fax: (351) 53 61 56 31

■ AFFILIATIONS: Centro de Estudos de Bioética, Coimbra, Portugal (*Member*).
■ TEACHING: specialized ethics, Faculty of Philosophy (Braga), special course, Ecole supérieure de biotechnologie (Porto).
■ WORKING ON: human rights [B3], ethical rules and principles [B47], philosophical ethics [C2], natural law [C24], double effect [C46], family members [D24], reproductive technologies [M26], abortion [M47], euthanasia [O15].

■ INTERESTED IN: bioethical issues [B96], biomedical research [F2], biological specimens procurement, blood transfusion, organ transplantation [I8], sperm banks [I11], organ procurement [I26], abortion [M47], terminal care [O7], euthanasia [O15].

P 8 **CERQUEIRA GONCALVES** Joaquim
Professor of philosophy, Faculdade de Letras de Lisboa, Universidade Portuguesa.

Conselho Nacional de Etica para as Ciências da Vida
Rua Prof. Gomes Teixeira
1300 Lisboa (Portugal)
✆ (351) 1-60 95 41

P 9 **COSTA** José Manuel Cardoso da
Lecturer, Faculty of Law, University of Coimbra. — Chairperson of the Constitutional Court of Portugal.

Faculté de droit de l'université de Coimbra
Rua Machado de Castro 173, 4° Dt
3000 Coimbra (Portugal)
✆ (351) 39-36 559

■ AFFILIATIONS: Centro de Estudos de Bioética, Coimbra, Portugal (*Member*).
■ PUBLICATIONS: yes.
■ WORKING ON: fundamental rights of the individuals [B2], constitutional law [K48], criminal code [K54], biotechnology — genetic engineering [Q20].
■ INTERESTED IN: Council of Europe [B25], European Convention on Human Rights [B36], European Convention of Bioethics [B97], constitutional amendments [K36], biotechnology — genetic engineering [Q20].

P 10 **FARIA** Rui
Head of the Department of Neurosurgery, Hospital of S. João.

Hospital de S. João
Asprela
4200 Porto (Portugal)
✆ (351) 2-48715

■ AFFILIATIONS: Centro de Estudos de Bioética, Coimbra, Portugal (*Member*).
■ WORKING ON: organ transplantation [I24], persistent vegetative state [N42], brain death [O6], withholding treatment [O10], prolongation of life [O13], euthanasia [O15].
■ INTERESTED IN: medical ethics [C52], advance directives [J18], living wills [J19].

P 11 **FORMOSINHO** Sebastião
Full Professor of Chemistry.

Departamento de Quimica
Universidade de Coimbra
3049 Coimbra cedex (Portugal)
✆ (351) 39-22 826

■ WORKING ON: education [B76].
■ INTERESTED IN: education [B76], scientific misconduct [E25], teleological ethics [C7], Christian ethics [C22], natural law [C24].

P 12 **FREITAS** Antonio
Chairperson of Internal Medicine Department, Hospital S. João, Porto. — Full Professor of Medicine, School of Medicine, Porto University.

Faculté de médecine de l'université de Porto
Rua Paulo da Gama n°420-1° C
4100 Porto (Portugal)
✆ (351) 2-61 80 987 Fax: (351) 2-48 55 12

■ WORKING ON: professional ethics [C51], health occupations [G61], diagnosis [I2], treatment [I30], physicians and researchers accountability [J23], cardiovascular diseases [N19].
■ INTERESTED IN: determination of death [O2], terminal care [O7], medical genetics [Q35].

P 13 **GAMEIRO** Aires

Head of Psychology and Health Pastoral Work Departments.

Instituto S. João de Deus
Casa de Saude do Telhal
2725 Mem Martins (Portugal)
✆ (351) 916 30 21 Fax: (351) 916 59 16

■ AFFILIATIONS: Secretariado de Humanização e pastoral, Mem Martins, Netherlands (*Member*). — Comissão Geral de Animação da Ordem Hospitaleira de S. João de Deus (Roma), Italy (*Member*). — Comissão Nacional da Pastoral da Saude (Lisboa), Lisboa, Portugal (*Member*).
■ TEACHING: Ongoing training sessions, lectures, conferences and seminars.
■ PUBLICATIONS: yes.
■ WORKING ON: human person [B12], obligations to society [B54], virtues [B64], health education [B77], values [C15], theology [C41], self concept [C83], pastoral care [C88], family members [D24], mental health [G4], mental institutions [G29], alternative therapies [I33], personhood [K3], mind [L20], aging [L50], suffering [N11], palliative care [O8], suicide [O19], substance dependence [P3], outpatient commitment [P18].
■ INTERESTED IN: mind [L20], substance dependence [P3], human person [B12], obligations to society [B54], values [C15], theology [C41], self concept [C83], pastoral care [C88], family members [D24], personhood [K3].

P 14 **LEANDRO** Armando

Judge, Court of Appeal. — Director of the Centro de Estudios Judiciarios (Education of the Magistrates).

Centro de Estudios Judiciarios
Largo do Limoeiro
1100 Lisboa (Portugal)
✆ (351) 1-87 47 13

■ TEACHING: bioethics and bioethics law to magistrates (artificial procreation, genetics experimentation on germ cells, transsexualism, euthanasia).
■ WORKING ON: fundamental rights of the individuals [B2], ethical rules and principles [B47], patients' rights [J2], jurisprudence — accountability [K8], legislation and law [K34].
■ INTERESTED IN: professional ethics [C51], social issues [D2], determination of death [O2], euthanasia [O15], teleological ethics [C7].

P 15 **LOPES-CARDOSO** Augusto

Barrister-at-law.

Rua Antonio Cardoso n°475 Apt 10
4100 Porto (Portugal)
✆ (351) 2-60 66 075 Fax: (351) 2-30 80 09

■ AFFILIATIONS: Centro de Estudos de Bioética, Coimbra, Portugal (*Member*).
■ PUBLICATIONS: yes.
■ WORKING ON: fundamental rights of the individuals [B2], physicians and researchers accountability [J23], legal personality [K2], jurisprudence — accountability [K8], legislation and law [K34], reproductive technologies [M26], communicable diseases [N27], acquired immunodeficiency syndrome [N35], HIV seropositivity [N36], terminal care [O7], euthanasia [O15], suicide [O19].
■ INTERESTED IN: fundamental rights of the individuals [B2], due process [B40], deontology [B59], physicians and researchers accountability [J23], legal personality [K2], jurisprudence — accountability [K8], legislation and law [K34], reproductive technologies [M26], communicable diseases [N27], acquired immunodeficiency syndrome [N35], HIV seropositivity [N36], terminal care [O7], euthanasia [O15], suicide [O19], sexuality and procreation [M2], reproductive organs and embryonic structures [M9], embryos [M10], fetuses [M11], ovum [M12], biomedical research [F2], human experimentation [F11].

P 16 **MARQUES** Adelino

Head of the Department of Nephrology, University Hospital of Coimbra, Professor of nephrology and Medical Deontology, Faculty of Medicine, University of Coimbra.

Universidade de Coimbra
R. Carolina Michaelis 78 -2°
3000 Coimbra (Portugal)
✆ (351) 39-40 39 39 Fax: (351) 39-23 907

■ AFFILIATIONS: Conselho Nacional de Etica e de Deontologia da Ordem dos Médicos (Conseil national d'éthique et de déontologie de l'Ordre des médecins), Lisboa, Portugal (*Member*). — Centro de Estudos de Bioética, Coimbra, Portugal (*Member*).
■ TEACHING: Deontology to undergraduates.
■ WORKING ON: ethics committees [B22], deontology [B59], code of deontology [B60].
■ INTERESTED IN: World Medical Assembly [B20], ethics committees [B22], Helsinki Declaration [B37], deontology [B59], code of deontology [B60], codes of biomedical ethics [B90], Jehovah's witnesses [C38], medical ethics [C52], transplantation [I22], renal dialysis [I32], terminal care [O7], euthanasia [O15].

P 17 **MARTINHO DA SILVA** Dra Paula

Barrister-at-law, Conseil de Lisbonne (Bar). — Member of the Conselho Nacional de Etica para as Ciências da Vida.

Avenida Luis Bivar 93 -8° Esq.
1000 Lisboa (Portugal)
✆ (351) 1-356 00 11 Fax: (351) 1-526 792

■ AFFILIATIONS: Conselho Nacional de Etica para as Ciências da Vida — CNE, Lisboa, Portugal (*Member*).
■ PUBLICATIONS: yes.
■ WORKING ON: review [A2], fundamental rights of the individuals [B2], privacy [B11], human person [B12], ethics committees [B22], ethical rules and principles [B47], ethics [B48], biological specimens procurement, blood transfusion, organ transplantation [I8], required request [I10], sperm banks [I11], donors [I14], organ donors [I15], donor cards [I16], anonymous donation [I17], transplantation [I22], transplant recipients [I23], organ transplantation [I24], organ donation [I25], organ procurement [I26], health care professionals, researchers and patient relationship [J1], patients' rights [J2], patient information [J4], patient access [J5], parental notification [J6], disclosure [J7], treatment refusal [J15], confidentiality [J20], physicians and researchers accountability [J23], inter-personal relationship [J27], physician patient relationship [J32], law, legislation and jurisprudence [K1], jurisprudence — accountability [K8], legitimacy [K21], legislation and law [K34], law [K56], law enforcement [K57], legal aspects [K58], bill [K60], stages of life — problems proper to childhood and elderly [L1], anatomy, physiology, development [L2], body parts and fluids [L4], aid children [L34], sexuality and procreation — unborn child [M1], sexuality and procreation [M2], embryos [M10], sperm [M14], procreation [M15], wish of children [M16], reproductive technologies [M26], ovum donors [M27], abortion on demand [M52], disease [N1], death and resuscitation [O1], genetics and applications — biotechnology [Q1].
■ INTERESTED IN: data bases [A3], bibliography [A4], information centers [A5], literature [A7], documentation [A8], publications [A9], editorial policies [A10], fundamental rights of the individuals [B2], human rights [B3], self determination [B4], women's rights [B6], privacy [B11], human person [B12], ethics committees [B22], CCNE — Comité Consultatif National d'Ethique (France) [B23], Council of Europe [B25], United Nations [B29], International Court [B30], CCPPRB — French committees for the protection of human subjects of biomedical research [B32], humanity heritage [B46], ethical rules and principles [B47], respect [B63], information — communication — media [B68], uncontrolled information [B88], bioethics [B89], codes of biomedical ethics [B90], bioethical issues [B96], European Convention of Bioethics [B97], Bioethics bill, 1992 (France) [B98], social issues [D2], legislation [D13], model legislation [D14], couple [D26], marital relationship [D29], family relationship [D31], parents [D32], mothers [D33], fathers [D34], females [D35], minority groups [D42], handicapped [D43], political issues [D56], EC — European Communities [D62], ECC — European Community Commission [D63], containment [E38], data protection [E52], biomedical research [F2], human genome project [F8], experimentation [F9], human experimentation [F11], biological specimens procurement, blood transfusion, organ transplantation [I8], required request [I10], donors [I14], organ donors [I15], transplantation [I22], organ transplantation [I24], health care professionals, researchers and patient relationship [J1], law, legislation and jurisprudence [K1], stages of life — problems proper to childhood and elderly [L1], sexuality and procreation — unborn child [M1], disease [N1], genetics and applications — biotechnology [Q1].

P 18 **MENDES** Jose Pinto

Head of the Department of Immuno-Allergology, Consultant, Department of Pneumology, University Hospital of Coimbra.

Rua Padre Americo, 62
3000 Coimbra (Portugal)
✆ (351) 39-723 924

■ AFFILIATIONS: Centro de Estudos de Bioética, Coimbra, Portugal (*Member*).
■ PUBLICATIONS: yes.

■ INTERESTED IN: health and biology mediatisation [B85], uncontrolled information [B88], biology and human future [C43], future generations [C44], medical ethics [C52], behavior [C73], social control [D3], biology [E5], genetics [E6], biomedical technologies [E28], clinical trials [F3], therapeutic research [F6], human experimentation [F11], patients' rights [J2], confidentiality [J20], inter-personal relationship [J27], sexuality and procreation [M2], reproductive technologies [M26], pregnancy and childbirth [M39], abortion [M47], fetal therapy [M57], determination of death [O2], terminal care [O7], euthanasia [O15], genetics and applications — biotechnology [Q1], biotechnology — genetic engineering [Q20], behavioral genetics [Q34], medical genetics [Q35], population genetics [Q43].

P 19 **NUNES** Rui
Assistant Professor.

Faculdade de Medicina
9925 Circunvalaçao
4200 Porto (Portugal)
✆ (351) 2-81 77 67 Fax: (351) 2-83 01 047

■ AFFILIATIONS: Centro de Estudos de Bioética, Coimbra, Portugal (*Member*).
■ TEACHING: Medicine.
■ WORKING ON: human genome project [F8], terminal care [O7], euthanasia [O15], active euthanasia [O16], involuntary euthanasia [O17], voluntary euthanasia [O18], genetic intervention [Q22], genome mapping [Q23], gene therapy [Q42], human genome [Q47].
■ INTERESTED IN: genetic counseling [Q36], genetic screening [Q37], eugenics [Q39], positive eugenics [Q41], gene therapy [Q42], carriers [Q46], human genome [Q47], ethics [B48], codes of ethics [B49], deontology [B59], code of deontology [B60], bioethics [B89], codes of biomedical ethics [B90], ethicists [B92], history of biomedical ethics [B94], bioethics movement [B95], European Convention of Bioethics [B97], deontological ethics [C6], values [C15], quality of life [C16], medical ethics [C52], EC — European Communities [D62], ECC — European Community Commission [D63], government and political systems [D64], state government [D67], biology [E5], genetics [E6], molecular biology [E8], biomedical technologies [E28], drug industry [E37], biological containment [E39], clinical trials [F3], therapeutic research [F6], human genome project [F8], animal experimentation [F10], human experimentation [F11], health legislation [G14], resource allocation [H4], costs and benefits [H10], medical secrecy [J22], privileged communication [J21], professional deontology [J25], physician patient relationship [J32], reproductive technologies [M26], death [O5], brain death [O6], terminal care [O7], euthanasia [O15], active euthanasia [O16], involuntary euthanasia [O17], voluntary euthanasia [O18], suicide [O19], attitudes to death [O20], resuscitation [O21], genetic intervention [Q22], genome mapping [Q23], cloning [Q24], clones [Q25], genetically modified organisms [Q28], transgenic animals [Q29], sex preselection [Q30], recombinant DNA research [Q31], DNA linkage [Q33], medical genetics [Q35].

P 20 **OLIVEIRA** Guilherme
Professeor, Faculty of Law, Coimbra, Research Fellow, founder of the Centre de droit biomédical de la faculté de droit.

Universidade de Coimbra
Faculté de droit
3049 Coimbra Corex (Portugal)
✆ (351) 39-22 113/4/5/6 Fax: (351) 39-23 353

■ AFFILIATIONS: Centre de droit biomédical, Coimbra, Portugal (*Chairperson*).
■ PUBLICATIONS: yes.
■ WORKING ON: human body commercialization [B13], reification [B14], CCNE advice (France) [B91], bioethics movement [B95], Bioethics bill, 1992 (France) [B98], patient information [J4], parental notification [J6], living wills [J19], medical law [K51], ovum donors [M27], semen donors [M28], AID [M32], host mothers [M35], embryo donation [M38].
■ INTERESTED IN: data bases [A3], documentation [A8], human genome project [F8], mandatory programs [G19], voluntary programs [G21], medical records [G50], prenatal diagnosis [I3], mass screening [I6], mandatory screening [I7], physician patient relationship [J32], legally incompetent person [K4], medical law [K51], aid children [L34], involuntary sterilization [M7], AID [M32], genome mapping [Q23], genetic counseling [Q36], genetic screening [Q37], gene therapy [Q42], human genome [Q47].

P 21 **PEDROSA** P. Léal
Chaplain, University of Coimbra. — Spiritual help for Catholic priests.

Monte Formoso Praceta Cidade de Salamanca 15- 5°
3000 Coimbra (Portugal)
✆ (351) 39-491 793

■ AFFILIATIONS: Centro de Estudos de Bioética (Coimbra) — CEB, Coimbra, Portugal (*Member*).
■ WORKING ON: respect of human dignity and human rights [B1], specific approaches to ethics [C1], health care professionals, researchers and patient relationship [J1], sexuality and procreation — unborn child [M1], death and resuscitation [O1].
■ INTERESTED IN: fundamental rights of the individuals [B2], dignity [B5], freedom [B10], human person [B12], ethical relativism [C10], conscience [C13], quality of life [C16], value of life [C18], patient information [J4], homosexuality [M22], selective abortion [M49], therapeutic abortion [M50], abortion on demand [M52], euthanasia [O15], active euthanasia [O16], involuntary euthanasia [O17], voluntary euthanasia [O18], attitudes to death [O20].

P 22 **PINTO MACHADO** Joaquim
Full professor of Anatomy. — Director of the Institute of Anatomy, Medical School, Porto.

Faculté de Médecine de Porto
Avenida da Boa Vista n°431
Alameda Professor Hernâni Monteiro
4100 Porto (Portugal)
✆ (351) 2-49 68 08

■ AFFILIATIONS: Conselho Nacional de Etica para as Ciências da Vida, Lisboa, Portugal (*Member*).
■ PUBLICATIONS: yes.
■ WORKING ON: review [A2], manifests and declarations concerning human rights [B33], competence [B51], professional competence [B52], obligations of society [B53], obligations to society [B54], code of deontology [B60], solidarity [B65], biomedical ethics education [B78], medical education [B80], internship and residency [B81], students [B82], curriculum [B83], universities [B84], codes of biomedical ethics [B90], values [C15], quality of life [C16], value of life [C18], biology and human future [C43], medical ethics [C52], biology [E5], clinical trials [F3], human genome project [F8], human experimentation [F11], family planning [G8], health care costs [H9], costs and benefits [H10], risks and benefices [H11], human body [L3], reproductive organs and embryonic structures [M9], sexuality [M20], abortion on demand [M52], drug abuse [P6].
■ INTERESTED IN: health care costs [H9], costs and benefits [H10], risks and benefices [H11], health care services and aged [L48], fetal therapy [M57], drug abuse [P6].

P 23 **PORTO** Armando
Head of the Department of Internal Medicine, HUC Coimbra.

Faculdade de Medicina
HUC
3000 Coimbra (Portugal)
✆ (351) 39-40 39 39 Fax: (351) 39-23 907

■ AFFILIATIONS: Centro de Estudos de Bioética, Coimbra, Portugal (*Member*).
■ TEACHING: yes.
■ PUBLICATIONS: yes.
■ WORKING ON: professional ethics [C51], treatment [I30], attitudes to death [O20].
■ INTERESTED IN: fundamental rights of the individuals [B2], professional ethics [C51], social issues [D2], methodology of research and experimentation [F15], treatment [I30], attitudes to death [O20].

P 24 **RAPOSO** Mario
Barrister-at-Law. — University teacher.

Cabinet d'avocat
Rua Rodrigo da Fonseca, 149-3°
1000 Lisboa (Portugal)
✆ (351) 1-388 72 50 Fax: (351) 387 47 76

■ AFFILIATIONS: Conselho Nacional de Etica para as Ciências da Vida, Lisboa, Portugal (*Chairperson*).
■ PUBLICATIONS: yes.
■ WORKING ON: malpractice [K19], negligence [K20], legal liability [K30], ghost surgery [K31], therapeutic risk [K32], wrongful life [K33].
■ INTERESTED IN: insurance [H5], health insurance [H6], national health insurance [H7], life insurance [H8], legal personality [K2], expert testimony [K18].

P 25 **RENAUD** Maria Isabel
Professor, Faculty of Social Sciences and Humanities, New University of Lisbon, Director of the Philosophy Department. — Visiting Professor, Faculty of Human Sciences, Universidade Catolica

Portuguesa (Lisbon). — Chairperson Conseil général de l'Institut national de défense du consommateur.

Universidade Católica Portuguesa, Universidade de Lisboa
Avenida Carolina de Michaelis Lote n°20-9 B
2795 Linda-A-Velha (Portugal)
✆ (351) -1-419 54 03 Fax: (351) -1-414 02 86

■ AFFILIATIONS: Centro de Estudos Bioética, Coimbra — CEB, Portugal (*Member*).
■ TEACHING: yes.
■ PUBLICATIONS: yes.
■ WORKING ON: women's rights [B6], human equality [B7], integrity [B8], justice [B9], freedom [B10], ethics [B48], respect [B63], virtues [B64], solidarity [B65], normative ethics [C5], deontological ethics [C6], teleological ethics [C7], morals [C12], professional ethics [C51], philosophy [C61], love [C80], trust [C81], personality [C82], self concept [C83].
■ INTERESTED IN: *idem.*

P 26 **RENAUD** Michel

Professor of philosophy, New University of Lisbon (Faculty of Social Sciences and Humanities) and Portuguese Catholic University of Lisbon (Faculty of Humanities).

Universidade Nova de Lisboa
Avenida Carolina Michaelis Lote n°20-9 B
2795 Linda-A-Velha (Portugal)
✆ (351) 1-419 54 03 Fax: (351) 1-414 02 86

■ AFFILIATIONS: Conseil national d'éthique pour les sciences de la vie, Portugal (*Member*). — Centro de Estudos de Bioética, Coimbra — CEB, Portugal (*Vice-Chairperson*).
■ TEACHING: yes.
■ PUBLICATIONS: yes.
■ WORKING ON: ethics [B48], codes of ethics [B49], moral obligations [B50], self determination [B4], freedom [B10], human person [B12], specific approaches to ethics [C1], Christian ethics [C22], Roman catholic ethics [C23], natural law [C24], medical ethics [C52], philosophy [C61], existentialism [C64], humanism [C66], mental processes [C69], motivation [C76], violence [D85], dehumanization [D86], nuclear warfare [D89], in vitro fertilization [M29].
■ INTERESTED IN: ethics committees [B22], CCNE — Comité Consultatif National d'Ethique (France) [B23], biomedical ethics education [B78], European Convention of Bioethics [B97], Bioethics bill, 1992 (France) [B98], homosexuality [M22], palliative care [O8], quality adjusted life years [O14], history of biomedical ethics [B94].

P 27 **ROBALO-CORDEIRO** Antonio

Full Professor. — Director of the University Clinic of Pneumology. — Director of the Centre of Pneumology (research unit).

University Hospital
Professor Mota-Pinto
3000 Coimbra (Portugal)
✆ (351) 39-36 262 Fax: (351) 39-23 907

■ AFFILIATIONS: Centro de Estudos de Bioética, Coimbra, Portugal (*Member*).
■ TEACHING: postgraduates and public conferences.
■ PUBLICATIONS: yes.
■ WORKING ON: aging [L50], occupational medicine [G65], pollution [E47].
■ INTERESTED IN: aging [L50], occupational medicine [G65], medical informatics [F21], fluoridation [E45], pollution [E47].

P 28 **RUEFF** José Alexandre

Professor of Genetics, Director of the Department of Genetics, New University of Lisbon (UNL).

Université nouvelle de Lisbonne (UNL) – Faculté des sciences médicales
Département de génétique
R. da Junqueira, 96
1300 Lisboa (Portugal)
✆ (351) 1-364 50 83 and 363 21 41 Fax: (351) 362 20 18 and 363 21 05

■ AFFILIATIONS: Conselho Nacional de Etica para as Ciências da Vida, Lisboa, Portugal (*Member*).

■ TEACHING: yes.
■ INTERESTED IN: biomedical research and experimentation [F1], health care professionals, researchers and patient relationship [J1], sexuality and procreation — unborn child [M1], genetics and applications — biotechnology [Q1], human genome project [F8], animal testing alternatives [F14], professional deontology [J25], investigator subject relationship [J29], human characteristics [L19], sexuality and procreation [M2], reproductive technologies [M26], fetal therapy [M57], cancer [N17], injuries, occupational diseases, intoxications [N46], genetic defects and hereditary diseases [Q2], biotechnology — genetic engineering [Q20], medical genetics [Q35].

P 29 **SANTOS** Agostinho de Almeida Médecine
Professor of Genetics, Medical School of Coimbra, Head of the Department of gynaecology, University Hospital, in charge of the unit of artificial procreation.

Faculté de médecine
Université de Coimbra
Quinta de S. Miguel Solum
3000 Coimbra (Portugal)
✆ (351) 39-71 20 52

■ AFFILIATIONS: Centro de Estudos de Bioética, Coimbra, Portugal (*Member*).
■ TEACHING: within courses in genetics, gynecology and human reproduction.
■ PUBLICATIONS: yes.

P 30 **SEBASTIÃO** Luis Miguel
Lecturer, University of Evora, Philosophy of Education.

Universidade de Evora
Departamento de Pedagogia e Educação Apto 94
7001 Evora (Portugal)
✆ (351) 66-25572 Fax: (351) 66-20 775

■ AFFILIATIONS: Centro de Estudos de Bioética, Coimbra, Portugal (*Member*).
■ PUBLICATIONS: yes.
■ WORKING ON: human person [B12], ethics [B48], codes of ethics [B49], evolution [C47], sociobiology [C63], behavior [C73].
■ INTERESTED IN: human person [B12], codes of ethics [B49], morality [B61], double effect [C46], primates [C50], sociobiology [C63], behavior [C73], behavioral genetics [Q34].

P 31 **SERRÃO** Daniel
Professor of Pathological Anatomy, Professor of Medical Deontology, Medical School of Porto. — Professor of Forensic Medicine, Faculty of Law, Catholic University of Porto. — Head of the Department of Pathology, Hospital S. João, Porto.

Lab. Anatomia Patologicã
Rua de S. Tomé 746
4200 Porto (Portugal)

■ AFFILIATIONS: Conselho Nacional de Etica para as Ciências da Vida, Lisboa, Portugal (*Member*).
■ TEACHING: Deontology and medical ethics for medical students (fourth year), ethics in forensic medicine for students in law (fifth year).
■ PUBLICATIONS: yes.
■ WORKING ON: ethics committees [B22], codes of ethics [B49], deontology [B59], biomedical ethics education [B78], codes of biomedical ethics [B90], ethicists [B92], bioethics movement [B95], European Convention of Bioethics [B97], biology and human future [C43], evolution [C47], speciesism [C48], medical ethics [C52], molecular biology [E8], research policy [E18], international research [E23], peer review [E63], organ transplantation [I24], consent to treatment [J8], physicians and researchers accountability [J23], human development [L12], human characteristics [L19], cancer [N17], autopsies [O3], euthanasia [O15].
■ INTERESTED IN: CCNE — Comité Consultatif National d'Ethique (France) [B23], code of deontology [B60], CCNE advice (France) [B91], Jewish ethics [C29], animal experimentation [F10], informed consent [J9], living wills [J19], blood donation [I20], excess embryos [M34], genome mapping [Q23].

P 32 **SOUSA FRANCO** António

Faculté de Droit
Universidade Católica Portugesa

Rua de S. Bernardo n° 50 - r/c Dto.
1200 Lisboa (Portugal)
✆ (351) 1-395 43 49

■ PUBLICATIONS: yes.

P 33 **TEIXEIRA-DIAS** José J. C.
Senior University Professor of Molecular Physical Chemistry, Director of Research Group on Molecular Physical Chemistry.

University of Coimbra
Chemistry Department
Paço das Escolas
3049 Coimbra (Portugal)
✆ (351) 39-265 41 Fax: (351) 39-26541

■ WORKING ON: research policy [E18], international research [E23], computers [E33], teaching methods [E59], technology assessment [E65].
■ INTERESTED IN: research policy [E18], drug industry [E37], social control of science [E64].

P 34 **TOSCANO RICO** José Manuel
Professor and Head of Department of Pharmacology, Faculty of Medicine, University of Lisbon.

Faculdade de medicina da Universidade de Lisboa
Instituto de Farmacologia
1600 Lisboa (Portugal)
✆ (351) 1-797 94 07 Fax: (351) 1-764 059

■ AFFILIATIONS: Conselho Nacional de Etica para as Ciências da Vida (*Member*).
■ WORKING ON: biology [E5], molecular biology [E8], investigators [E15], research institutes [E17], biomedical technologies [E28], radiation [E48], nontherapeutic research [F5], cognitive research [F7], animal experimentation [F10], animal testing alternatives [F14].
■ INTERESTED IN: animal care committees [B21], ethics committees [B22], Council of Europe [B25], Helsinki Declaration [B37], biomedical ethics education [B78], curriculum [B83], medical ethics [C52], nursing ethics [C53], research policy [E18], research design [E19], drug industry [E37], human experimentation [F11], research subjects [F12], healthy volunteers [F13], control groups [F17], random selection [F18], selection of subjects [F19], medical informatics [F21], tissue banking [I12], informed consent [J9], presumed consent [J10], medical secrecy [J22].

Spain

Area: 504 782 km²
Population: 39.7m
Gross domestic product: Ptas 57.1trn; US$ 510bn
Gross domestic product per head: $ 12,850
Gross domestic product: 1991: 2.7%
1992: 3.4%

Organisations

O E 1 **ALIANZA EVANGELICA ESPANOLA (ASSOCIÉE À L'ALLIANCE ÉVANGÉLIQUE EUROPÉENNE)**
Valencia, 2, 5, 6
Apartado 20095
08080 Barcelona (Spain)

O E 2 **ASSOCIACIO BIOÉTICA DE L'ACADEMIA DE CIENCIES MEDIQUES DE CATALUNYA I BALEARS**
Passeig de la Bonanova 47
08017 Barcelona (Spain)
✆ (34) 3-212 38 95

O E 3 **CATEDRA DE BIOETICA – UNIVERSIDAD PONTIFICIA COMILLAS**
Calle Universidad Comillas 7
28049 Madrid (Spain)
✆ (34) 1-734 66 00 Fax: (34) 1-734 45 70

O E 4 **COMITÉ ÉTICO DEL HOSPITAL CLINICO UNIVERSITARIO**
Avda. Blasco Ibànez 17
46010 Valencia (Spain)
✆ (34) 6-38 32 600 Fax: (34) 6-36 02 144

O E 5 **INSTITUT BORJA DE BIOÉTICA**
Calle Llasseres
08190 Sant Cugat del Valles (Barcelona) (Spain)
✆ (34) 3-674 47 66 Fax: (34) 3-674 79 80

■ DATE OF FOUNDATION: 1975.
■ EXECUTIVE BOARD: Dr. Narcis Jubany (*Chairperson*), Dr. Moisès Broggi (*Vice-Chairperson*), Dr. Joaquin Plaza (*Treasurer*), Sr Jordi Abel (*Secretary*),Dr. Francesc Able (Director), Dr. Lluis Campos (*Member*), Sra Dolores Cesari (*Member*), Dr. Victor Conill (*Member*), José Luis Fonseca (*Member*), Dr. Diego Gracia Guillèn (*Member*), Carmen Lleida Basiana (*Member*)
■ TEACHING: Schools of theology (Barcelona, Roma). — Schools of Medicine (Barcelona, Lerida). — Hospital Ethics Committees (Hospital Sant Joan de Deu; San Rafael; Virgen de Montserrat; San Raffaele (Milano).
■ PUBLICATIONS: Horitzons de bioetica. — Regular contribution in the journal "Labor Hospitalaria".
■ DOCUMENTATION CENTRE: open to the public.
■ COLLOQUIUMS, SYMPOSIUMS: this Institute has been instrumental in the organisation of 19 international conferences from 1980 until now, within the premises of the FIUC/IFCU. In the last 5 years, participation in 40 international conferences contributing 30 major papers, organisation of 30 courses (intensive courses of 40 hours). In the last three years the members of the Institute gave one hundred lectures at University level.
■ DECISIONS, ADVICE: in collaboration with Hospital Sant Joan De Deu: Protocolo de asistencia inmediata al niño con mielomeningocele. — Criterios de muerte cerebral en el niño. — Diagnostico prenatal de los defectos congenitos. — Protocolo de Actuacion ante pacientes testigos de Jehova.
■ WORKING ON: human person [B12], ethics committees [B22], code of deontology [B60], interdisciplinary communication [B70], information dissemination [B71], Roman catholic ethics [C23], medical ethics [C52], obstetrics and gynecology [G66], pediatrics [G68], prenatal diagnosis [I3], biological specimens procurement, blood transfusion, organ transplantation [I8], patients' rights [J2], sexuality and procreation [M2], reproductive technologies [M26], abortion [M47], death and resuscitation [O1], genetics and applications — biotechnology [Q1].

O E 6 **INSTITUTO SUPERIOR DE CIENCIAS MORALES**
Calle Felix Boix 13
28036 Madrid (Spain)
✆ (34) 1-345 36 00 and 345 36 01 and 350 82 18 Fax: (34) 1-345 86 79

0 E 7 **UNIVERSIDAD DE NAVARRA – FACULTAD DE MEDICINA**
Departamento de Bioética
Apartado 273
31080 Pamploma (Spain)
✆ (34) 48-25 21 50

Individuals

E 1 **ABEL** Francesc
Director of the Institut Borja de Bioetica.

Institut Borja de Bioetica
Calle Llasseres 30
Sant Cugat del Vallès
08190 Barcelona (Spain)
✆ (34) 3-674 47 66 Fax: (34) 3-674 79 80

■ AFFILIATIONS : Sant Cugat del Valles
■ TEACHING: Hospital Lectures. — Lectures at the Institute.
■ PUBLICATIONS: yes.
■ WORKING ON: ethics committees [B22], interdisciplinary communication [B70], bioethical issues [B96], Christian ethics [C22], resource allocation [H4], guidelines [K55], selective abortion [M49], terminal care [O7], genetic counseling [Q36].
■ INTERESTED IN: ethics committees [B22], resource allocation [H4], audiovisual aids [B74], data protection [E52].

E 2 **ANGULO** Fernando
Head of the Department of Psychiatry for children and adolescents, San Juan de Dios Hospital, Barcelona. — Director of the Program for Mental Health (Conselleria de Sanitat-General de Catalunya).

Hospital San Juan de Dios
Carretera de Esplugas S/H
08034 Barcelona (Spain)
✆ (34) 3-280 40 00 Fax: (34) 3-203 39 59

■ PUBLICATIONS: yes.
■ WORKING ON: organization of health care; facilities, manpower and services; health occupations [G1], health economics, population characteristics [H1], health care, medical acts [I1], health care professionals, researchers and patient relationship [J1], stages of life problems proper to childhood and elderly [L1], sexuality and procreation — unborn child [M1], neurosciences — psychiatry [P1].
■ INTERESTED IN: maternal health [G3], mental health [G4], residential facilities [G26], health care, medical acts [I1], sexuality and procreation — unborn child [M1].

E 3 **BERNAD PEREZ** Luisa
Expert in Forensic Medicine (evaluation of physical damage).

Catedra de Medicina Legal — Facultad de Medicina
Universidad de Zaragoza
Calle Domingo Miral
50009 Zaragoza (Spain)
✆ (34) 76-56 72 57

■ TEACHING: nursing ethics.
■ WORKING ON: deontology [B59], nursing education [B79], nursing ethics [C53], patient information [J4], consent to treatment [J8], treatment refusal [J15], confidentiality [J20], privileged communication [J21], medical secrecy [J22], expert evaluation [K15], malpractice [K19], negligence [K20], legal liability [K30], abortion on demand [M52], terminal care [O7], euthanasia [O15].
■ INTERESTED IN: deontology [B59], nursing education [B79], nursing ethics [C53], patient information [J4], disclosure [J7], consent to treatment [J8], treatment refusal [J15], confidentiality [J20], privileged communication [J21], medical secrecy [J22], professional deontology [J25], nurse patient relationship

[J30], expert evaluation [K15], malpractice [K19], negligence [K20], economic value of life [K28], beginning of life [L24], in vitro fertilization [M29], artificial insemination [M31], therapeutic abortion [M50], abortion on demand [M52], terminal care [O7], withholding treatment [O10], allowing to die [O11], euthanasia [O15].

E 4 **BILLOCH BARCELO** Miquel

Executive Secretary (Hospital del Espiritu Santo).

Hospital del Espiritu Santo
Provenza 304, 1°, 1a
08008 Barcelona (Spain)
✆ (34) 3-2155878 and 2155882 Fax: (34) 3-2154390

E 5 **BROGGI-VALLES** Moisès

Member of the Royal Medical Academy of Barcelona (Chairperson). — Former chairman of the Comision Deontologica del Col. de Médicos de Barcelona. — Deputy Chairman of the Institut Borja de Bioetica.

Institut Borja de Bioetica
Carrer Putxet 51
8023 Barcelona (Spain)
✆ (34) 3-212 03 74

■ PUBLICATIONS: yes.

■ WORKING ON: medical ethics [C52], North-South relationships [D60], international organizations [D61], war [D87], nuclear warfare [D89], quality of environment [E41], ecology [E43], health services research [F4], family planning [G8], population growth [H20], elderly [H23], legal liability [K30], allowing to die [O11], quality adjusted life years [O14], drug abuse [P6].

E 6 **BUISAN** Lidia

Head of the Department of Anaesthetics and Intensive care. — Member of the Deontology Committee, Collegio de Médicos, Barcelona.

Hospital Cruz Rosa
L'Hospital del Llobregat
4 José Molins 29-41
08906 Barcelona (Spain)
✆ (34) 3-440 75 00

■ AFFILIATIONS: Institut Borja de Bioética, Barcelona, Spain (*Member of the Board*).

■ PUBLICATIONS: yes.

■ WORKING ON: ethics committees [B22], professional deontology organs [B26], patient advocacy [B42], codes of ethics [B49], code of deontology [B60], Red Cross [B67], computerized file [E35], safety [E51], data protection [E52], medical informatics [F21], anesthesia [I31], confidentiality [J20], medical secrecy [J22], law, legislation and jurisprudence [K1], technical expertise [K16], wrongful life [K33], legislation and law [K34], authorization [K37], contracts [K39], medical law [K51], criminal code [K54], law [K56], case studies [N4], suffering [N11], pain [N12], coma [N41], persistent vegetative state [N42], death and resuscitation [O1], death [O5], brain death [O6], terminal care [O7], palliative care [O8], life-sustaining treatment [O9], withholding treatment [O10], allowing to die [O11], terminally ill [O12], prolongation of life [O13], quality adjusted life years [O14], euthanasia [O15], active euthanasia [O16], involuntary euthanasia [O17], voluntary euthanasia [O18], suicide [O19], attitudes to death [O20], resuscitation [O21].

■ INTERESTED IN: respect of human dignity and human rights [B1], fundamental rights of the individuals [B2], self determination [B4], institutions involved in the protection of rights [B17], ethics committees [B22], CCNE — Comité Consultatif National d'Ethique (France) [B23], College of Physicians [B27], International Court [B30], codes of biomedical ethics [B90], ethical review [B93], Bioethics bill, 1992 (France) [B98], normative ethics [C5], deontological ethics [C6], situational ethics [C9], ethical relativism [C10], quality of life [C16], natural law [C24], eastern orthodox ethics [C26], Protestant ethics [C27], Islamic ethics [C28], Jewish ethics [C29], Jehovah's witnesses [C38], professional ethics [C51], medical ethics [C52], nursing ethics [C53], formal social control [D10], legislation [D13], model legislation [D14], torture [D94], biomedical technologies [E28], computerized file [E35], quality of environment [E41], safety [E51], data protection [E52], methods [E53], decision making [E54], decision analysis [E55], [E56], evaluation [E61], technology assessment [E65], biomedical research [F2], human experimentation [F11], selection of subjects [F19], medical informatics [F21], health legislation [G14], institutional policies [G15], institutional obligations [G16], residential facilities [G26], intensive care units [G33], hospices [G34].

E 7 **CAMPOS** Luis
Professor of Obstetrics and Gynaecology (retired). — Member of the Board of Trustees of the Institut Borja de Bioetica.

Institut Borja de Bioetica
Carrer Llaseres 30
08190 Sant Cugat (Spain)
✆ (34) 3-67 411 50 Fax: (34) 3-67 479 80

■ WORKING ON: obstetrics and gynaecology [G66], beginning of life [L24], sexuality and procreation [M2], brain death [O6], terminal care [O7], euthanasia [O15], genetic counseling [Q36].
■ INTERESTED IN: diagnosis [I2], beginning of life [L24], sexuality and procreation [M2], brain death [O6], terminal care [O7], euthanasia [O15], genetic counseling [Q36].

E 8 **CESARI I ALIBERCH**

Av. Portal de l'Angel, 7, 1r 4a A,
08002 Barcelona (Spain)
✆ (34) 3-301 72 09 Fax: (34) 3-674 79 80

■ AFFILIATIONS: Institut Borja de Bioetica, Barcelona, Spain (*Member, assistant*).
■ TEACHING: sexual pedagogy to teachers (secondary school and university).
■ PUBLICATIONS: yes.
■ WORKING ON: stages of life — problems proper to childhood and elderly [L1], human development [L12], adolescents [L14], mind [L20], child development [L23], child and family [L31], sexuality and procreation — unborn child [M1], sexuality and procreation [M2], sexuality [M20], sexual behavior [M21].

E 9 **CORTES** Higini
Director of Education, Nursing Department.

Hospital Clinico y Provincial de Barcelona
Via Broelle, 170
08036 Barcelona (Spain)
✆ (34) 3-454 60 00 Fax: (34) 3-454 66 91

■ AFFILIATIONS: Alianza Evangélica Española (associée à l'Alliance Evangélique Européenne) — AEE, Barcelona, Spain (*Member of the bioethics department*).
■ TEACHING: Collaborating Professor in the course "Master en Gestion y Administracion de Enfermeria", Nursing School of the University of Barcelona.
■ PUBLICATIONS: yes.
■ WORKING ON: review [A2], bibliography [A4], literature [A7], biomedical ethics education [B78], nursing education [B79], Protestant ethics [C27], religion and sects [C30], scriptural interpretation [C40], theology [C41], nursing ethics [C53], patient care [G53], hospitalisation [G46], nurse patient relationship [J30]death [O5], terminal care [O7], attitudes to death [O20].
■ INTERESTED IN: data bases [A3], bibliography [A4], information centers [A5], information — communication — media [B68], bioethics [B89], religion and religious ethics [C19], professional ethics [C51], ethics and humanities [C54], research [E14], teaching methods [E59], biomedical research [F2], health care [G45], inter-personal relationship [J27], aged — problems related with aging [L45], death and resuscitation [O1].

E 10 **CORTINA** Adela
Full Professor of Ethics, Political and Legal Philosophy.

Universidad de Valencia (Fac. Filosofia)
Avda. Blasco Ibànez, 21
46010 Valencia (Spain)
✆ (34) 6-38 64 420 Fax: (34) 6-36 96 023

■ AFFILIATIONS: Comité ético del Hospital Clinico Universitario, Valencia, Espagne (*Member*).
■ TEACHING: Participation en el Master de Bioética del Departamento de Historia de la Medicina de la Facultad de Medicina en la Universidad Complutense de Madrid, dirigido por el Prof. Diego Gracia. — Participacion en la signatura de Bioética en las Licenciatures de Medicina y Filosofia en la Universidad de Valencia.
■ PUBLICATIONS: yes.
■ WORKING ON: human rights [B3], self determination [B4], justice [B9], freedom [B10], human person

[B12], ethics committees [B22], ethics [B48], morality [B61], solidarity [B65], philosophical ethics [C2], ethics and humanities [C54], political issues [D56], democracy [D82], genetics [E6], molecular biology [E8], clinical trials [F3], human genome project [F8], legal personality [K2], terminal care [O7], euthanasia [O15], biotechnology — genetic engineering [Q20].
■ INTERESTED IN: human rights [B3], self determination [B4], justice [B9], freedom [B10], human person [B12], ethics committees [B22], biomedical ethics education [B78], European Convention of Bioethics [B97], Bioethics bill, 1992 (France) [B98], religion and sects [C30], North-South relationships [D60], democracy [D82], genetics [E6], molecular biology [E8], clinical trials [F3], human genome project [F8], wrongful death [K12], terminal care [O7], euthanasia [O15], biotechnology — genetic engineering [Q20].

E 11 **CUYAS** Manuel

Professor Emeritus of Deontology, Medical School Barcelona. — Professor of Theology, Faculty of Theology of Catalonia (Barcelona). — Visiting Professor at Pontificia Università Gregoriana (Rome).

Institut Borja de Bioética
C. Llasseres 30
08190 Sant Cujat del Valles (Barcelona) (Spain)
✆ (34) 3-674 11 50/674 56 29 Fax: (34) 3-674 79 80

■ AFFILIATIONS: Institut Borja de Bioética, Sant Cujat Del Vallès (Barcelona), Spain (*Member*).
■ TEACHING: Institut Borja de Bioética. — Faculty of Theology of Catalonia (undergraduates). — Pontificia Università Gregoriana (undergraduates and postgraduates). — Istituto Scientifico Ospedale Sant Raffaele, dipartimento di Medicina e Scienze Umane (Milano).
■ PUBLICATIONS: yes.
■ WORKING ON: dignity [B5], ethics committees [B22], deontology [B59], biomedical ethics education [B78], professional ethics [C51], marital relationship [D29], patient information [J4], informed consent [J9], treatment refusal [J15], medical secrecy [J22], professional deontology [J25], beginning of life [L24], sexuality and procreation — unborn child [M1], euthanasia [O15].
■ INTERESTED IN: dignity [B5], biomedical ethics education [B78], professional ethics [C51], patient information [J4], medical secrecy [J22], professional deontology [J25], beginning of life [L24], sexuality and procreation — unborn child [M1], euthanasia [O15].

E 12 **DE LA FUENTE HONTAÑON** Carmen

Physician of the National Health System (specialises in internal medicine). — Regional Coordinator of Continual Medical Education. — Professor of Medical Ethics and Deontology, Faculty of Medicine in Valladolid.

Colegio Oficial de Médicos
Pasion 13-2°
47001 Valladolid (Spain)
✆ (34) 83-351703 and 355488 Fax: (34) 83-350254

■ AFFILIATIONS: Sociedad bioética de Castilla-Leon, Valladolid, Spain (*Founding member*). — Sociedad cuidados paliativos, Madrid, Spain (*Founding member*). — Asociacion médica mundial — AMM (*Member*).
■ TEACHING: responsible for a course in ethics and deontology at the Faculty of Medicine, Valladolid, and a seminar on the ethical aspects of paliative cares. Participates in a seminar on the ethics of nursing care.
■ PUBLICATIONS: yes.
■ WORKING ON: human person [B12], College of Physicians [B27], codes of ethics [B49], biomedical ethics education [B78].
■ INTERESTED IN: *idem, plus:* bioethics [B89], specific approaches to ethics [C1], professional ethics [C51], organization of health care; facilities, manpower and services; health occupations [G1], health care professionals, researchers and patient relationship [J1]

E 13 **EGOZCUE** J.

Professor of Cell biology, Head of Department of Biology and Physiology.

Universidad Autonoma de Barcelona
Bellaterra
08193 Barcelona (Spain)
✆ (34) 3-581 16 60 Fax: (34) 3-581 22 95

■ AFFILIATIONS: European Society of Human Reproduction and Embryology, Brussels, Belgium (*Member (member of executive committee 1984-90)*).

■ TEACHING: Symposia, courses.
■ PUBLICATIONS: yes.
■ WORKING ON: deontology [B59], codes of biomedical ethics [B90], evolution [C47], primates [C50], medical ethics [C52], genetics [E6], molecular biology [E8], biomedical technologies [E28], freezing [E29], radiation [E48], nontherapeutic research [F5], animal experimentation [F10], animal testing alternatives [F14], health legislation [G14], prenatal diagnosis [I3], amniocentesis [I4], chorionic villus sampling [I5], sperm banks [I11], tissue banking [I12], beginning of life [L24], immunologic contraception [M5], reproductive organs and embryonic structures [M9], embryos [M10], ovum [M12], sperm [M14], procreation [M15], reproduction [M17], fertility [M18], infertility [M19], in vitro fertilization [M29], excess embryos [M34], abortion [M47], fetal development [M54], fetal therapy [M57], cancer [N17], congenital defects [N23], genetic defects and hereditary diseases [Q2], genetic defects [Q3], chromosomal disorders [Q4], XYY karyotype [Q5], down's syndrome [Q6], hereditary diseases [Q18], sex linked defects [Q19], biotechnology [Q21], genetic intervention [Q22], sex determination [Q26], hybrids [Q27], transgenic animals [Q29], sex preselection [Q30], recombinant DNA research [Q31].
■ INTERESTED IN: review [A2], human body commercialization [B13], traffic of organs [B15], Council of Europe [B25], deontology [B59], medical education [B80], codes of biomedical ethics [B90], European Convention of Bioethics [B97], medical ethics [C52], health legislation [G14], living wills [J19], wrongful life [K33], euthanasia [O15].

E 14 **ELIZARI BASTERRA** Francisco Javier

Director of the Instituto Superior de Ciencias Morales (Universidad Comillas, Madrid).

Instituto Superior de Ciencias Morales
Calle Felix Boix 13
28036 Madrid (Spain)
✆ (34) 1-345 36 00 and 345 36 01 and 350 82 18 Fax: (34) 1-345 86 79

■ AFFILIATIONS: Instituto Superior de Ciencias Morales, Madrid, Spain (*Director*).
■ TEACHING: Faculty of Theology at the Universitdad Camillas (Madrid).
■ PUBLICATIONS: yes.
■ WORKING ON: ethics committees [B22], Roman catholic ethics [C23], Roman catholicism [C31], theology [C41], patient information [J4], informed consent [J9], advance directives [J18], living wills [J19], medical secrecy [J22], sexuality and procreation — unborn child [M1], reproductive technologies [M26], abortion [M47], determination of death [O2], terminal care [O7], euthanasia [O15].
■ INTERESTED IN: ethics committees [B22], patient advocacy [B42], CCNE advice (France) [B91], European Convention of Bioethics [B97], Bioethics bill, 1992 (France) [B98], sociology of medicine [D55], random selection [F18], selection of subjects [F19], placebos [I36], advance directives [J18], living wills [J19], attitudes to death [O20].

E 15 **ESCUDE I CASALS** Jordi M

Teacher of Fundamental Theology, (Facultat de Teologia de Catalunya) Catalonia.

Instituto Borja de Bioetica
30 Calle Llasseres, Sant Cugat del Valles,
08190 Barcelona (Spain)
✆ (34) 3-674 47 66 Fax: (34) 3-674 79 80

■ AFFILIATIONS: Institut Borja de Bioètica, Spain (*Member*).
■ TEACHING: Lectures in bioethics. — Fundamental theology.
■ PUBLICATIONS: yes.
■ WORKING ON: morals [C12], Christian ethics [C22], Roman Catholic ethics [C23].
■ INTERESTED IN: morals [C12], Christian ethics [C22].

E 16 **FONT** Jordi

Research Fellow and Director of the Fundacio Vidal i Barraquer.

Fundacio Vidal i Barraquer
Rivadeneyra 6
08002 Barcelona (Spain)
✆ (34) 3-412 38 40

■ AFFILIATIONS: Associacio Bioética de l'Academia de Ciencies Mèdiques de Catalunya i Balears, Barcelona, Spain (*Member*).
■ WORKING ON: psychology [C68], mental processes [C69], single persons [D25], couple [D26], investigators [E15], health legislation [G14], psychiatry [G69], physician patient relationship [J32],

professional patient relationship [J33], privileged communication [J21], medical secrecy [J22], mind [L20], homosexuality [M22], homosexuals [M23], pain [N12], psychiatric diagnosis [P29], suffering [N11], group therapy [P32], psychotherapy [P30].

■ INTERESTED IN: review [A2], book review [A6], psychology [C68], family relationship [D31], handicapped [D43], minority groups [D42], research team [E16], research policy [E18], mental health [G4], health services [G25], residential facilities [G26], intensive care units [G33], psychiatry [G69], costs and benefits [H10], risks and benefices [H11], health care professionals, researchers and patient relationship [J1], mind [L20], homosexuality [M22], homosexuals [M23], transsexualism [M24], sexuality [M20], suffering [N11], pain [N12], withholding treatment [O10], death [O5], allowing to die [O11], quality adjusted life years [O14], suicide [O19], neurosciences — psychiatry [P1].

E 17 **GAFO** F-Javier

Teacher of Moral Theology (Moral de la Persona) in the Faculty of Theology at the Universidad Pontificia Comillas, Madrid.

Calle Universidad Comillas, 7
Canto Blanco
28049 Madrid (Spain)
✆ (34) 1-734 45 70

■ AFFILIATIONS: Catedra de Bioética, Madrid, Spain (*Director*).

■ TEACHING: Annual bioethics course, concrete bioethical themes (Assisted Procreation, Genetic Engineering, Drugs, AIDS, Ecology).

■ PUBLICATIONS: yes.

■ WORKING ON: prostitution [M25], mifepristone [M53], fetal therapy [M57], terminal care [O7], resuscitation [O21], biotechnology — genetic engineering [Q20], human genome [Q47].

■ INTERESTED IN: respect of human dignity and human rights [B1], religion and religious ethics [C19], theology [C41], population [H19], elderly [H23], spousal consent [J12], parent child relationship [J34].

E 18 **GARCIA-CONDE** Javier

Professor of Medicine at the University of Valencia (Spain). — Head of the Haematology and Medical Oncology service. — Physician in chief of the Hospital Clinico Universitario de Valencia.

Hospital Clinico Universitario
Avd Blasco Ibanez 17
46010 Valencia (Spain)
✆ (34) 6-386 2625 Fax: (34) 6-386 4767

■ TEACHING: Doctoral courses. — Courses for specialists in oncology, Committee of bioethics in the Hospital Clinico Universitario of Valencia.

■ PUBLICATIONS: yes.

■ WORKING ON: cancer [N17], euthanasia [O15], active euthanasia [O16], involuntary euthanasia [O17], voluntary euthanasia [O18], suicide [O19], attitudes to death [O20], ethics committees [B22].

■ INTERESTED IN: cancer [N17], euthanasia [O15], active euthanasia [O16], involuntary euthanasia [O17], voluntary euthanasia [O18], suicide [O19], attitudes to death [O20], human rights [B3], dignity [B5], justice [B9], freedom [B10], education [B76], health education [B77], biomedical ethics education [B78], nursing education [B79], medical education [B80], students [B82], ethics committees [B22], transplantation [I22].

E 19 **GASULL** Maria

Professor, Nursing School of Sant Pau, Free University of Barcelona.

E. U. I. Hospital Sta Creu i Sant Pau
Sant Antoni Ma Claret, 167
08025 Barcelona (Spain)
✆ (34) 3-235 89 87 Fax: (34) 3-456 0190

■ AFFILIATIONS: Institut Borja de Bioética, Sant Cugat del Vallès, Spain (*Scientific collaborator*).

■ TEACHING: nursing ethics.

■ PUBLICATIONS: yes.

■ WORKING ON: nursing education [B79], nursing ethics [C53], maternal health [G3], palliative care [O8].

■ INTERESTED IN: human rights [B3], self determination [B4], dignity [B5], communication [B69], humanism [C66], motivation [C76], counseling [C87], maternal welfare [G10], maternal life [G11], hospices [G34], nurses [G39], nurse midwives [G43], patients' rights [J2], patient information [J4], patient access [J5], disclosure [J7], consent to treatment [J8], patient participation [J16], privileged communication [J21], nurse patient relationship [J30], physician nurse relationship [J31], allowing to die [O11], terminally ill [O12], patient care [G53].

E 20 **GUARDIA I CANELA** Josep Delfi
Chairperson of the Academia de Jurisprudencia i Legislacio de Catalunya.

Academia de Jurisprudencia i Legislacio de Catalunya
Rambla de Catalunya, 101, 3r. 1a
08008 Barcelona (Spain)
✆ (34) 3-215 16 39 Fax: (34) 3-335 81 84

■ WORKING ON: legitimacy [K21], accountability [K29], legal liability [K30], civil laws [K43], civil code [K44], European law [K49].
■ INTERESTED IN: *idem.*

E 21 **HERRANZ RODRIGUEZ** Gonzalo
Professor of Medical Ethics Medical School, University of Navarra. — Chairperson, Central Ethics Committee, Spanish Medical Association (Consejo General de Colegios Médicos de España).

Facultad de Medicina, Depart. de Bioética,
Universidad de Navarra
Apartado 273
31080 Pamplona (Spain)
✆ (34) 48-25 21 50 Fax: (34) 48-17 55 00

■ TEACHING: Undergraduate (fifth year students) of Medical Ethics. — Postgraduate students of Medical Research Ethics.
■ PUBLICATIONS: yes.
■ WORKING ON: codes of ethics [B49], code of deontology [B60], clinical trials [F3], therapeutic research [F6], medical informatics [F21], inter-personal relationship [J27], reproductive technologies [M26], death and resuscitation [O1].

E 22 **LACADENA** Juan Ramon
Head of Genetics Department, Faculty of Biological Sciences.

Universidad Complutense
Facultad de Ciencias Biologicas
Departamento de Genética
Ciudad Universataria
28040 Madrid (Spain)
✆ (34) 1-39 41 641 Fax: (34) 1-39 41 638

■ PUBLICATIONS: yes.
■ WORKING ON: genetics [E6], beginning of life [L24], reproductive technologies [M26], abortion [M47], genetics and applications — biotechnology [Q1], genetic defects and hereditary diseases [Q2], biotechnology — genetic engineering [Q20], behavioral genetics [Q34], medical genetics [Q35], population genetics [Q43].
■ INTERESTED IN: *idem.*

E 23 **LOPEZ AZPITARTE** Eduardo
Professor of Moral Theology, Faculty of Theology, University of Granada. — Director of the Teachers Residence, Faculty of Theology.

Facultad de Teologia
Apartado 2002
18080 Granada (Spain)
✆ (34) 58-29 45 59

■ AFFILIATIONS: Hospital Virgen de las Nieves, Granada (*Member of the Ethics committee)*
■ TEACHING: Bioethics in foreign countries (South America and Africa).
■ PUBLICATIONS: yes.
■ WORKING ON: ethical rules and principles [B47], biomedical ethics education [B78], bioethics [B89], philosophical ethics [C2], Roman Catholic ethics [C23], medical ethics [C52], war [D87], torture [D94], patients' rights [J2], confidentiality [J20], aged [L49], sexuality and procreation [M2], reproductive technologies [M26], abortion [M47], terminal care [O7], euthanasia [O15], biotechnology — genetic engineering [Q20].
■ INTERESTED IN: *idem.*

E 24 **MASIA** Juan

Professor of the Faculty of Philosophy. — Visiting professor of the Faculty of Theology.

Universidad Comillas
Calle Commillas, n. 3
28049 Madrid (Spain)
✆ (34) 1-734 39 50 Fax: (34) 1-734 45 70

■ TEACHING: Seminars (2nd cycle) in the Faculty of Theology.
■ PUBLICATIONS: yes.
■ WORKING ON: ethical analysis [C3], quality of life [C16], biology and human future [C43], cultural pluralism [D51].
■ INTERESTED IN: potentiality of personhood [K5], embryos [M10], fetuses [M11], euthanasia [O15], brain death [O6].

E 25 **MONTAYA** Eladio

Professor of Human Physiology, Medical School, University of Alcalá de Henares, Madrid. — Head of the Department of Quality of Life and Natural Resources, Secretaria General del Plan Nacional de I+D, Comision Interministerial de Ciencia y Tecnologia.

Comision Interministerial de Ciencia y Tecnologia
Rosario Pino 14-16,
28020 Madrid (Spain)
✆ (34) 1-579583 25 Fax: (34) 1-571 57 81

■ PUBLICATIONS: yes.
■ INTERESTED IN: nontherapeutic research [F5], therapeutic research [F6], human genome project [F8], reproductive technologies [M26], fetal therapy [M57], genetic defects and hereditary diseases [Q2], biotechnology — genetic engineering [Q20].

E 26 **NUNEZ CUBERO** Dra Ma Pilar

Gynaecologist, Assistant Director of the Gynaecology and Obstetrics Department, Hospital del Espiritu Santo. — Member of the Board, Institut Borja de Bioética. — Head of the department of family planning, Association catholique internationale des services à la jeunesse féminine (ACISJF).

Institut Borja de Bioética
Santa Amalia 9-13, 1° 1a,
08031 Barcelona (Spain)
✆ (34) 3-674 11 50 Fax: (34) 3-674 41 86

■ AFFILIATIONS: Institut Borja de Bioética, Sant Cugat del Vallès, Barcelona (*Member of the Board*).
■ TEACHING: Lectures and debates organised by the Institut Borja de Bioética.
■ PUBLICATIONS: yes.
■ WORKING ON: bibliography [A4], documentation [A8], publications [A9], dignity [B5], human person [B12], Amnesty International [B18], ethics committees [B22], professional deontology organs [B26], College of Physicians [B27], patient advocacy [B42], ethics [B48], respect [B63], interdisciplinary communication [B70], education [B76], codes of biomedical ethics [B90], values [C15], religious sciences [C39], theology [C41], double effect [C46], medical ethics [C52], socioeconomic factors [D18], marital relationship [D29], ethnic groups [D36], disadvantaged [D46], international aspects [D59], dehumanization [D86], genetics [E6], biomedical technologies [E28], safety [E51], teaching methods [E59], human genome project [F8], human experimentation [F11], debriefing [F16], maternal health [G3], family planning [G8], maternal welfare [G10], hospital [G28], private hospitals [G31], patient care team [G38], physicians [G40], hospitalisation [G46], patient care [G53], obstetrics and gynecology [G66], prenatal diagnosis [I3], patient information [J4], physician patient relationship [J32], contraception [M3], fertility [M18], sexuality [M20], childbirth [M40].
■ INTERESTED IN: fundamental rights of the individuals [B2], ethical rules and principles [B47], biomedical ethics education [B78], bioethics [B89], quality of life [C16], medical ethics [C52], socioeconomic factors [D18], ethnic groups [D36], North-South relationships [D60], genetics [E6], biomedical technologies [E28], human genome project [F8], human experimentation [F11], maternal health [G3], hospices [G34], selection for treatment [G52], prenatal diagnosis [I3], patient information [J4], child donation [K22], child's interest [K24], adoption [L32], unwanted children [L35], exploited children [L40], child abuse [L41], wish of children [M16], reproductive technologies [M26], pregnancy in adolescence [M45], prenatal injuries [N50], palliative care [O8], genetic defects and hereditary diseases [Q2], genetic counseling [Q36].

E 27 **PALACIOS** Marcelo

Physician, Director of the Policlinica del Instituto Social de la Marina, Gijon, Spain, (non active). —MP — Member of the Council of Europe. — Chairperson of the Spanish parliamentary committee for artificial procreation. — Member of the ESLA (genetics, legal and social aspects) E.C. programme "Analysis of the Human Genome".

Congreso de los Diputados
Calle Floridablanca, 1
28071 Madrid (Spain)
✆ (34) 1-429 51 93 Fax: (34) 1-429 96 27

■ AFFILIATIONS: Assembly of the Council of Europe, Stasbourg Cedex, France (*Chairperson of Bioethics subcommittee*).
■ TEACHING: in several Spanish Universities — Conferences and publications.
■ PUBLICATIONS: yes.
■ WORKING ON: fundamental rights of the individuals [B2], World Medical Assembly [B20], ethics committees [B22], CCNE — Comité Consultatif National d'Ethique (France) [B23], Council of Europe [B25], European Convention on Human Rights [B36], health education [B77], codes of biomedical ethics [B90], European Convention of Bioethics [B97], evolution [C47], biological life [C49], primates [C50], biomedical research [F2], health legislation [G14], tissue donation [I13], required request [I10], transplantation [I22], informed consent [J9], living wills [J19], legal personality [K2], malpractice [K19], child's interest [K24], preconception injuries [K27], therapeutic risk [K32], legal rights [K45], law [K56], law enforcement [K57], legal aspects [K58], legal obligation [K59], conscience clause [K63], reproductive technologies [M26], embryo donation [M38], illegal abortion [M48], pregnant women [M42], biotechnology — genetic engineering [Q20], behavioral genetics [Q34], medical genetics [Q35].
■ INTERESTED IN: codes of ethics [B49], family members [D24], single persons [D25], couple [D26], cohabitation [D27], married persons [D28], marital relationship [D29], technology [E27], biomedical technologies [E28], freezing [E29], preservation [E30], cryonic suspension [E31]children's rights [L25], child and family [L31], aid children [L34], unwanted children [L35], reproductive organs and embryonic structures [M9], embryos [M10], fetuses [M11], ovum [M12], placentas [M13], sperm [M14], procreation [M15], wish of children [M16], reproduction [M17], fertility [M18], infertility [M19], abortion on demand [M52], iatrogenic disease [N3], determination of death [O2], withholding treatment [O10], prolongation of life [O13], euthanasia [O15], active euthanasia [O16], involuntary euthanasia [O17], voluntary euthanasia [O18], population genetics [Q43], genetic identity [Q44], gene pool [Q45], carriers [Q46], human genome [Q47].

E 28 **PIGA RIVERO** Antonio

Full Professor of Forensic Medicine, Health and Bioethics Law.

Universidad de Alcala de Ilenares
Facultad de Medicina
Carretera Madrid-Barcelona KM. 33'600
28871 Madrid (Spain)
✆ (34) 1-885 45 32 Fax: (34) 1-885 45 44

■ AFFILIATIONS: Association Internationale Droit, Ethique et Science (*Member of the Board*).
■ TEACHING: Medicine, Health and Nursing care.
■ PUBLICATIONS: yes.
■ WORKING ON: human rights [B3], ethical analysis [C3], medical ethics [C52], health legislation [G14], medical law [K51], child abuse [L41].
■ INTERESTED IN: health legislation [G14], codes of biomedical ethics [B90].

E 28b **QUINTANA TRIAS** Octavi

President of the Comité directeur pour la bioéthique (CDBI), Conseil de l'Europe. — Member of the Group on Ethical, Social, and Legal Aspects of Human Genome Analysis (ESLA).

Insalud
Alcalá, 56
28071 Madrid (Espagne)
✆ (34) 1 420 08 14 Fax : (34) 1 429 22 56

■ THEMES DE TRAVAUX : ethics committees [B22], Council of Europe [B25], European Convention of Bioethics [B97], biomedical technologies [E28], technology assessment [E65], human experimentation [F11], health policy [G13], resource allocation [H4], health care costs [H9], prenatal diagnosis [I3], reproductive technologies [M26], terminal care [O7].

■ CENTRES D'INTÉRET : technology assessment [E65], health policy [G13], medical evaluation [G51], resource allocation [H4], health care costs [H9], soutien financier [H18], health care services and aged [L48].

E 29 **RIVERO HERNANDEZ** Francisco

Universidad de Barcelona
Avenida Diagonal, 684 (Zona Pedralbes)
08017 Barcelona (Spain)
✆ (34) 3-280 28 28 Fax: (34) 3-280 01 34

■ PUBLICATIONS: yes.
■ WORKING ON: dignity [B5], freedom [B10], child's interest [K24], child and family [L31], artificial insemination [M31].
■ INTERESTED IN: dignity [B5], integrity [B8], human person [B12], Bioethics bill, 1992 (France) [B98], donors [I14], transplantation [I22], legitimacy [K21], child donation [K22], child's interest [K24], child and family [L31], aid children [L34], in vitro fertilization [M29], artificial insemination [M31], host mothers [M35], privacy [B11].

E 30 **ROMEO CASABONA** Carlos Maria
Professor of Criminal Law. — Dean.

Universitad de La Laguna
Camino de la hornera, S/N
38071 La Laguna, Tenerife (Spain)
✆ (34) 22-256 840 Fax: (34) 22-259 245

■ AFFILIATIONS: Association mondiale de droit médical (*Member*).
■ TEACHING: postgraduates of several universities. — Master of Bioethics, Universidad Complutense (Madrid).
■ PUBLICATIONS: yes.
■ WORKING ON: respect of human dignity and human rights [B1], fundamental rights of the individuals [B2], human rights [B3], self determination [B4], dignity [B5], privacy [B11], human body commercialization [B13], traffic of organs [B15], Council of Europe [B25], killing [D90], justifiable killing [D91], sex offenses [D92], rape [D93], torture [D94], genetics [E6], research [E14], technology [E27], biomedical technologies [E28], freezing [E29], preservation [E30], computers [E33], records [E34], computerized file [E35], biomedical research and experimentation [F1], biomedical research [F2], clinical trials [F3], nontherapeutic research [F5], therapeutic research [F6], human genome project [F8], human experimentation [F11], research subjects [F12], healthy volunteers [F13], health legislation [G14], nurses [G39], medical records [G50], prenatal diagnosis [I3], amniocentesis [I4], mandatory screening [I7], biological specimens procurement, blood transfusion, organ transplantation [I8], drugs [I34], placebos [I36], health care professionals, researchers and patient relationship [J1], medical secrecy [J22], law, legislation and jurisprudence [K1], legal personality [K2], minors [K6], jurisprudence — accountability [K8], fraud [K13], malpractice [K19], negligence [K20], injuries [K26], preconception injuries [K27], economic value of life [K28], accountability [K29], medical law [K51], criminal law [K53], criminal code [K54], conscience clause [K63], contraception [M3], sterilization [M6], involuntary sterilization [M7], volontary sterilization [M8], reproductive technologies [M26], abortion [M47], fetal therapy [M57], anencephaly [N25], acquired immunodeficiency syndrome [N35], HIV seropositivity [N36], coma [N41], death [O5], brain death [O6], terminal care [O7], mentally handicapped [P7], mentally retarded [P8], mentally ill [P9], dangerousness [P15], outpatient commitment [P18], institutionalized persons [P23], psychosurgery [P38], genetic defects and hereditary diseases [Q2], biotechnology — genetic engineering [Q20].
■ INTERESTED IN: *idem, plus:* information sources, data bases [A1].

E 31 **SERRAT MORE** Dolores
Full Professor Forensic Medicine.

Facultad de Medicina
Universidad de Zaragoza, Domingo Miral S/M
Alfonso I n° 26 3° C
50009 Zaragoza (Spain)
✆ (34) 76-56 72 57 Fax: (34) 76-56 80 92

■ TEACHING: Students of Medicine (undergraduates and postgraduate), Faculty of Medicine of the Universidad de Zaragoza.
■ WORKING ON: deontology [B59], medical ethics [C52], acquired immunodeficiency syndrome [N35].
■ INTERESTED IN: health care professionals, researchers and patient relationship [J1], patients' rights

[J2], confidentiality [J20], physicians and researchers accountability [J23], law, legislation and jurisprudence [K1], illegal abortion [M48], abortion on demand [M52], acquired immunodeficiency syndrome [N35], HIV seropositivity [N36].

E 32 **VEGA GUTIERREZ** Javier

Professor of Forensic Medicine and Toxicology, University of Valladolid.

Dept. de Medicina Legal, Universidad de Valladolid
Calle Ranion y Cajal 7
47005 Valladolid (Spain)
✆ (34) 83-30 71 52 or 42 30 00 Fax: (34) 83-42 30 65

■ TEACHING: undergraduate Medicine (6th course). — Doctorate: Forensic Medicine and Toxicology; Deontology and Medical Law.
■ PUBLICATIONS: yes.
■ WORKING ON: respect of human dignity and human rights [B1], health policy [G13], patients' rights [J2], confidentiality [J20], physicians and researchers accountability [J23], law, legislation and jurisprudence [K1], sexuality and procreation [M2], reproductive technologies [M26], abortion [M47], cancer [N17], injuries, occupational diseases, intoxications [N46], determination of death [O2], terminal care [O7], euthanasia [O15], suicide [O19], attitudes to death [O20], resuscitation [O21], behavioral and mental disorders [P2], genetic defects and hereditary diseases [Q2].
■ INTERESTED IN: respect of human dignity and human rights [B1], law, legislation and jurisprudence [K1], sexuality and procreation — unborn child [M1], death and resuscitation [O1], neurosciences — psychiatry [P1].

E 33 **VIDAL** Marciano

Profesor de Universidad (U. Pontificia Comillas, Madrid; Academia Alfonsiana, Roma). — Director del Instituto Universitario "Matrimonio y familia" (Madrid).

Universidad Pontificia Comillas
Canto Blanco
28049 Madrid (Spain)
✆ (34) 1-734 39 50 Fax: (34) 1-734 45 70

■ AFFILIATIONS: Instituto superior de ciencias morales, master en bioéetica, Madrid, Spain (*Member*).
■ TEACHING: Fundamentacion de la bioética. — La bioética y la religion, sobre todo la religion catolica (theological bioethics).
■ PUBLICATIONS: yes.
■ WORKING ON: ethical rules and principles [B47], bioethics [B89], religion and religious ethics [C19].
■ INTERESTED IN: information sources, data bases [A1], ethical rules and principles [B47], bioethics [B89], religion and religious ethics [C19], sexuality and procreation — unborn child [M1], resuscitation [O21], genetics and applications — biotechnology [Q1].

E 34 **ZARRALUQUI** L.

Barrister-at-law, Writer, University Professor.

Pintor Rosales 82
28008 Madrid (Spain)
✆ (34) 1-449 39 14 Fax: (34) 1-549 36 22

■ PUBLICATIONS: yes.
■ WORKING ON: fundamental rights of the individuals [B2], bioethics [B89], cosmetic surgery [I38], host mothers [M35], convenience medicine [G57].
■ INTERESTED IN: cosmetic surgery [I38], legislation and law [K34], host mothers [M35].

United Kingdom

Area : 244 100 km²
Population : 57.7m
Gross domestic product : £ 608bn ; US$ 1.02trn
Gross domestic product per head : 17,710
Gross domestic product growth : 1991 : -1.8%
1992 : 2.4%

Organisations

O GB 1 **BOARD FOR SOCIAL RESPONSIBILITY – CHURCH OF ENGLAND**
Church House
Great Smith Street
SWIP 3NZ London (United Kingdom)
✆ (44) 71-222 9011 or 71 387 4499 Fax: (44) 71-233 2576 and 71-383 6233

O GB 2 **CENTRE FOR CONTEMPORARY ETHICS – KEELE UNIVERSITY**
University of Keele
Department of Philosophy
Keele
ST 5 5BG Staffs (United Kingdom)
✆ (44) 782-583 304 Fax: (44) 782-583 399

O GB 3 **CENTRE OF MEDICAL LAW AND ETHICS**
King's College Strand
WC 2 2LS London (United Kingdom)
✆ (44) 71-873 2382 Fax: (44) 71-873 2465

O GB 4 **CENTRE FOR BUSINESS AND PROFESSIONAL ETHICS** (CBPE)
The University of Leeds
LS2 9JT Leeds (United Kingdom)
✆ (44) 532 333 260 Fax: (44) 532 333 265

■ OTHER FIELDS OF ACTIVITIES: Professional ethics: Business ethics. — Legal ethics. — Media ethics. — Teachings of ethics.
■ AFFILIATION: a centre of activity with components from the Departments of Philosophy, Medicine, Law, Economics, Business Studies, and Theology.
■ DATE OF FOUNDATION: 1989.
■ EXECUTIVE BOARD: Jennifer Jackson (*Director*), Gwyneth Pitt (*Deputy Director*), Professor Peter Moizer (*Executive*), Professor Gerald Richards (*Executive*), Dr. Haddon Wilmer (*Executive*).
■ TEACHING: postgraduate: M.A. in Health CareEthics, Ph.D. in Health CareEthics. — Undergraduate: third year course on Medical Ethics.
■ PUBLICATIONS: collections of papers on philosophical ethics in reproductive ethics. — Philosophical ethics in reproductive medicine (ed. David Bromham, Maureen Dalton, Jennifer Jackson and Peter Milligan).
■ COLLOQUIUMS, SYMPOSIUMS: First International Conference on Philosophical Ethics in Reproductive Medicine 18-22 April 1988. — Second International Conference on Philosophical Ethics in Reproductive Medicine 1991. — Third International Conference on Philosophical Ethics in Reproductive Medicine 1994.
■ WORKING ON: ethical analysis [C3], medical ethics [C52], nursing ethics [C53], human rights [B3], integrity [B8], ethics committees [B22], experimentation [F9], beginning of life [L24], aged — problems related to aging [L45], biotechnology — genetic engineering [Q20].

O GB 5 **CENTRE FOR HEALTH CARE LAW**
University of Leicester
LE1 7RH Leicester (United Kingdom)
✆ (44) 533-522 522

■ OTHER FIELDS OF ACTIVITIES: Medical law. — Teaching and research.
■ AFFILIATION: Research Unit within the Faculty of Law, Leicester University.
■ DATE OF FOUNDATION: 1988.
■ EXECUTIVE BOARD: John Finch (*Co-director*), Dr. John Keown (*Co-director*).
■ TEACHING: teaching of medical ethics to undergraduate lawyers.
■ PUBLICATIONS: examining euthanasia (an edition arising from a Conference posted by the Centre in 1991), (Cambridge University Press, forthcoming).

■ COLLOQUIUMS, SYMPOSIUMS: conference on euthanasia, 26/10/91.
■ DECISIONS, ADVICE: The Centre's advice is often sought by practising lawyers and health care professionals, as well as by Members of Parliament and the media.
■ WORKING ON: human experimentation [F11], tissue donation [I13], blood transfusions [I19], organ transplantation [I24], bone marrow transplantation [I29], personhood [K3], malpractice [K19], medical law [K51], therapeutic abortion [M50], anencephaly [N25], persistent vegetative state [N42], active euthanasia [O16], involuntary euthanasia [O17], voluntary euthanasia [O18], outpatient commitment [P18], group therapy [P32].

O GB 6 CENTRE FOR HEALTH ECONOMICS

University of York
Heslington
YO1 5DD York (United Kingdom)
✆ (44) 904 433 646 Fax: (44) 904 433 644

■ OTHER FIELDS OF ACTIVITIES: health economics. — Determinants of health. — Technology assessment. — Measurement and valuation of health. — Research and teaching.
■ AFFILIATION: the CHE is part of the Institute for Research into the Social Sciences, an Institute of the University of York.
■ DATE OF FOUNDATION: 1983.
■ EXECUTIVE BOARD: Professor Alan Maynard (*Director*), Ken Wright (*Deputy director*).

O GB 7 CENTRE FOR PHILOSOPHY AND HEALTH CARE

University College Swansea
SA2 8PP Swansea (United Kingdom)
✆ (44) 792 - 205 678

O GB 8 CHURCH OF SCOTLAND — BOARD OF SOCIAL RESPONSIBILITY

29 Milton Road East
EH15 2NN Edinburgh (United Kingdom)
✆ (44) 31 225 5722 Fax: (44) 31 225 7867

■ OTHER FIELDS OF ACTIVITIES: Social interests section: study groups and commitees covering the moral, ethical and social problems; produces reports for the Church of Scotland and publishes ensuing literature. — Social work section: Elderly services: 43 residential homes, 4 specifically caring for those suffering from dementia; also provides community services for elderly people. — Community services: 23 residential establishments providing services for people suffering from alcohol and drug addiction, HIV injection, mental handicap, mental illness, residents in probation hostels, single homeless, residential prostitutes, single day centres for lifestyle, two day centres for mental illness, 1 resource worker HIV/AIDS, project worker for women and children. — Staff training: staff team training members of staff of board of social responsability establishments; supervising students on placement from Universities and Colleges and supervising staff studying for Scottish vocational qualification.
■ AFFILIATION: the social interests section of the board of social responsibility .
■ OFFICIALLY REGULATED ORGANISATION: 1984 Deliverance 1 (page 24): Renaming of the Department on Social Responsibility as the Board on Social Responsability. — 1975 Deliverance 2 (pages 23,25): Establishment of the Social and Moral Welfare Board as the Department on Social Responsability. — 1963 Deliverance 15 (page 354): The amalgamation of the Committee on Temperance and Morals with the Committee on Social Service to form The Social Service and Moral Welfare Board .
■ EXECUTIVE BOARD: Ian D. Baillie (*Director of Social Work*), David J. Kellock (*Deputy Director of Community Services*), Kristine J. Gibbs (*Social Interests Officer*).
■ PUBLICATIONS: Abortion in debate 1987. — Lifestyle survey 1987. — Report on child abuse 1990. — Study facts on child abuse 1991. — Report on human transplants 1990. — Report on ritual abuse 1992. — Report on family matters 1992. — The future of the family — A crisis of commitment 1992. — Quorum press ISBN 0 715204517. — Quorum press ISBN 0 86153090X.— St Andrew's press ISBN 0861531574.
■ DOCUMENTATION CENTRE: internal.
■ COLLOQUIUMS, SYMPOSIUMS: annual three day conference on "The Ministry of Healing" - October or November (1989, 1990, 1991, 1992). — Advanced level colloquium on "The Ministry of Healing" - April 1991.
■ WORKING ON: respect of human dignity and human rights [B1], fundamental rights of the individuals [B2], human equality [B7], human person [B12], human body commercialization [B13], ethics committees [B22], patient advocacy [B42], common good [B45], ethical rules and principles [B47], ethics [B48], codes of ethics [B49], moral obligations [B50], obligations to society [B54], moral policy [B55], morality [B61], respect [B63], virtues [B64], information — communication — media [B68],

communication [B69], interdisciplinary communication [B70], mass media [B72], biomedical ethics education [B78], bioethics [B89], bioethical issues [B96], specific approaches to ethics [C1], philosophical ethics [C2], morals [C12], moral development [C14], values [C15], quality of life [C16], social worth [C17], value of life [C18], religion and religious ethics [C19], religious ethics [C21], Christian ethics [C22], Protestant ethics [C27], scriptural interpretation [C40], theology [C41], professional ethics [C51], social sciences [C89], social and political issues [D1], social issues [D2], legislation [D13], model legislation [D14], government regulation [D17], family members [D24], single persons [D25], couple [D26], cohabitation [D27], married persons [D28], marital relationship [D29], sibling [D30], family relationship [D31], parents [D32], mothers [D33], fathers [D34], females [D35], males [D45], social problems [D52], public policy [D77], violence [D85], sex offenses [D92], research [E14], research policy [E18], research design [E19], drug industry [E37], organization of health care; facilities, manpower and services; health occupations [G1], health policy [G13], social workers [G44], amniocentesis [I4], biological specimens procurement, blood transfusion, organ transplantation [I8], donors [I14], organ donors [I15], donor cards [I16], anonymous donation [I17], transplantation [I22], transplant recipients [I23], organ transplantation [I24], organ donation [I25], fetal tissue transplantation [I27], intracerebral fetal tissue transplantation [I28], health care professionals, researchers and patient relationship [J1], patients' rights [J2], living wills [J19], confidentiality [J20], professional patient relationship [J33], law, legislation and jurisprudence [K1], personhood [K3], child's interest [K24], health legislation [K52], stages of life — problems proper to childhood and elderly [L1], anatomy, physiology, development [L2], human development [L12], age [L13], adolescents [L14], adults [L15], childhood difficulties, diseases, protection [L21], child development [L23], child and family [L31], children and media [L36].

■ INTERESTED IN: respect of human dignity and human rights [B1], fundamental rights of the individuals [B2], ethics committees [B22], ethical rules and principles [B47], codes of ethics [B49], obligations to society [B54], moral policy [B55], morality [B61], information — communication — media [B68], mass media [B72], biomedical ethics education [B78], bioethics [B89], bioethical issues [B96], specific approaches to ethics [C1], religion and religious ethics [C19], religious ethics [C21], Christian ethics [C22], Roman catholic ethics [C23], eastern orthodox ethics [C26], cohabitation [D27], married persons [D28], marital relationship [D29], family relationship [D31], parents [D32], organization of health care; facilities, manpower and services; health occupations [G1], family planning [G8], health legislation [G14], social workers [G44], health care, medical acts [I1], biological specimens procurement, blood transfusion, organ transplantation [I8], health care professionals, researchers and patient relationship [J1], patients' rights [J2], confidentiality [J20], law, legislation and jurisprudence [K1], personhood [K3], legally incompetent person [K4], child's interest [K24], European law [K49], international law [K50], health legislation [K52], stages of life — problems proper to childhood and elderly [L1], children's rights [L25], children and media [L36], child abuse [L41], aged — problems related with aging [L45], aging [L50], old person abuse [L51], sexuality and procreation — unborn child [M1], sexuality and procreation [M2], sexuality [M20], sexual behavior [M21], homosexuality [M22], homosexuals [M23], transsexualism [M24], prostitution [M25], in vitro fertilization [M29], pregnancy in adolescence [M45], fetal development [M54], acquired immunodeficiency syndrome [N35], persistent vegetative state [N42], death and resuscitation [O1], determination of death [O2], terminal care [O7], euthanasia [O15], active euthanasia [O16], involuntary euthanasia [O17], voluntary euthanasia [O18], suicide [O19], attitudes to death [O20], alcohol abuse [P4], smoking [P5], drug abuse [P6], psychoactive drugs [P33], heroin [P34], LSD [P35], biotechnology — genetic engineering [Q20].

O GB 9 COLLEGE OF OPHTHALMOLOGISTS

Bramber Court
2, Bramber Road
W14 9PQ London (United Kingdom)
✆ (44) 71 385 6281 Fax: (44) 71 381 1799

■ OTHER FIELDS OF ACTIVITIES: Education. — Examinations. — Professional standards. — Scientific.
■ DATE OF FOUNDATION: April 1988.
■ EXECUTIVE BOARD: Peter Wright (*Chairperson*), Peter Watson (*Senior Vice-Chairperson*), Andrew Elkington (*Vice-président*), Brian Martin (*Vice-Chairperson*), Patrick Holmes Sellors (*Vice-Chairperson*), Paul Hunter (*Honorary Secretary*), Sidney Davidson (*Honorary Treasurer*), Richard Porter, David Hopkins, Nicholas Galloway
■ TEACHING: yes.
■ PUBLICATIONS: yes.
■ DOCUMENTATION CENTRE: internal.
■ COLLOQUIUMS, SYMPOSIUMS: annual scientific congress (ophthalmologists).
■ WORKING ON: publications [A9], College of Physicians [B27], professional competence [B52], communication [B69], interdisciplinary communication [B70], information dissemination [B71],

education [B76], health education [B77], medical education [B80], internship and residency [B81], medical ethics [C52], medical evaluation [G51], ophthalmology [G67], age [L13], health care services and aged [L48], diabetes [N22], eye diseases [N38].
■ INTERESTED IN: College of Physicians [B27], communication [B69], interdisciplinary communication [B70], information dissemination [B71], education [B76], medical education [B80], ophthalmology [G67], diabetes [N22], eye diseases [N38].

O GB 10 DUNDEE MEDICAL GROUP

Ninewells Hospital and Medical School
Department of Biochemical Medecine
DD1 9SY Dundee Tayside (United Kingdom)
✆ (44) 382- 644 620

O GB 11 FACULTY OF LAW AND CENTRE FOR SOCIAL ETHICS AND POLICY – UNIVERSITY OF MANCHESTER

Mansfields Cooper Building
Oxford Road
M13 9PL Manchester (United Kingdom)
✆ (44) 61-275 3414

O GB 12 FACULTY OF OCCUPATIONAL MEDICINE

6 St Andrew's Place
Regent's Park
NW1 4LE London (United Kingdom)
✆ (44) 71 - 487 3414 Fax: (44) 71 - 487 5218

O GB 13 FACULTY OF PHARMACEUTICAL MEDICINE

6 St Andrew's Place
Regent's Park
NW1 4LE London (United Kingdom)
✆ (44) 71 224 0343 Fax: (44) 71 224 5381

■ OTHER FIELDS OF ACTIVITIES: the Faculty of Pharmaceutical Medicine concerns itself with all aspects of the setting, maintenance and improvement of standards in all facets of Pharmaceutical Medicine.
■ AFFILIATION: the Faculty of Pharmaceutical Medicine operates under the aegis of the Royal college of Physicians of the United Kingdom.
■ DATE OF FOUNDATION: 1989.
■ EXECUTIVE BOARD: Dr. John George Domenet (*Vice-Chairperson*), Dr. Richard Rondel (*Registrar*), Dr. Felicity Gabbay (*Academic Registrar*), Dr. Richard Bax (*Treasurer*).
■ TEACHING: the setting of the syllabus and the administration of the examinations for the granting of Associateship and Membership of the Faculty.

O GB 14 FACULTY OF PUBLIC HEALTH MEDICINE

4, St Andrew's Place
Regent's Park
NW 1 4LE London (United Kingdom)
✆ (44) 71- 935 0243 Fax: (44) 71 - 224 6973

■ OTHER FIELDS OF ACTIVITIES: Quality standards of public health medicine practice. — Training of medical practitioners in the principles and practice of public health medicine. — Development of health promotion. — Encouragement of research in these fields.
■ DATE OF FOUNDATION: 1972.
■ EXECUTIVE BOARD: Dr. John Michael O'Brien (*Chairperson*), Professor Roderic Keith Griffiths (*Vice-chairperson*), Dr. June Madge Crown (*Registrar*), Dr. Peter Henderson Gentle (*Assistant registrar*), Dr. David Lyess Miller (*Academic registrar*), Dr. James Montgomery Dunlop (*Treasurer*).
■ TEACHING: yes.

O GB 15 GUILD OF CATHOLIC DOCTORS

60 Grove End Road
NW 8 9NH London (United Kingdom)
✆ (44) 71 - 266 4246

O GB 16 IMPERIAL COLLEGE OF SCIENCE, TECHNOLOGY AND MEDICINE

Continuing Education Centre
Imperial College
SW7 2AZ London (United Kingdom)
✆ (44) 71 - 225 8666 Fax: (44) 71 - 225 8668

■ OTHER FIELDS OF ACTIVITIES: post experience education in all areas of science, engineering and medicine.
■ AFFILIATION: a Centre within Imperial college.
■ DATE OF FOUNDATION: 1906.
■ EXECUTIVE BOARD: Dr. Mervyn Evan Jones (*Director*).
■ TEACHING: annual short course in medical ethics.

O GB 17 INSTITUTE OF LAW AND ETHICS IN MEDICINE

University of Glasgow
Stair Building
G12 8QQ Glasgow (United Kingdom)
✆ (44) 041 330 5577 Fax: (44) 041 330 4698

■ AFFILIATION: multi-disciplinary Centre within the University of Glasgow.
■ DATE OF FOUNDATION: 1985.
■ EXECUTIVE BOARD: Professor Sheila Mc Lean (*Director*), Professor Bryan Jennett (*Executive Director*), Dr. Mary Gilhooly (*Research co-ordinator*), Professor Robin Downie (*Executive member*), Dr. Edith Hillan (*Executive member*).
■ TEACHING :the Institute offers a part-time degree of M. Philosophy in Law and Ethics in Medicine over 2 years. Supervision is also available for the degrees of LL. M; and Ph. D.

O GB 18 INSTITUTE OF MEDICAL ETHICS (IME)

Department of Medicine – Royal Infirmary
Lauriston Place
EH3 9YW Edinburgh (United Kingdom)
✆ (44) 31 229 2477 Fax: (44) 31 225 6485

■ DATE OF FOUNDATION: 1964.
■ EXECUTIVE BOARD: Professor Thomas Oppé (*Chairperson*), Professor Richard West (*General Secretary*), Dr. Raanan Gillon (*Editor*), Dr. Kenneth Boyd (*Director of Research*), Sir Douglas Black (*Chairperson*).
■ TEACHING: staff and associates teach medical and nursing ethics in their local medical and nursing colleges. — The Institute conducts enquiries into bioethics teaching in the UK.
■ PUBLICATIONS: Journal of Medical Ethics, Quarterly. — Other publications.
■ DECISIONS, ADVICE: medical and nursing ethics in publications.
■ WORKING ON: ethical rules and principles [B47], bioethics [B89], philosophical ethics [C2], professional ethics [C51], ethics and humanities [C54], biomedical research [F2], experimentation [F9], health policy [G13], health facilities [G22], health care [G45], family practice [G60], health occupations [G61], biological specimens procurement, blood transfusion, organ transplantation [I8], treatment [I30], patients' rights [J2], confidentiality [J20], childhood difficulties, diseases, protection [L21], aged — problems related with aging [L45], sexuality and procreation — unborn child [M1], communicable diseases [N27], death and resuscitation [O1], biotechnology — genetic engineering [Q20].

O GB 19 INSTITUTE OF MEDICAL ETHICS — MEDICAL POSTGRADUATE DEPARTMENT (IME)

Canynge Hall
Whiteladies Road
BS8 2PR Bristol (United Kingdom)
✆ (44) (0272) 303030 Fax: (44) (0272) 733929

■ DATE OF FOUNDATION: 1963.
■ EXECUTIVE BOARD: Professor Sir Douglas Black (*Chairperson*), Professor Thomas Oppé (*Chairperson*), Professor Richard West (*General Secretary*).
■ PUBLICATIONS: Journal of Medical Ethics (Jointly with BMJ). — Papers and books on ethical issues.
■ COLLOQUIUMS, SYMPOSIUMS: International Meeting of Medical Ethics, London September 1993.
■ WORKING ON: Philosophical ethics [C2], ethical analysis [C3], values [C15], quality of life [C16], social worth [C17], value of life [C18], medical ethics [C52], nursing ethics [C53], biomedical technologies

[E28], clinical trials [F3], health services research [F4], nontherapeutic research [F5], animal experimentation [F10], human experimentation [F11], healthy volunteers [F13], health legislation [G14], hospital [G28], hospices [G34], nurses [G39], physicians [G40], health care delivery [G48], medical records [G50], selection for treatment [G52], patient care [G53], diagnosis [I2], patients' rights [J2], legislation and law [K34], stages of life — problems proper to childhood and elderly [L1], sexuality and procreation — unborn child [M1].

O GB 21 **MEDICAL ETHICS COMMITTEE AND BRITISH MEDICAL ASSOCIATION**
BMA House
Tavistock Square
WCIH 9JP London (United Kingdom)
✆ (44) 071 387 4499 Fax: (44) 071-383 6233

O GB 22 **OXFORD PRACTICE SKILLS PROJECT**
John Radcliffe Hospital
Headington
OX 3 9 DU Oxford (United Kingdom)
✆ (44) 865 221 972 Fax: (44) 865 750 750

O GB 23 **ROYAL COLLEGE OF GENERAL PRACTITIONERS**
14 Prince's Gate
Hyde Park
SW7 1PU London (United Kingdom)
✆ (44) 71 - 581 32 32 Fax: (44) 71 - 225 30 47

O GB 24 **ROYAL COLLEGE OF OBSTETRICIANS AND GYNAECOLOGISTS**
27 Sussex Place
Regent's Park
NW 1 4RG London (United Kingdom)
✆ (44) 71 - 262 5425 Fax: (44) 71 - 723 0575

O GB 25 **ROYAL COLLEGE OF PATHOLOGISTS**
2 Carlton House Terrace
SW1 Y5AF London (United Kingdom)
✆ (44) 71 - 930 5861 Fax: (44) 71 - 321 0523

O GB 26 **ROYAL COLLEGE OF PHYSICIANS**
11 St Andrew's Place
NW1 4LE London (United Kingdom)
✆ (44) 71 - 935 1174 Fax: (44) 71 - 487 52 18

O GB 27 **ROYAL COLLEGE OF PHYSICIANS AND SURGEONS OF GLASGOW**
234-242 St Vincent's Street
G2 5RJ Glasgow (United Kingdom)
✆ (44) 41 - 221 6072 Fax: (44) 41 - 221 1804

O GB 28 **ROYAL COLLEGE OF PHYSICIANS OF EDINBURGH**
9 Queen Street
EH2 1QJ Edinburgh (United Kingdom)
✆ (44) 31 - 225 7324 Fax: (44) 31 - 220 3939

O GB 29 **ROYAL COLLEGE OF PSYCHIATRISTS**
17 Belgrave Square
SW1X 8PG London (United Kingdom)
✆ (44) 71 235 2351 Fax: (44) 71 245 1231

■ OTHER FIELDS OF ACTIVITIES: Royal Medical college Concerned with educational standards (postgraduate), organisation of the recognised postgraduate qualifying examination in psychiatry, promotion of study and research in psychiatry, publication of the British Journal of Psychiatry. The Royal College of Psychiatrists is also concerned with general professional issues.
■ OFFICIALLY REGULATED ORGANISATION: college Bye-laws and regulations.
■ DATE OF FOUNDATION: 16 June 1971.
■ EXECUTIVE BOARD: Professor A.C.P. Sims (*Chairperson*), Professor I.. Kolvin (*Vice-chairperson*), Professor AH. Crisp (*Vice-chairperson*), Dr. W.D. Boyd (*Treasurer*), Professor A.M.G. Gath (*Registrar*),

Dr. F. Caldicott (*Dean*).
■ TEACHING: lectures and courses for Junior Psychiatrists.
■ PUBLICATIONS: some publications in the British Journal of Psychiatry.
■ DOCUMENTATION CENTRE: internal.
■ COLLOQUIUMS, SYMPOSIUMS: regular courses held twice a year for Junior Psychiatrists, venues vary.
■ DECISIONS, ADVICE: the Royal College of Psychiatrists produces guidelines on many aspects of psychiatric care which are available to its membership.
■ WORKING ON: review [A2], data bases [A3], bibliography [A4], literature [A7], documentation [A8], codes of ethics [B49], professional competence [B52].

O GB 30 ROYAL COLLEGE OF RADIOLOGISTS

38 Portland Place
W1N 3DG London (United Kingdom)
✆ (44) 71 - 636 4432 / 3 Fax: (44) 71 - 323 3100

O GB 31 ROYAL COLLEGE OF SURGEONS OF EDINBURGH

18, Nicholson Street
EH8 9DW Edinburgh (United Kingdom)
✆ (44) 31 - 556 6206 Fax: (44) 31 - 557 6406

O GB 32 THE LINACRE CENTRE FOR HEALTH CARE ETHICS

60 Grove End Road
St. John's Wood
NW8 9 NH London (United Kingdom)
✆ (44) 71- 289 36 25 Fax: (44) 71- 266 48 13

O GB 33 THE LINACRE CENTRE FOR THE HEALTH CARE ETHICS

60 Grove End Road
St. John's Wood
NW8 9NH London (United Kingdom)
✆ (44) 71 289 3625

■ DATE OF FOUNDATION: May 1977.
■ EXECUTIVE BOARD: Luke Gormally (*Director*), Agneta Sutton (*Deputy director*), Ailsa Connolly (*Secretary*), Dr. Helen Watt (*Research Fellow*), The Rev. Dr. John Berry (*Associate Research Fellow*), Dr. Teresa Iglesias (*Honorary Associate Fellow*).
■ TEACHING: intensive courses for doctors, nurses, hospital chaplains.
■ PUBLICATIONS: series of studies in bioethics.
■ DECISIONS, ADVICE: consultants to various church bodies (especially the Catholic Bishops' Joint Committee on bioethical issues) and to catholic hospitals.
■ WORKING ON: dignity [B5], integrity [B8], justice [B9], human person [B12], respect [B63], virtues [B64], solidarity [B65], biomedical ethics education [B78], value of life [C18], Roman catholic ethics [C23], natural law [C24], double effect [C46], medical ethics [C52], nursing ethics [C53], killing [D90], justifiable killing [D91], human genome project [F8], health care delivery [G48], selection for treatment [G52], resource allocation [H4], health care costs [H9], elderly [H23], fetal tissue transplantation [I27], treatment refusal [J15], advance directives [J18], handicapped children [L38], infanticide [L42], health care services and aged [L48], death [O5], brain death [O6], life-sustaining treatment [O9], withholding treatment [O10], allowing to die [O11], terminally ill [O12], prolongation of life [O13], quality adjusted life years [O14], euthanasia [O15], genetic counseling [Q36], genetic screening [Q37], preimplantation genetic diagnosis [Q38], eugenics [Q39], gene therapy [Q42], human genome [Q47].
■ INTERESTED IN: human genome project [F8], health care delivery [G48], selection for treatment [G52], resource allocation [H4], health care costs [H9], health care services and aged [L48], death [O5], brain death [O6], genetic screening [Q37], preimplantation genetic diagnosis [Q38], eugenics [Q39], gene therapy [Q42], human genome [Q47].

O GB 34 THE MEDICAL RESEARCH COUNCIL (MRC)

20, Park Crescent
W1N 4AL London (United Kingdom)
✆ (44) 71 636 5422 Fax: (44) 71 436 6179

■ OTHER FIELDS OF ACTIVITIES: The MRC'S aim is to promote research which has as its ultimate objective the maintenance and improvement if human health. It currently supports research in the following areas: mental health, AIDS, Cancer, disorders of the body's various systems (e.g. nervous systems),

UNITED KINGDOM (UK)

the effectiveness of different types of health care, environmental influences on health, infections diseases and inherited diseases.
■ OFFICIALLY REGULATED ORGANISATION: the Royal Charter of the Medical Research Council.
■ DATE OF FOUNDATION: 1913.
■ EXECUTIVE BOARD: Sir David Plastow (*Chairperson*), Professor IV Allen (*Neuropathology, Belfast*), Professor D.T. Baird (*Obstetrics & Gynaecology, Edinburgh*), R.P. Bauman (*Chief Executive, Smithkline Beecham*), Professor C.L. Berry (*Morbid Anatomy, London*), Professor A.M. Breckenridge (*Clinical Pharmacology, Liverpool*), Dr. K.C. Calman (*Chief Medical Officer, Department of Health*), Sir Michael Carlisle, Dr.. J.T. Carter (*Health and Safety Executive*), Dr. P. Doyle (*ICI*)

O GB 35 THE NUFFIELD COUNCIL ON BIOETHICS

28 Bedford Square
WC1B 3EG London (United Kingdom)
✆ (44) 71 631 05 66 Fax: (44) 71 323 48 77

■ DATE OF FOUNDATION: 1991.
■ EXECUTIVE BOARD: Rt Hon Sir Patrick Nairne (*Chairperson*), Beverly Anderson, Margaret Auld, Professor Canon Gordon Dunstan (*Reverend*), Professor John Gurdon, Professor Eve Johnstone (*Psych.*), Professor Ian Kennedy, Dr. Anne McLaren, Caroline Miles, Dr. Brian Newbould
■ PUBLICATIONS: Studies (forthcoming in 1993) on medical & scientific uses of human tissue & genetic screening.
■ DOCUMENTATION CENTRE: internal.
■ COLLOQUIUMS, SYMPOSIUMS: as yet no programme developed.
■ WORKING ON: genetic defects and hereditary diseases [Q2], sex linked defects [Q19], DNA fingerprinting [Q32], medical genetics [Q35], genetic counseling [Q36], genetic screening [Q37], preimplantation genetic diagnosis [Q38], eugenics [Q39], negative eugenics [Q40], positive eugenics [Q41], patients' rights [J2], diagnosis [I2], fetal tissue transplantation [I27], intracerebral fetal tissue transplantation [I28], bone marrow transplantation [I29].
■ INTERESTED IN: genetics and applications — biotechnology [Q1], patients' rights [J2], diagnosis [I2], biological specimens procurement, blood transfusion, organ transplantation [I8], respect of human dignity and human rights [B1], ethical rules and principles [B47], bioethics [B89], political activity [D57], congenital defects [N23].

O GB 36 THE ROYAL COLLEGE OF PSYCHIATRISTS

University of Oxford - Philosophy group
Warneford Hospital
OX3 7JX Oxford (United Kingdom)
✆ (44) 235 2351 Fax: (44) 245 1231

O GB 37 UNIT FOR THE STUDY OF HEALTH CARE ETHICS

Department of General Practice, Univ. of LIverpool
Whelan Building, P.O. Box 147
L 69 3BX Liverpool (United Kingdom)
✆ (44) 51 - 794 5608 Fax: (44) 51 794 5604

■ OTHER FIELDS OF ACTIVITIES: limited interest in other areas of social ethics.
■ AFFILIATION: University of Liverpool. — Faculty of Medicine/Department of General Practice.
■ DATE OF FOUNDATION: 1988.
■ EXECUTIVE BOARD: Dr. Heather Draper (*Head of Unit*), Dr. Peter Campion, Professor Ian Stanley.
■ TEACHING: Responsible for all undergraduate medical students ethics training. — Runs an M.Sc. in the Ethics of Health Care. — Contributes to courses elsewhere outside the University on request.

Individuals

GB 1 ALDERSON Priscilla

Secretary, Consumers for Ethics in Research (forum for research subjects and Health Service users and all others concerned with Health Research).

Consumers for Ethics in Research (CERES)
PO Box 1365
N16 OBW London (United Kingdom)

■ TEACHING: Multi disciplinary conferences.
■ PUBLICATIONS: yes.

GB 2 **BAILLIE** Ian D.
Director of Social Work, Board of Social Responsibility, Social Interests Section (Church of Scotland).

Board of Social Responsibility
Church of Scotland
121, George Street
EH2 4YN Edinburgh (United Kingdom)
✆ (44) 31 - 225 5722 Fax: (44) 31 - 225 7867

■ TEACHING: The Training Section educates care staff concerning the professional, social, moral and ethical issues in which they work.
■ PUBLICATIONS: yes.
■ WORKING ON: literature [A7], documentation [A8], publications [A9], respect of human dignity and human rights [B1], fundamental rights of the individuals [B2], self determination [B4], dignity [B5], women's rights [B6], human equality [B7], human person [B12], human body commercialization [B13], traffic of organs [B15], ethics committees [B22], patient advocacy [B42], common good [B45], ethical rules and principles [B47], ethics [B48], codes of ethics [B49], moral obligations [B50], professional competence [B52], obligations to society [B54], moral policy [B55], morality [B61], respect [B63], virtues [B64], information — communication — media [B68], communication [B69], interdisciplinary communication [B70], information dissemination [B71], mass media [B72], audiovisual aids [B74], education [B76], health education [B77], biomedical ethics education [B78], bioethics [B89], codes of biomedical ethics [B90], bioethical issues [B96], specific approaches to ethics [C1], philosophical ethics [C2], ethical analysis [C3], morals [C12], conscience [C13], moral development [C14], values [C15], quality of life [C16], social worth [C17], value of life [C18], religion and religious ethics [C19], clergy [C20], religious ethics [C21].
■ INTERESTED IN: literature [A7], documentation [A8], respect of human dignity and human rights [B1], fundamental rights of the individuals [B2], ethics committees [B22], ethical rules and principles [B47], codes of ethics [B49], professional competence [B52], obligations to society [B54], moral policy [B55], morality [B61], information — communication — media [B68], mass media [B72], biomedical ethics education [B78], bioethics [B89], codes of biomedical ethics [B90], bioethical issues [B96], quality of life [C16], social worth [C17], value of life [C18], religion and religious ethics [C19], religious ethics [C21], Christian ethics [C22], Roman catholic ethics [C23], eastern orthodox ethics [C26], Protestant ethics [C27], social sciences [C89], social and political issues [D1], social issues [D2], legislation [D13], government regulation [D17], family members [D24], single persons [D25], couple [D26], cohabitation [D27], married persons [D28], marital relationship [D29], sibling [D30], family relationship [D31], parents [D32], mothers [D33], fathers [D34], females [D35], handicapped [D43], males [D45], social problems [D52], community services [D78], research [E14], research policy [E18], research design [E19].

GB 3 **BRAZIER** Margaret
Professor of Law and Legal Studies. — Director of the Centre for Social Ethics and Policy, University of Manchester.

Faculty of Law and Legal Studies — Centre for Social Ethics and Policy
University of Manchester
Mansfields Cooper Bulding, Oxford Road
M13 9PL Manchester (United Kingdom)
✆ (44) 61 - 275 3593 Fax: (44) 61 - 275 3579

■ TEACHING: Teaching M.A. in Health Care Ethics, University of Manchester. — Lecturing to Health Care and Legal Professionals.
■ PUBLICATIONS: yes.
■ WORKING ON: human rights [B3], self determination [B4], women's rights [B6], human body commercialization [B13], reification [B14], animal care committees [B21], European Convention on Human Rights [B36], codes of ethics [B49], biomedical research [F2], patients' rights [J2], battery [K10].
■ INTERESTED IN: *idem.*

GB 4 **BRUCE** Donald
Director of the Society, Religion and Technology Project, working with an advisory committee and reporting to the funding body (the Church of Scotland) through the National Mission Department.

Society, Religion and Technology Project
John Knox House
45 High Street
EH1 1SR Edinburgh (United Kingdom)
✆ (44) 31 556 2953 Fax: (44) 31 556 7478

■ PUBLICATIONS: yes.

GB 5 **CALLAGHAN** Brendan
Principal of Heythrop College, University of London.

Heythrop College
11-13 Cavendish Sq.
WOM 0AN London (United Kingdom)
✆ (44) (71) 580 69 41 Fax: (44) (71) 580 50 31

■ AFFILIATIONS: Institute of Medical Ethics, Bristol, United Kingdom (*Member of Governing Body*). — West Lambeth Area Health Authority Ethics Committee, London, United Kingdom (*Member of Governing Body*).
■ PUBLICATIONS: yes.

GB 6 **CALMAN** K.C.
Chief Medical Officer, Department of Health.

585 Anniesland Road
G13 1UX Glasgow (United Kingdom)

■ TEACHING: Medical students, doctors, public health workers.
■ PUBLICATIONS: yes.
■ WORKING ON: health care, medical acts [I1], health care professionals, researchers and patient relationship [J1].
■ INTERESTED IN: *idem.*

GB 7 **CHADWICK** Ruth
Lecturer in Philosophy. — Director of the Centre for Applied Ethics.

University of Wales College, Cardiff
PO Box 94
CF1 3XB Cardiff (United Kingdom)
✆ (44) 222-874 025 Fax: (44) 222-874 242

■ AFFILIATIONS: European Society for the Philosophy of Medicine and Health Care — ESPMH, Maastricht, Netherlands (*Member*).
■ TEACHING: Lectures at University of Wales College of Medicine. — Lectures to Philosophy students at University of Wales, Cardiff. — Study for nurses.
■ PUBLICATIONS: yes.

GB 8 **COCE** Anthony
Clinician.

Worcester Royal Infirmary
Castle Street
WRI 3 AS Worcester (United Kingdom)
✆ (44) 0905 76 33 33

■ AFFILIATIONS: Linacre Centre, London, United Kingdom (*Governor*).
■ TEACHING: Symposia Courses in Bioethics.
■ PUBLICATIONS: yes.

GB 9 **COTTEY** Alan
Lecturer in Physics, University of East Anglia.

School of Physics
University of East Anglia
NR4 7TJ Norwich (United Kingdom)
✆ (44) 603 592 596 Fax: (44) 603 259 513

■ TEACHING: Coordinates Higher Education course on Science, Values and Ethics.
■ WORKING ON: science [E2].
■ INTERESTED IN: review [A2], data bases [A3], bibliography [A4], nuclear warfare [D89], science [E2].

GB 10 **DANCY** Jonathan
Professor of Philosophy.

University of Keele
Department of Philosophy
Keele
ST 5 5BG Staffs (United Kingdom)
✆ (44) 782 - 62 11 11 Fax: (44) 782 - 613847

■ AFFILIATIONS: Centre for Contemporary Ethics, Keele University, Staffs ST5 5BG, United Kingdom (*Member*).
■ WORKING ON: specific approaches to ethics [C1], values [C15], value of life [C18], consequences [C45], double effect [C46], philosophy [C61], humanism [C66], judgement [C71], intention [C75], motivation [C76], killing [D90], justifiable killing [D91].
■ INTERESTED IN: situational ethics [C9], motivation [C76], judgement [C71], consequences [C45].

GB 11 **DE RAEVE** Anne-Louise
Macmillan Lecturer in Nursing Ethics, Centre for Philosophy and Health Care, University College of Swansea.

Centre for Philosophy and Health Care
University College of Swansea
SA2 8PP Swansea (United Kingdom)
✆ (44) 792 29 56 11 et 792 29 56 12 Fax: (44) 792 29 57 69

■ TEACHING: Participating in teaching the MA degree course run by the department. — Teaching nurses elsewhere on a peripatetic and sessional basis, usually as a response to requests.
■ WORKING ON: integrity [B8], codes of ethics [B49], altruism [B56], ethical analysis [C3], morals [C12], conscience [C13], nursing ethics [C53], emotions [C79], love [C80], trust [C81], counseling [C87], pastoral care [C88], teaching methods [E59], hospital [G28], hospices [G34], nurses [G39], patient care [G53], patient information [J4], patient participation [J16], nurse patient relationship [J30], physician nurse relationship [J31], cancer [N17], palliative care [O8], withholding treatment [O10], allowing to die [O11], terminally ill [O12].
■ INTERESTED IN: *idem.*

GB 12 **DOWNIE** R.S.
Executive Director, Institute of Law and Ethics in Medicine. — Professor of Moral Philosophy.

University of Glasgow
Department of Philosophy
G12 8 QQ Glasgow (United Kingdom)
✆ (44) 41 - 339 8855

■ AFFILIATIONS: Institute of Law and Ethics in Medicine — ILEM, Glasgow, United Kingdom
■ TEACHING: Degree courses (medical undergraduate, nursing undergraduate, medical postgraduate, nursing postgraduate).
■ PUBLICATIONS: yes.
■ WORKING ON: ethical rules and principles [B47], education [B76], health education [B77], biomedical ethics education [B78], nursing education [B79], medical education [B80], history of biomedical ethics [B94], specific approaches to ethics [C1], professional ethics [C51], nursing ethics [C53], ethics and humanities [C54], social sciences [C89], inter-personal relationship [J27], parent child relationship [J34].
■ INTERESTED IN: *idem.*

GB 13 **DOYAL** Len
Senior lecturer in Medical Ethics.

St Bartholomew and London Hospital
Medical College
Turner Street
E1 2AD London (United Kingdom)
✆ (44) 71 601 88 88 Fax: (44) 71 251 87 24

■ TEACHING: The sole Senior Lecturer in Medical Ethics in the UK employed by a Medical School. Saint-Bartholomew's and the London Hospital Medical Colleges have the most exclusive programme of

teaching Medical Ethics and Law in the UK.
■ PUBLICATIONS: yes.
■ WORKING ON: information sources, data bases [A1], information centers [A5], book review [A6], documentation [A8], publications [A9], literature [A7], respect of human dignity and human rights [B1], protection of rights — involved institutions [B16], ethical rules and principles [B47], information — communication — media [B68], bioethics [B89], philosophical ethics [C2], professional ethics [C51], medical ethics [C52], ethics and humanities [C54], social and political issues [D1], life sciences [E4], containment [E38], biomedical research [F2], methodology of research and experimentation [F15], health [G2], health policy [G13], health facilities [G22], organization of health care [G35], health care [G45], immunization [G59], economics [H2], diagnosis [I2], treatment [I30], patients' rights [J2], confidentiality [J20], inter-personal relationship [J27], legal personality [K2], jurisprudence — accountability [K8], legislation and law [K34].
■ INTERESTED IN: anatomy, physiology, development [L2], childhood difficulties, diseases, protection [L21], aged — problems related with aging [L45], sexuality and procreation — unborn child [M1], reproductive technologies [M26], pregnancy and childbirth [M39], abortion [M47], fetal development [M54], common symptomatology [N2], cancer [N17], cardiovascular diseases [N19], diabetes [N22], congenital defects [N23], communicable diseases [N27], eye diseases [N38], nervous system diseases [N40], injuries, occupational diseases, intoxications [N46], death and resuscitation [O1], neurosciences — psychiatry [P1], genetics and applications — biotechnology [Q1].

GB 14 **DUNSTAN** G.R.

Professor Emeritus of Moral and Social Theology, University of London. — Honorary Research Fellow, University of Exeter.

9, Maryfield Av.
Pennsylvania
Exeter EX4 6 JN Devon (United Kingdom)
✆ (44) 392 - 214 691

■ AFFILIATIONS: Nuffield Council on Bioethics, London, United Kingdom (*Member*).
■ TEACHING: Training courses for medical doctors, nursing and administrative members of the Health Services, community of specialist interest groups (the clergy, etc.).
■ PUBLICATIONS: yes.

GB 15 **EVANS** Donald

Director of the Centre for Philosophy and Health Care.

Centre for Philosophy and Health Care
University College, Swansea
SA2 8 PP Swansea (United Kingdom)
✆ (44) 792 29 56 12 Fax: (44) 792 29 57 69

■ TEACHING: On masters, doctoral degrees and Diploma in Ethics of Health Care. — Diploma in Health Care Law. — Training conferences for members of research ethics committees. — Health module for management post-graduates. — Professional development courses for general practitioners and nurses.
■ PUBLICATIONS: yes.

GB 16 **EVANS** Martyn

University Lecturer. — Sub-Dean, Faculty of Health Care Studies.

Centre for Philosophy and Health Care
University College of Swansea
Singleton Park
SA2 8 PP Swansea (United Kingdom)
✆ (44) 792 - 20 56 78
■ AFFILIATIONS: International Programme in Bioethics Education and Research, Nijmegen, Netherlands (*Scholar*).
■ TEACHING: Postgraduate degrees of University of Wales, taught and supervised by the Centre for Philosophy and Health Care. — European Seminar in Bioethics, taught by International Programme in Bioethics Education and Research.
■ PUBLICATIONS: yes.
■ WORKING ON: human person [B12], human body commercialization [B13], traffic of organs [B15], moral obligations [B50], altruism [B56], wedge argument [C4], deontological ethics [C6], metaethics [C11], clinical trials [F3], nontherapeutic research [F5], therapeutic research [F6], human experimentation [F11], research subjects [F12], control groups [F17], random selection [F18], group of vulnerable subjects [F20], resource allocation [H4], costs and benefits [H10], risks and benefices

[H11], tissue donation [I13], donors [I14], organ donors [I15], donor cards [I16], anonymous donation [I17], transplantation [I22], transplant recipients [I23], organ transplantation [I24], organ donation [I25], organ procurement [I26], fetal tissue transplantation [I27], intracerebral fetal tissue transplantation [I28], informed consent [J9], parental consent [J13], treatment refusal [J15], living wills [J19], life extension [L47], aging [L50], death [O5], life-sustaining treatment [O9], withholding treatment [O10].
■ INTERESTED IN: audiovisual aids [B74], self concept [C83], egoism [C84], mentally ill [P9], psychotic disorders [P10], dementia [P11], schizophrenia [P12], psychiatric diagnosis [P29].

GB 17 **FARSIDES** Callope

Lecturer in Philosophy. — Director of the Centre for Contemporary Ethical Studies, Keele University. — Teaching and Research in Health Care Ethics. — Member of Local Research Ethics Committee (Deputy Chairperson).

University of Keele
Department of Philosophy
Keele
ST5 5 BG Staffs (United Kingdom)
✆ (44) 782 - 62 11 11 Fax: (44) 782 - 61 38 47

■ AFFILIATIONS: United Kingdom Forum on Health Care Ethics and Law, London, United Kingdom (*Committee Member*).
■ TEACHING: MA Medical Ethics (Course Supervisor). — Medical Ethics for MBA Health Executive Programme, Centre for Health Planning and Management, Keele University. — Diploma/MA Primary Case, Department of Postgraduate Medicine, Keele University.
■ PUBLICATIONS: yes.
■ WORKING ON: women's rights [B6], human body commercialization [B13], coercion [D5], hospices [G34], ecology [E43], patient compliance [G49], obstetrics and gynecology [G66], resource allocation [H4], consent to treatment [J8], parental consent [J13], advance directives [J18], personhood [K3], HIV seropositivity [N36], palliative care [O8], euthanasia [O15], biotechnology [Q21].
■ INTERESTED IN: reproductive technologies [M26], multiple pregnancy [M46], fetal therapy [M57], HIV seropositivity [N36], medical genetics [Q35], gene therapy [Q42], genetics and applications — biotechnology [Q1], acquired immunodeficiency syndrome [N35].

GB 18 **FINNIS** John

Professor of Law and Legal Philosophy (Oxford University).

University College
Highstreet
OX1 4BH Oxford (United Kingdom)
✆ (44) 865 276 602 Fax: (44) 865 276 675

■ AFFILIATIONS: Linacre Centre for Health Care Ethics, London, United Kingdom (*Vice-Chairperson*).
■ PUBLICATIONS: yes.
■ WORKING ON: ethics [B48], moral policy [B55], fundamental rights of the individuals [B2], philosophical ethics [C2], deontological ethics [C6], Roman catholic ethics [C23], double effect [C46], intention [C75], advance directives [J18], wrongful life [K33], in vitro fertilization [M29], euthanasia [O15].
■ INTERESTED IN: GIFT (Gamete Intrafallopian Transfer) [M37], deontological ethics [C6], double effect [C46], intention [C75], virus diseases [N37], euthanasia [O15].

GB 19 **FRITH** Lucy

Lecturer in Social Ethics, St Martin's College. — Teaching. — Researching.

St Martin's College
HA1 3JD Lancaster (United Kingdom)
✆ (44) 05 24 63 446

■ TEACHING: Medical Ethics teaching at St Martin's to B.A. Honours students, student nurses, medical and health care professionals.
■ WORKING ON: reproductive technologies [M26], pregnancy and childbirth [M39], fetal development [M54], methodology of research and experimentation [F15].
■ INTERESTED IN: sexuality and procreation [M2], reproductive technologies [M26], pregnancy and childbirth [M39], fetal development [M54].

GB 20 **FULFORD** K.W.M.

Research Fellow, Green College Oxford. — Director, Oxford Practice Skills Project, University of Oxford Medical School.

University of Oxford
Dept. of Psychiatry
Warneford Hospital
OX3 7JX Oxford (United Kingdom)
✆ (44) 865 22 64 82 Fax: (44) 865 22 19 72

■ AFFILIATIONS: Royal College of Psychiatrists (Philosophy Section), London, United Kingdom (*Chairperson*).
■ TEACHING: Teaching Medical Ethics (clinical medical students and post-graduate students). — Lecturing to postgraduates on ethical and concrete aspects of psychiatry.
■ PUBLICATIONS: yes.
■ WORKING ON: review [A2], book review [A6], publications [A9], respect of human dignity and human rights [B1], competence [B51], biomedical ethics education [B78], ethical analysis [C3], metaethics [C11], values [C15], professional ethics [C51], determinism [C62], psychology [C68], mental processes [C69], intention [C75], motivation [C76], normality [C78], dissent [D58], psychiatry [G69], consent to treatment [J8], personhood [K3], legally incompetent person [K4], potentiality of personhood [K5], psychotic disorders [P10], outpatient commitment [P18], psychiatric diagnosis [P29].
■ INTERESTED IN: information sources, data bases [A1], respect of human dignity and human rights [B1], competence [B51], biomedical ethics education [B78], ethical analysis [C3], metaethics [C11], values [C15], professional ethics [C51], determinism [C62], psychology [C68], mental processes [C69], intention [C75], motivation [C76], normality [C78], dissent [D58], psychiatry [G69], consent to treatment [J8], personhood [K3], legally incompetent person [K4], potentiality of personhood [K5], psychotic disorders [P10], outpatient commitment [P18], psychiatric diagnosis [P29].

GB 21 **GILLON** Raanan

Director, Imperial College Health Service, London University. — Visiting Professor of Medical Ethics, Saint Mary's Hospital Medical School/Imperial College. — Editor (Journal of Medical Ethics).

Imperial College
14, Prince's Gardens
SW7 1 NA London (United Kingdom)
✆ (44) 071 584 6301 Fax: (44) 071 823 8681

■ AFFILIATIONS: Institute of Medical Ethics, Bristol, United Kingdom (*Editor, member of governing body*).
■ TEACHING: Teaching Clinical Medical Students at St Mary's Hospital Medical School (London University). — One week intensive course in Medical Ethics for medical and nursing teachers and members of Ethics Committees at Imperial College. — Lectures in a variety of invited contexts.
■ PUBLICATIONS: yes.
■ WORKING ON: medical ethics [C52], human person [B12], biomedical ethics education [B78], bioethical issues [B96], resource allocation [H4], consent to treatment [J8], advance directives [J18], living wills [J19], physician's role [J26], personhood [K3], human characteristics [L19], mind [L20], involuntary sterilization [M7], mother fetus relationship [M43], persistent vegetative state [N42], death [O5], brain death [O6], withholding treatment [O10], euthanasia [O15], gene therapy [Q42].
■ INTERESTED IN: information sources, data bases [A1], respect of human dignity and human rights [B1], protection of rights — involved institutions [B16], ethical rules and principles [B47], bioethics [B89], philosophical ethics [C2], religion and religious ethics [C19], philosophy of biology [C42], professional ethics [C51], ethics and humanities [C54], coercion [D5], punishment [D8], prisoners [D12], scarcity [D21], violence [D85], scientific misconduct [E25], technology [E27], occupational medicine [E46], health hazards [E49], methods [E53], evaluation [E61], biomedical research [F2], experimentation [F9], methodology of research and experimentation [F15], maternal health [G3], mental health [G4], health legislation [G14], mandatory programs [G19], artificial organs [G24], health services [G25], intensive care units [G33], hospices [G34], WHO — World Health Organization [G36], selection for treatment [G52], patient care [G53], medicine [G63], sports medicine [G64], occupational medicine [G65], economics [H2], resource allocation [H4], population control [H22], elderly [H23], prenatal diagnosis [I3], mandatory screening [I7], required request [I10], donors [I14], organ donors [I15], donor cards [I16], anonymous donation [I17], transplantation [I22], alternative therapies [I33], extraordinary treatment [I39], health care professionals, researchers and patient relationship [J1], law, legislation and jurisprudence [K1], stages of life — problems proper to childhood and elderly [L1], sexuality and procreation — unborn child [M1], disease [N1], death and resuscitation [O1], neurosciences — psychiatry [P1], genetics and applications — biotechnology [Q1].

GB 22 **GORMALLY** Luke

Director of the Linacre Centre for Health Care Ethics.

The Linacre Centre for Health Care Ethics
60, Grove End Road
St. John's Wood
NW8 9 NH London (United Kingdom)
✆ (44) 71- 266 4813 et 790 7681 Fax: (44) 71 - 266 2316

■ TEACHING: Intensive Courses organised by the Linacre Centre. — Courses organised by hospitals and postgraduate medical centres.
■ PUBLICATIONS: yes.
■ WORKING ON: dignity [B5], integrity [B8], justice [B9], human person [B12], common good [B45], respect [B63], virtues [B64], biomedical ethics education [B78], natural law [C24], double effect [C46], cultural pluralism [D51], killing [D90], health care delivery [G48], resource allocation [H4], beginning of life [L24], embryos [M10], brain death [O6], prolongation of life [O13], euthanasia [O15].
■ INTERESTED IN: health care delivery [G48], resource allocation [H4], beginning of life [L24], embryos [M10], brain death [O6], prolongation of life [O13], euthanasia [O15].

GB 23 **GUNNING** Jennifer

Head of Assessment Branch at the Agricultural and Food Research Council's Central Office.

Agricultural and Food Research Council
North Star Avenue
SN2 1UH Swindon (United Kingdom)
✆ (44) -793- 413250 Fax: (44) -793- 413201

■ PUBLICATIONS: yes.
■ WORKING ON: ethicists [B92], ethical review [B93], bioethical issues [B96], life sciences [E4], research policy [E18], research design [E19], peer review [E63], social control of science [E64], technology assessment [E65], animal experimentation [F10], human experimentation [F11], healthy volunteers [F13], animal testing alternatives [F14], animal organs [L18], biotechnology — genetic engineering [Q20], transgenic animals [Q29].
■ INTERESTED IN: bioethical issues [B96], European Convention of Bioethics [B97], Bioethics bill, 1992 (France) [B98], life sciences [E4], research policy [E18], scientific misconduct [E25], peer review [E63], social control of science [E64], animal experimentation [F10], human experimentation [F11], healthy volunteers [F13], animal testing alternatives [F14], health care delivery [G48], selection for treatment [G52], health care costs [H9], costs and benefits [H10], risks and benefices [H11], biological specimens procurement, blood transfusion, organ transplantation [I8], fetal tissue transplantation [I27], reproductive technologies [M26], embryo donation [M38], abortion [M47], mifepristone [M53], biotechnology — genetic engineering [Q20], transgenic animals [Q29], preimplantation genetic diagnosis [Q38], eugenics [Q39].

GB 24 **HARRIS** John

Professor of Applied Philosophy Research Director, Centre of Social Ethics and Policy.

Centre of Social Ethics and Policy
University of Manchester
Oxford Road
M13 9PL Manchester (United Kingdom)
✆ (44) 61 - 275 3519

■ AFFILIATIONS: International Association of Bioethics, Victoria, Australia (*Director*).
■ TEACHING: Masters degree Programme Health Care Ethics, research students (M.A., Ph.D.).
■ PUBLICATIONS: yes.

GB 25 **HILL** Donald

Visiting Tutor in Medical Ethics (St Mary's Hospital School). —Assistant Editor (Journal of Applied Philosophy).

St Mary's Hospital Medical School
W7 London (United Kingdom)
✆ (44) 071 723 1252 Fax: (44) 071 724 7349

■ AFFILIATIONS: Society for Applied Philosophy (*Workshop Co-ordinator and Executive Commitee Member*).
■ PUBLICATIONS: yes.
■ WORKING ON: respect of human dignity and human rights [B1], ethical rules and principles [B47], ethics [B48], bioethics [B89], ethicists [B92], bioethical issues [B96], specific approaches to ethics [C1], philosophical ethics [C2], professional ethics [C51], medical ethics [C52], philosophy [C61], health care professionals, researchers and patient relationship [J1], physician's role [J26], physician patient relationship [J32], death and resuscitation [O1].
■ INTERESTED IN: bioethics [B89], ethicists [B92], bioethical issues [B96], medical ethics [C52], withholding treatment [O10], suicide [O19], brain death [O6].

GB 26 HOPE Tony

Project Leader, Oxford Practiced Skills Project.

Oxford Practiced Skills Project – University of Oxford Medical School
John Radcliffe Hospital
Headington
OX 3 9 DU Oxford (United Kingdom)
✆ (44) 865 - 22 19 72 Fax: (44) 865 - 75 07 50

■ TEACHING: The Central purpose of the OPSP is to develop and evaluate a teaching programme for clinical medical students in Ethics, Communication Skills and the Law.
■ PUBLICATIONS: yes.

GB 27 HORNETT Stuart Ian

Barrister-at-law.

Centre for Health Care
16, Links Avenue, Gidea Park
RM2 GND Essex (United Kingdom)
✆ (44) 0708- 74 28 15 Fax: (44) 533 522 200

■ AFFILIATIONS: Centre for Health Care Law (University of Leicester), Leicester, United Kingdom (*Associate Researcher*).
■ PUBLICATIONS: yes.
■ WORKING ON: Council of Europe [B25], public advocacy [B41], personhood [K3], legally incompetent person [K4], potentiality of personhood [K5], negligence [K20], therapeutic risk [K32], European law [K49], international law [K50], health legislation [K52], property rights [K46], court decision [K40], medical law [K51].

GB 28 JACKSON Jennifer

Lecturer in Philosophy, Department of Philosophy, University of Leeds. — Director of the Centre for Business and Professional Ethics.

University of Leeds
LS2 9JT Leeds (United Kingdom)
✆ (44) 532-33 32 80 Fax: (44) 532-33 32 65

■ AFFILIATIONS: Centre for Business and Professional Ethics (CBPE), United Kingdom (*Director*). — UK Forum on Health Care Ethics and Law, Cardiff, United Kingdom (*Executive Committee Member*).
■ TEACHING: Conducts an M.A. course in Health Care Ethics. — Teaches undergraduates in Medicine and in Arts and Law. — Teaches short courses (e.g. for clinical pharmacists, nurses).
■ PUBLICATIONS: yes.
■ WORKING ON: human rights [B3], self determination [B4], ethics committees [B22], moral obligations [B50], competence [B51], paternalism [B62], virtues [B64], codes of biomedical ethics [B90], bioethical issues [B96], ethical analysis [C3], trust [C81], consent to treatment [J8], informed consent [J9], presumed consent [J10], third party consent [J11], parental consent [J13], treatment refusal [J15], deception [K14], withholding treatment [O10], allowing to die [O11], prolongation of life [O13], voluntary euthanasia [O18], attitudes to death [O20], involontary commitment [P21].
■ INTERESTED IN: ethics committees [B22], virtues [B64], trust [C81], treatment refusal [J15], deception [K14], allowing to die [O11], withholding treatment [O10].

GB 29 KENNEDY I.M.

Director, Centre of Medical Law and Ethics, King's College, London. — Professor of Medical Law and Ethics, King's College, London.

Centre of Medical Law and Ethics
King's College Strand
WC 2 2LS London (United Kingdom)
✆ (44) 71 873 2382 Fax: (44) 71 873 24 65

■ AFFILIATIONS: Centre of Medical Law and Ethics, London, United Kingdom (*Director*).
■ TEACHING: Teaching university undergraduates and postgraduate students, supervising, presenting seminars for professionals engaged in health care.
■ PUBLICATIONS: yes.

GB 30 **KENNET** Wayland
Honorary Secretary of the Parliamentary and Scientific Commitee, Vice Chairperson of the Parliamentary Office of Science and Technology (British Parliament).

The Parlementiary Office of Science and Technology
House of Lords
SW1 London (United Kingdom)
✆ (44) 71 219 31 41 Fax: (44) 71 219 59 79

■ AFFILIATIONS: House of Lords (*Member*). — Centre for Medical Law and Ethics, Strand, United Kingdom (*Member*).

GB 31 **KEOWN** John
Lecturer in Law, specializing in Medical Law and Ethics, University of Leicester.

Centre for Health Care Law
Faculty of Law
University of Leicester
LEI 7 RH Leceister (United Kingdom)
✆ (44) 533 522 366 Fax: (44) 533 525 023

■ AFFILIATIONS: Forum for Health Care Ethics and the Law (Manchester), United Kingdom (*Member*).
■ TEACHING: Teaching undergraduate lawyers about medical law and ethics.
■ PUBLICATIONS: yes.
■ WORKING ON: biological specimens procurement, blood transfusion, organ transplantation [I8], law, legislation and jurisprudence [K1], sexuality and procreation — unborn child [M1], terminal care [O7], outpatient commitment [P18].
■ INTERESTED IN: *idem.*

GB 32 **LAMB** David
Reader in Philosophy (lectures).

University of Manchester
Oxford Road
M13 9PL Manchester (United Kingdom)
✆ (44) 61 275 20 00

■ TEACHING: Undergraduate courses and post-graduate (Ph.D.), supervision, University of Manchester.
■ PUBLICATIONS: yes.
■ WORKING ON: death and resuscitation [O1], biological specimens procurement, blood transfusion, organ transplantation [I8].
■ INTERESTED IN: death and resuscitation [O1], determination of death [O2], death [O5], brain death [O6], life-sustaining treatment [O9], withholding treatment [O10], allowing to die [O11], euthanasia [O15], active euthanasia [O16], involuntary euthanasia [O17], prolongation of life [O13], biological specimens procurement, blood transfusion, organ transplantation [I8].

GB 33 **MACMILLAN** Maureen S.
Research Fellow.

Edinburgh Medical Group
Nursing Studies Research Unit
12 Buccleuch Place
EH8 Edinburgh (United Kingdom)
✆ (44) 31 650 4276 Fax: (44) 31 667 7938

■ AFFILIATIONS: Nursing Studies Research Unit, Edinburgh, United Kingdom (*Member*).
■ TEACHING: Ethics to medical students and nursing students.
■ PUBLICATIONS: yes.
■ WORKING ON: respect of human dignity and human rights [B1], ethics committees [B22], patient advocacy [B42], ethical rules and principles [B47], quality of life [C16], nursing ethics [C53], research design [E19], research policy [E18], methods [E53], group of vulnerable subjects [F20], patients' rights [J2], aged — problems related with aging [L45], terminal care [O7].
■ INTERESTED IN: patient advocacy [B42], nursing ethics [C53], research design [E19], selection of subjects [F19], group of vulnerable subjects [F20], teaching methods [E59], health care services and aged [L48], aged [L49], terminal care [O7].

GB 34 MAHONEY Jack

Professor of Moral and Social Theology, University of London, King's College.

King's College
University of London Strand
WC2R 2LS London (United Kingdom)
✆ (44) 71 - 873 2587 Fax: (44) 71 873 2265

■ AFFILIATIONS: International Study Group on Bioethics of the International Federation of Catholic Universities — FIUC (*Member*).
■ TEACHING: Courses on bioethics to undergraduate and graduate students of King's College London, and directing postgraduate research students.
■ PUBLICATIONS: yes.
■ WORKING ON: respect of human dignity and human rights [B1], ethics [B48], philosophical ethics [C2], religion and religious ethics [C19], resource allocation [H4], reproductive technologies [M26], biotechnology — genetic engineering [Q20].
■ INTERESTED IN: bioethics [B89], philosophical ethics [C2], religion and religious ethics [C19], drug industry [E37], biomedical research [F2], resource allocation [H4], biological specimens procurement, blood transfusion, organ transplantation [I8], beginning of life [L24], children's rights [L25], life extension [L47], reproductive technologies [M26], abortion [M47], congenital defects [N23], persistent vegetative state [N42], euthanasia [O15], biotechnology — genetic engineering [Q20].

GB 35 MASON J.Kenyon

Professor (Emeritus) of Forensic Medicine. — Honorary Fellow, Faculty of Law.

Faculty of Law (Room 412)
University of Edinburgh
Old College
EH8 9YL Edinburgh (United Kingdom)
✆ (44) 31 650 20 51 Fax: (44) 31 667 49 02

■ AFFILIATIONS: UK Forum on Health Care Ethics and the Law, Cardiff, United Kingdom (*Member*).
■ TEACHING: Undergraduate Honours Courses in Medical Jurisprudence.
■ PUBLICATIONS: yes.
■ WORKING ON: medical ethics [C52], patient advocacy [B42], Roman catholic ethics [C23], sex offenses [D92], state government [D67], child abuse [L41], consent to treatment [J8], treatment refusal [J15], confidentiality [J20], human body commercialization [B13], traffic of organs [B15], selection for treatment [G52], determination of death [O2], terminal care [O7], euthanasia [O15], resource allocation [H4], acquired immunodeficiency syndrome [N35].
■ INTERESTED IN: sexuality and procreation [M2], abortion [M47], organ transplantation [I24], fetal tissue transplantation [I27], intracerebral fetal tissue transplantation [I28], euthanasia [O15], terminal care [O7], consent to treatment [J8].

GB 36 MAYNARD Alan

Director of the Centre for Health Economics.

Centre for Health Economics
University of York
Heslington
YO 1 5 DD York (United Kingdom)
✆ (44) 904 - 433 646 Fax: (44) 904 - 433 644

■ WORKING ON: economics [H2], developing countries [H21], health care, medical acts [I1], patients' rights [J2], disease [N1], terminal care [O7].
■ INTERESTED IN: *idem*.

GB 37 **MC NAUGHTON** David

Lecturer in Philosophy.

University of Keele
Department of Philosophy
Keele
ST5 5BG Staffs (United Kingdom)
✆ (44) 782 583406 Fax: (44) 782 583399

■ TEACHING: Teacher on a postgraduate course in Medical Ethics run jointly by Departments of Law and Philosophy at Keele University.
■ PUBLICATIONS: yes.
■ WORKING ON: human rights [B3], self determination [B4], integrity [B8], freedom [B10], human person [B12], codes of ethics [B49], moral obligations [B50], deontology [B59], paternalism [B62], virtues [B64], ethical analysis [C3], deontological ethics [C6], utilitarianism [C8], situational ethics [C9], metaethics [C11], morals [C12], conscience [C13], values [C15], quality of life [C16], value of life [C18], future generations [C44], consequences [C45], speciesism [C48], medical ethics [C52], philosophy [C61], animal experimentation [F10], human experimentation [F11], research subjects [F12], healthy volunteers [F13], medical staff [G41], consent to treatment [J8], informed consent [J9], presumed consent [J10], third party consent [J11], spousal consent [J12], parental consent [J13], advance directives [J18], living wills [J19], death [O5], brain death [O6], withholding treatment [O10], allowing to die [O11], quality adjusted life years [O14], active euthanasia [O16], voluntary euthanasia [O18], mentally ill [P9].
■ INTERESTED IN: *idem*.

GB 38 **MORTON** David B.

Director of Biomedical Services. — Professor of Biomedical Science and Ethics.

University of Birmingham Medical School
Department of Biomedical Science and Ethics
Edgbaston
B 15 2 TT Birmingham (United Kingdom)
✆ (44) 21 - 414 3616 Fax: (44) 21 - 414 6979

■ AFFILIATIONS: Institute of Medical Ethics, Royal College of Veterinary Surgery, United Kingdom (*Member*).
■ TEACHING: Teaching medical, veterinary and biology undergraduates.
■ PUBLICATIONS: yes.
■ WORKING ON: nontherapeutic research [F5], therapeutic research [F6], animal experimentation [F10], animal testing alternatives [F14].
■ INTERESTED IN: ethical analysis [C3], biology and human future [C43], biological life [C49], primates [C50], biomedical research [F2], nontherapeutic research [F5], therapeutic research [F6], animal experimentation [F10], human experimentation [F11], healthy volunteers [F13], animal testing alternatives [F14], law [K56].

GB 39 **NICHOLSON** Richard H.

Editor, Bulletin of Medical Ethics. — Bioethics publishing and teaching.

Bulletin of Medical Ethics
31, Corsica Street
N5 1LA London (United Kingdom)
✆ (44) (71) 354-4252 Fax: (44) (71) 704-2874

■ AFFILIATIONS: Bioethics Publications, London, United Kingdom (*Chairperson*). — European Association of Centres of Medical Ethics — EACME (*Member*).
■ TEACHING: Teaching Medical Ethics at all levels.
■ PUBLICATIONS: yes.

GB 40 **PARRY** JIm

Senior Lecturer in Philosophy, Leeds University.

Centre for Business and Professional Ethics
University of Leeds
LS 2 9 JT Leeds (United Kingdom)
✆ (44) 532 33 32 60 Fax: (44) 532 33 32 65

■ TEACHING: Teaching on MA course in Health Care Ethics, first year undergraduate in Practical Ethics.

GB 41 **POKINGHORNE** John
Head of a College at Cambridge University.

Lodge Queen's College
CB3 9ET Cambridge (United Kingdom)
✆ (44) 223 335 511 Fax: (44) 223 335 522

■ AFFILIATIONS: Board for Social Responsibility Church of England, London, United Kingdom (*Member*). — Medical Ethics Committee, British Medical Association, London, United Kingdom
■ PUBLICATIONS: yes.
■ WORKING ON: Christian ethics [C22], theology [C41], ecology [E43], embryos [M10], fetuses [M11], fetal therapy [M57].
■ INTERESTED IN: *idem.*

GB 42 **SHAPIRO** David
Executive Secretary, Nuffield Council on Bioethics.

Nuffield Council on Bioethics
The Nuffield Foundation
28 Bedford Square
WCI B 3EG London (United Kingdom)
✆ (44) (71) 631-0566 Fax: (44) (71) 323-4877

■ WORKING ON: social and political issues [D1], genetic defects and hereditary diseases [Q2], DNA fingerprinting [Q32], diagnosis [I2], fetal tissue transplantation [I27], patients' rights [J2].
■ INTERESTED IN: respect of human dignity and human rights [B1], ethical rules and principles [B47], bioethics [B89], social and political issues [D1], diagnosis [I2], biological specimens procurement, blood transfusion, organ transplantation [I8], patients' rights [J2], congenital defects [N23], genetics and applications — biotechnology [Q1].

GB 43 **SHOTTER** Edward
Dean of Rochester.

Dean of Rochester
The Deanery Rochester
MEI IT9 Kent (United Kingdom)
✆ (44) 634 844 023

■ PUBLICATIONS: yes.
■ WORKING ON: morals [C12], religion and religious ethics [C19], clergy [C20], religious ethics [C21], Christian ethics [C22], Roman catholic ethics [C23], natural law [C24], death and resuscitation [O1], determination of death [O2], death [O5], industry [E36].
■ INTERESTED IN: information sources, data bases [A1], ethical rules and principles [B47], ethics [B48], morality [B61], information — communication — media [B68], biomedical ethics education [B78], nursing education [B79], medical education [B80], bioethics [B89], specific approaches to ethics [C1], philosophical ethics [C2], morals [C12], religion and religious ethics [C19], clergy [C20], religious ethics [C21], Christian ethics [C22], Roman catholic ethics [C23], natural law [C24], eastern orthodox ethics [C26], Protestant ethics [C27], Islamic ethics [C28], Jewish ethics [C29], theology [C41], professional ethics [C51], medical ethics [C52], nursing ethics [C53], counseling [C87], pastoral care [C88], social sciences [C89], social and political issues [D1], family members [D24], single persons [D25], couple [D26], cohabitation [D27], married persons [D28], marital relationship [D29], sibling [D30], family relationship [D31], parents [D32].

GB 44 **SHULMAN** Nisson E.
Rabbi of Saint John's Wood Synagogue. — Director of the Department of Medical Ethics of the Chief Rabbis Rabbinic Cabinet. — Coordinator of the Centre for Medical Ethics of The Jewish College, London.

St. John's Wood Synagogue
37- 41 Grove End Road
NW 8 9 NA London (United Kingdom)
✆ (44) 71 286 6333 Fax: (44) 266 2123

■ AFFILIATIONS: Centre for Medical Ethics of The Jewish College, London, United Kingdom (*Coordinator*).
■ TEACHING: Lecturing in selected areas of Bioethics at monthly lectures for the Service Medical Training in Bioethics, The Jewish College.
■ PUBLICATIONS: yes.

GB 45 **SMITH** Lewis L.
Director, Medical Research Council (MRC) Toxicology Unit.

MRC Toxicology Unit – Medical Council Research Laboratories
Woodmansterne Road
Carshalton, Surrey
SM5 4EF Carshalton (United Kingdom)
✆ (44) 81 643 8000 Fax: (44) 81 642 6538

■ AFFILIATIONS: Interdisciplinary Research Centre, University of Leicester (*Associate Director*).
■ PUBLICATIONS: yes.
■ WORKING ON: health services research [F4], nontherapeutic research [F5], animal experimentation [F10], research subjects [F12], animal testing alternatives [F14], methodology of research and experimentation [F15], control groups [F17], group of vulnerable subjects [F20], medical informatics [F21].
■ INTERESTED IN: *idem*.

GB 46 **SUTTON** Agneta
Deputy Director of the Linacre Centre of Ethics.

Linacre Centre for Health Care Ethics
60, Grove End Road
St. John's Wood
NW 8 9 NH London (United Kingdom)
✆ (44) 71 - 289 3625

■ TEACHING: Courses for student nurses, courses for hospital chaplains.
■ PUBLICATIONS: yes.
■ WORKING ON: dignity [B5], integrity [B8], justice [B9], human person [B12], ethics [B48], respect [B63], quality of life [C16], Roman catholicism [C31], medical ethics [C52], family relationship [D31], parents [D32], mothers [D33], handicapped [D43], genetics [E6], prenatal diagnosis [I3], amniocentesis [I4], chorionic villus sampling [I5], mass screening [I6], mandatory screening [I7], personhood [K3], minors [K6], wrongful life [K33], beginning of life [L24], children's rights [L25], child and family [L31], reproduction [M17], fertility [M18], reproductive technologies [M26], viability [M56], genetic defects [Q3], chromosomal disorders [Q4], biotechnology [Q21], medical genetics [Q35], human genome [Q47].
■ INTERESTED IN: *idem*.

GB 47 **WALSH** David B.
Medical Consultant, Chemical Pathologist.

Ninewells Hospital and Medical School
Department of Biochemical Medecine
DD1 9SY Dundee Tayside (United Kingdom)
✆ (44) 382 - 644 620

■ AFFILIATIONS: Dundee Medical Group, United Kingdom (*Honorary Secretary*).
■ TEACHING: MB ChB course in Dundee University.
■ WORKING ON: respect of human dignity and human rights [B1], specific approaches to ethics [C1], social and political issues [D1], science, technology, methods [E1], biomedical research and experimentation [F1], organization of health care; facilities, manpower and services; health occupations [G1], health economics, population characteristics [H1], health care, medical acts [I1], health care professionals, researchers and patient relationship [J1], law, legislation and jurisprudence [K1], stages of life — problems proper to childhood and elderly [L1], cancer [N17], communicable diseases [N27], kidney diseases [N39], injuries, occupational diseases, intoxications [N46], behavioral and mental disorders [P2], psychiatric technics [P24], genetic defects and hereditary diseases [Q2], biotechnology — genetic engineering [Q20], medical genetics [Q35], population genetics [Q43].
■ INTERESTED IN: medical ethics [C52], nursing ethics [C53].

GB 48 **WELLS** Frank O.
Director, Medical Affairs ABPI (providing an interface between the medical profession and the pharmaceutical industry).

The Association of the British Pharmaceutical Industry (ABPI)
12, Whitehall
SW 1 A 2 DY London (United Kingdom)
✆ (44) 71 - 930 32 90

■ AFFILIATIONS: Medical Ethics Committee, British Medical Association, London, United Kingdom (*Member*).

■ WORKING ON: codes of ethics [B49], professional competence [B52], information dissemination [B71], codes of biomedical ethics [B90], bioethical issues [B96], quality of life [C16], medical ethics [C52], EC — European Communities [D62], research [E14], investigators [E15], research team [E16], research policy [E18], industrial research [E22], international research [E23], scientific misconduct [E25], drug industry [E37], biomedical research [F2], clinical trials [F3], experimentation [F9], human experimentation [F11], research subjects [F12], healthy volunteers [F13], methodology of research and experimentation [F15], selection of subjects [F19], medical informatics [F21], family practice [G60], medicine [G63], drugs [I34], placebos [I36], patients' rights [J2], patient information [J4], consent to treatment [J8], informed consent [J9], consent forms [J14], confidentiality [J20], privileged communication [J21], medical secrecy [J22], physician's role [J26], medical law [K51], criminal law [K53].

■ INTERESTED IN: codes of ethics [B49], professional competence [B52], information dissemination [B71], codes of biomedical ethics [B90], history of biomedical ethics [B94], quality of life [C16], medical ethics [C52], EC — European Communities [D62], research [E14], investigators [E15], research team [E16], research institutes [E17], research policy [E18], industrial research [E22], military research [E24], scientific misconduct [E25], drug industry [E37], biomedical research and experimentation [F1], biomedical research [F2], clinical trials [F3], experimentation [F9], animal experimentation [F10], human experimentation [F11], research subjects [F12], healthy volunteers [F13], animal testing alternatives [F14], methodology of research and experimentation [F15], selection of subjects [F19], group of vulnerable subjects [F20], medical informatics [F21], family practice [G60], medicine [G63], drugs [I34], placebos [I36], patients' rights [J2], patient association [J3], patient information [J4], disclosure [J7], consent to treatment [J8], informed consent [J9], presumed consent [J10], consent forms [J14], advance directives [J18], confidentiality [J20], privileged communication [J21], medical secrecy [J22], physician's role [J26], medical law [K51], criminal law [K53].

GB 49 WILLIAMS C.J.

Specialist in Medical Oncology. — Coordinator of undergraduate course for medical students on Law and Ethics in Medicine.

CR.C. Medical Oncology Unit
Southampton General Hospital
SO 9 4 XY Southampton (United Kingdom)
✆ (44) 703 - 796 184 Fax: (44) 703 - 846 713

■ TEACHING: Coordinator of Course on Law and Ethics in Medicine for medical students.

■ PUBLICATIONS: yes.

■ WORKING ON: human rights [B3], Amnesty International [B18], ethics committees [B22], International Charter of Human Rights [B34], European Convention on Human Rights [B36], Helsinki Declaration [B37], patient advocacy [B42], ethics [B48], professional competence [B52], health education [B77], internship and residency [B81], students [B82], codes of biomedical ethics [B90], bioethical issues [B96], ethical analysis [C3], utilitarianism [C8], quality of life [C16], social worth [C17], medical ethics [C52], nursing ethics [C53], emotions [C79], French Senate [D73], handicapped [D43], physically handicapped [D44], males [D45], disadvantaged [D46], research [E14], investigators [E15], research design [E19], pilot projects [E20], drug industry [E37], decision making [E54], teaching methods [E59], clinical trials [F3], health services research [F4], nontherapeutic research [F5], random selection [F18], resource allocation [H4], costs and benefits [H10], placebos [I36].

■ INTERESTED IN: ethics committees [B22], patient advocacy [B42], bioethical issues [B96], quality of life [C16], medical ethics [C52], emotions [C79], research [E14], drug industry [E37], costs and benefits [H10], patients' rights [J2], physician patient relationship [J32], cancer [N17], terminally ill [O12].

Complementary list

This complementary list has been established by the Commission of European Communities, DG XII.

Organisations

Germany

[LC 1 D] **ARBEITSKREIS BERUFSBILD UND SELBSVERSTANDNIS IN DER BIOLOGIE**
Dachtelstrasse, 20 ◊ 7406 Mossingen (Germany)

[LC 2 D] **DEUTSCHER GEWERKSCHAFTSBUND, BUNDERSVORSTAND, ABTEILUNG TECHNOLOGIE, HUMANISIERUNG DER ARBEIT**
Hans-Böckler-Strasse ◊ 4000 Düsseldorf (Germany)

[LC 3 D] **FORSCHUNGSSTELLE ETHIK UND RECHT IN DER MEDIZIN**
Stefan Meier-Strasse,26 ◊ 7800 Fribourg, i. Br. (Germany)

Belgium

[LC 4 B] **CENTRE D'ÉTUDES BIOÉTHIQUES**
UCL 43/4534, Promenade de l' Alma, 51 ◊ 1200 Brussels (Belgium)

[LC 5 B] **CENTRE FOR BIOETHICS**
Université libre de Brussels, Koopliendenstraat, 11 ◊ 1000 Brussels (Belgium)

[LC 6 B] **CENTRUM VOOR BIO-MEDISCHE ETHIEK EN RECHT**
Kapucijnenoev, 35 ◊ 3000 Louvain (Belgium)

Spain

[LC 7 E] **CATEDRA DE HISTORIA DE LA MEDICINA**
Universidad Complutense, Ciudad Universitaria ◊ 28040 Madrid (Spain)

France

[LC 8 F] **CENTRE D'ÉTUDES ET DE RÉFLEXIONS BIOÉTHIQUES** (PROFESSEUR D. FOLSCHEID)
Université de Rennes, Institut de Philosophie, Avenue du Général-Leclerc ◊ 35042 Rennes Cedex (France)

[LC 9 F] **CLUB DE RÉFLEXION ÉTHIQUE "MÉDECINE, SANTÉ ET SOCIÉTÉ"** (E. MAHEN)
3, rue de Sevigné ◊ 75004 Paris (France)

[LC 10 F] **GROUPE DE RECHERCHES PSYCHANALYTIQUES ET SOCIOLOGIQUES EN SANTÉ PUBLIQUE** (P. PINEL)
Hôpital Necker-Enfants-Malades, 149, rue de Sèvres ◊ 75743 Paris Cedex (France)

[LC 11 F] **GROUPE INTERNATIONAL D'ÉTUDES BIOÉTHIQUES** (F. ABEL)
Fédération internationale des universités catholiques, 78, rue de Sèvres, ◊ 75341 Paris Cedex 07 (France)

Italy

[LC 12 I] **CENTRO DE BIOETICA UNIVERSITÀ CATOLICA DEL SACRO CUORE**
Largo Francesco Vito 1 ◊ 00168 Rome (Italy)

[LC 13 I] **FONDAZIONE LANZA**
Via Dante, 55 ◊ 35139 Padoue (Italy)

Luxembourg

[LC 14 L] **INSTITUT INTERNATIONAL DE BIOÉTHIQUE**
B.P. 419 ◊ 2014 Luxembourg (Luxembourg)

Netherlands

[LC 15 NL] **CENTRE FOR STUDIES OF SCIENCE, TECHNOLOGY AND SOCIETY**
Research Group on Genetics and Technology, University of Twente, P.O. Box 217 ◊ 7500 AE Enschede (Netherlands)

[LC 16 NL] **DEPARTMENT OF BIOETHICS OF THE NATIONAL FEDERATION OF HEALTH CARE INSTITUTIONS**
P.O. Box 9696 ◊ 3506 GR Utrecht (Netherlands)

[LC 17 NL] **GEZONDHEIDSRAAD** (CONSEIL NATIONAL DE SANTÉ)
Research Group on Genetics and Technology, University of Twente, P.O. Box 90517 ◊ 2509 LM The Hague (Netherlands)

United Kingdom

[LC 18 GB] **BRITISH MEDICAL ASSOCIATION ETHICAL COMMITTEE**
BMA House, Tavistock Square ◊ W1N 3AF London (United Kingdom)

Individuals

Germany

[LC 19 D] **AMMON** Ursula
Landesinstitut Sozialforschungsstelle, Rheinlanddamm, 199 ◊ 4600 Dortmund 1 (Germany)

[LC 20 D] **BEIER** (Prof. Dr.) Henning
RWTM Aachen-Medizinische Fakultät, Pauwelsstrasse ◊ 5100 Aachen (Germany)

[LC 21 D] **BELCHAUS** Gunter
Bundesministerium der Justiz, Heinemannstrasse, 6 ◊ 5300 Bonn 2 (Germany)

[LC 22 D] **DEUTSCH** E. *[Membre CDBI]*
Platz der Göttinger Sieben 6 ◊ 3400 Göttingen (Germany)

[LC 23 D] **DIEDRICH** (Prof. Dr.) K.
Universitäts-Frauenklinik, Sigmund-Freud-Strasse 25 ◊ 5300 Bonn 1 (Germany)

[LC 24 D] **FEIDEN** Karl
Federal Ministry of Health, Deutschherrenstr. 87 ◊ 5300 Bonn 2 (Germany)

[LC 25 D] **FIEBERG** M. G. *[Membre CDBI]*
Bundesjustizministerium, Hennemannstrasse 6 ◊ 5300 Bonn 2 (Germany)

[LC 26 D] **HEPP** (Prof. Dr.) H.
Ludwig-Maximilian-Universität, Frauenklinik, Marchioninistrasse 15 ◊ 8000 Munich 70 (Germany)

[LC 27 D] **HESS** (Dr.) B.
Max-Planck-Institut für Ernährungsphysiologie, Rheinlanddamm 201 ◊ 4600 Dortmund (Germany)

[LC 28 D] **KETTNER** (Dr.) Mathias
J.W. Goethe-Universität, Postfach 111932 ◊ 6000 Francfort 11 (Germany)

[LC 29 D] **KREBS** (Prof. Dr.) D.
Universitäts-Frauenklinik, Sigmund-Freud-Strasse 25 ◊ 5300 Bonn (Germany)

[LC 30 D] **LANGE** (Dr.) P.
Bundesministerium für Forschung und Technologie, Heinemannstrasse 2 ◊ 5300 Bonn 2 (Germany)

[LC 31 D] **MIETH** (Prof. Dr.) Dietmar
Katholisch-Theologisches Seminar, Abteilung für Theologische Ethik (Chairman, Centre for Ethics in the Sciences and Humanities), Eberhard-Karls-Universität ◊ 7400 Tübingen (Germany)

[LC 32 D] **NIEMER** (Dr.) Ursula *[Membre CDBI]*
Bundesministerium für Gesundheit, Deutscheherrenstr. 87 ◊ 5300 Bonn (Germany)

[LC 33 D] **OTT** Konrad
Graduiertenkolleg "Ethik in den Wissenchaften", Liebermeisterstrasse 16 ◊ 7400 Tübingen (Germany)

[LC 34 D] **WINNACKER** (Prof. Dr.) E. L.
Laboratorium für Molekulare Biologie, Genzentrum-Am-Kopferspitz ◊ 8033 Martinsried (Germany)

[LC 35 D] **WUERMELING** (Prof. Dr.) H. B.
Institüt für Rechtsmedizin, Universitätstrasse 22 ◊ 8520 Erlangen (Germany)

[LC 36 D] **ZERRES** (Prof. Dr.) Klaus
Institüt für Humangenetik der Universität Bonn,Wilhelmstr. 31 ◊ 5300 Bonn (Germany)

Belgium

[LC 37 B] **DEMEYERE** Frank
Centre for Bioethics, VUB, Koopliedenstraat 11 ◊ 1000 Brussels (Belgium)

[LC 38 B] **DIERKENS** Raphaël *[Secrétaire général]*
World Association for Medical Law, Apotheckstraat 5 ◊ Gand (Belgium)

[LC 39 B] **DONY** Jeanne
Ministère de la Santé, Léopold III laan, 51 ◊ 1200 Brussels (Belgium)

[LC 40 B] **GOFFIOUL** Fernand
Université de Liège, 11, avenue du Luxembourg ◊ 4020 Liège (Belgium)

[LC 41 B] **MALHERBE** Jean-François
Centre d'études bioéthiques UCL, Promenade de l'Alma, 51 ◊ 1200 Brussels (Belgium)

[LC 42 B] **MEIRE** Philippe
Ministère de la Justice, Place Polaert, 3 ◊ 1000 Brussels (Belgium)

[LC 43 B] **NUYTS** Marleen
Ministère de la Justice, Place Polaert, 3 ◊ 1000 Brussels (Belgium)

[LC 44 B] **VAN ELDER** Annick *[Membre CDBI]*
Cabinet du Ministre de la Justice, Place Polaert, 3 ◊ 1000 Brussels (Belgium)

[LC 45 B] **VAN STEENDAM** Guido
Commission d'éthique biomédicale, Craenendonck, 15 ◊ 3000 Leuven (Belgium)

[LC 46 B] **VAN STREITEGHEM** A.
Centre for Reproductive Medicine, VUJ, Laarbeeklaan, 101 ◊ 1090 Brussels (Belgium)

Denmark

[LC 47 DK] **KELSTRUP** Jørgen
Central Research Committee of Denmark, 62 Øterbrogade ◊ 100 Copenhagen (Denmark)

[LC 48 DK] **LAURITZEN** (Prof. Dr.) J. C.
Université de Copenhagen, Département d'obstétrique et gynécologie, Blegdamsvej, 9 ◊ Allboard Copenhagen (Denmark)

[LC 49 DK] **MICHELSEN** Nils
Danish Medical Association, Trondhjemsgade, 9 ◊ 2100 Copenhagen (Denmark)

[LC 50 DK] **PEDERSEN** *[Membre CDBI]*
Ministry of Health, Herluf Trolles Gade, 11 ◊ 1052 Copenhagen K (Denmark)

[LC 51 DK] **PELLE** Hedvig
Danish Medical Association, Trondhjemsgade, 9 ◊ 2100 Copenhagen (Denmark)

[LC 52 DK] **SEJER LARSEN** Else Marie *[Présidente]*
Conseil national d'éthique, Ravnsborgadde, 2-4 ◊ 2200 Copenhagen-N (Denmark)

[LC 53 DK] **TÄCKHOLM** Henrik
Ministry of Health, Herluf Trolles Gade, 11 ◊ 1052 Copenhagen-K (Denmark)

[LC 54 DK] **WACHENFELD** Margaret
Danish Centre of Human Rights, Studiestræde, 39 ◊ 1455 Copenhagen (Denmark)

Spain

[LC 55 E] **CORBELLA** Jacint
Catedra de Medicina Legal, Universidad de Barcelona, Av. Juan XXIII ◊ 08028 Barcelone (Spain)

[LC 56 E] **GRACIA** Diego *[Profesor de Historia de la Medicina]*
Calle Maldonado,48-7° B ◊ 28006 Madrid (Spain)

[LC 57 E] **PERIS TUSER** Mercè
Institut Català de la Salut, Gran Via de les Cortes Catalanes, 587 ◊ 08007 Barcelone (Spain)

[LC 58 E] **QUINTANA TRIAS** Octavi *[Président CDBI]*
Ministerio de Sanidad, Paseo del Prado, 18-20 ◊ 28004 Madrid (Spain)

[LC 59 E] **VIEDMA** Antonia
Institut Català de la Salut, Gran Via de les Cortes Catalanes, 587 ◊ 08007 Barcelona (Spain)

France

[LC 60 F] **AMBROSELLI** Claire
Institut supérieur de la santé et de la recherche médicale (INSERM), 101, rue de Tolbiac ◊ 73564 Paris Cedex 13 (France)

[LC 61 F] **BOTREAU-ROUSSEL** (Dr.) Pierre
Ministère de la Santé, 1, place Fontenoy ◊ 75700 Paris (France)

[LC 62 F] **COTTE** (Prof.) Louis
Faculté de Médecine, place Saint-Jacques ◊ 25000 Besançon (France)

[LC 63 F] **DAVID** Georges
19, rue Gazan ◊ 75014 Paris (France)

[LC 64 F] **EDELMAN** (Avocat) Bernard
61-63, rue Hallé ◊ 75014 Paris (France)

[LC 65 F] **GHIATI** Claude
Inserm-Ima Hôpital de Bicêtre ◊ 94270 Kremlin-Bicêtre (France)

[LC 66 F] **GINSBURGER**
Association "Ethique et Société", 15, impasse Pommier ◊ 69003 Lyon (France)

[LC 67 F] **GROMB** (Prof.) Claude
136, avenue de Canéjan ◊ 33600 Pessac (France)

[LC 68 F] **HERMITTE** (Juriste) Marie-Angèle
63, rue Hallé ◊ 75014 Paris (France)

[LC 69 F] **HUMEAU** (Prof.) Claude
Groupe d'études de la fécondation en France (GEFF), Laboratoire de fécondation in vitro, 13, avenue du Professeur-Granet ◊ 34000 Montpellier (France)

[LC 70 F] **JACQUOT** (Dr.) Sophie
Direction générale de la santé, 1, place Fontenoy ◊ 75007 Paris (France)

[LC 71 F] **LABRUSSE-RIOU** (Prof.) *[Professeur de droit]*
Université de Paris-I Panthéon-Sorbonne ◊ Paris (France)

[LC 72 F] **LEROY-GISSENGER**
Ministère de la Justice, 13, place Vendôme ◊ 75001 Paris (France)

[LC 73 F] **LHUGUENOT** Marie-Helène
Comité consultatif national d'éthique (CCNE), secrétariat général, 101, rue de Tolbiac ◊ 75654 Paris cedex 13 (France)

[LC 74 F] **MANTZ** (Prof.) J.
Centre hospitalier régional, Comité d'éthique, 4, rue Kirschleger ◊ 67085 Strasbourg Cedex (France)

[LC 75 F] **MAZEN** M. J.
Association française de droit de la santé, Faculté de droit, 4, boulevard Gabriel ◊ 2100 Dijon (France)

[LC 76 F] **MEYER** Philippe
Hôpital Necker, Département de pharmacologie, 161, rue de Sèvres ◊ 75015 Paris (France)

[LC 77 F] **PIERSON** (Prof.) M.
Hôpital d'enfants, CHU de Nancy Comité d'éthique ◊ 54500 Vandœuvre (France)

[LC 78 F] **PROUST** (Dr.) Bernard
CHU Comité d'éthique, 1, rue Germont ◊ 76031 Rouen (France)

[LC 79 F] **RENÉ** Louis
Président du Conseil national de l'Ordre des médecins, 60, boulevard Latour-Maubourg ◊ 75007 Paris (France)

[LC 80 F] **SUDRE** (Prof.) Frédéric *[Directeur de l'IRDHIC]*
Université de Montpellier-I, 9, rue Saint-Louis ◊ 34000 Montpellier (France)

[LC 81 F] **TESTART** (Dr.) Jacques
Hôpital Antoine-Béclère, 157, rue de la Porte-de-Trivaux ◊ 91241 Clamart (France)

[LC 82 F] **TORRELI** (Prof.) Maurice
Villa la Renaissance, rue Jussien-Prolongée ◊ 0600 Nice (France)

[LC 83 F] **VULLIET-TAVERNIER** Sophie
Commission nationale de l'informatique et des libertés (CNIL), 2, rue Saint-Guillaume ◊ 75007 Paris (France)

Ireland

[LC 84 IRL] **DOOLEY** (Reverend Prof. Dr.) Maurice
St Patrick's College ◊ Thurles (Ireland)

[LC 85 IRL] **HENRY** (Dr.) May
Postgraduate Medical Board, Corrigan House, 12 Burlington Road ◊ Dublin 4 (Ireland)

[LC 86 IRL] **SHORTALL** Maurice
Spiritan Missionaires, Holy Ghost Missionary College, Kimmage Manor ◊ Dublin 12 (Ireland)

Italy

[LC 87 I] **ANTICO** Leonardo
Centre for Bioethics, Catholic University, Via Cavalese ◊ 00135 Rome (Italy)

[LC 88 I] **BELLINO** Francesco
Facoltà Magisterio, Via Tenente de Liguori,13 ◊ 70126 Bari (Italy)

[LC 89 I] **BERTAZZONI** (Prof.) Umberto
Université de Pavie, Département de génétique, Via S. Epifanio, 14 ◊ 27100 Pavie (Italy)

[LC 90 I] **BOMBELLI** (Dr.) Fernandino *[Gynécologue]*
Ospedale San Raffaele, Via Olgettina, 60 ◊ 20132 Milan (Italy)

[LC 91 I] **BONO** (Dr.) Angiole
Istituto Sieroterapico, Via Della Signora, 2 ◊ 20122 Milan (Italy)

[LC 92 I] **BORDIGNON** (Dr.) Claudio
Ospedale San Raffael, Laboratorio Hematologia, Via Olgettina, 60 ◊ 20132 Milan (Italy)

[LC 93 I] **CORRADO** Viafora *[Coordinateur du Projet "éthique et médecine"]*
Fondazione Lanza, Via Dante, 55 ◊ 35139 Padova (Italy)

[LC 94 I] **COSMI** (Prof.) Ermelando
Istituto Ostetrica e Ginecologica, Via Regina Elene, 3254 ◊ 00161 Rome (Italy)

[LC 95 I] **CROSIGNANI** (Prof. Dr.) P. G. *[Director]*
Clinica Ostetrica e Ginecologica, Università degli Studi, Via M. Melloni, 52 ◊ 20129 Milan (Italy)

[LC 96 I] **DIPPOLITO** (Dr.) Fiorenza *[Direttore Diu Protezione Materno-Infantelo, (Membre CDBI)]*
Ministero Sanità, D.G. Med. Soc.11, Ple. Industria, 20 ◊ 00144 Rome (Italy)

[LC 97 I] **IACCARINO** (Prof. Dr.) M.
International Institute of Genetics and Biophysics — CNR Via Marconi,10 ◊ 80125 Naples (Italy)

[LC 98 I] **MAZZONI** (Prof.) Aldo
Policlinico S. Orsola, Via Massarenti, 9 ◊ 40138 Bologna (Italy)

[LC 99 I] **MILANI-COMPARETTI** Marco *[Directeur scientifique, "Istituto Internazionale, Studi Etico-Giuridici sulla nuova biologia"]*
Villa Eolia, Via Cappuccini ◊ 99057 Milazzo (Messina) (Italy)

[LC 100 I] **POLIZZI** (Prof.) Francesco
Via Deitizii, 10 ◊ 00185 Rome (Italy)

[LC 101 I] **PUGLISI** (Prof.) Salvatore *[Membre CBDI]*
Via San Godenzo, 169 ◊ 00189 Rome (Italy)

[LC 102 I] **ROGNONI** (Prof. Dr.) Giancarla
Istituto Medicina Legale, Université de Pavie ◊ 27100 Pavie (Italy)

[LC 103 I] **ROMANO** (Prof. Dr.) Carlo
Commissione Etica, Università di Napoli Istituto Medicina Legale, Policlinico, Via S. Pansini, 5 ◊ 80131 Naples (Italy)

[LC 104 I] **SGARAMELLA** (Prof. Dr.) V.
Université de Pavie, Département de génétique, Via S. Epifanio, 14 ◊ 27100 Pavie (Italy)

[LC 105 I] **ZANOBIO** (Prof.) Bruno
Istituto Storia Medicina, Via Albricci, 9 ◊ 20122 Milan (Italy)

[LC 106 I] **ZUCCO** (Dr.) Maria-Flavia
Istituto Tecnologie Biomediche CNR, Via G.B. Morgagni, 30/E ◊ 00161 Rome (Italy)

Luxembourg

[LC 107 L] **BAULER** (Prof.) André
Avenue de la Gare, 14 ◊ 2919 Luxembourg (Luxembourg)

[LC 108 L] **BETZ** Arsène *[Président de la Commission nationale d'éthique, et membre CDBI]*
4, rue Jos Tockert ◊ 2620 Luxembourg (Luxembourg)

[LC 109 L] **MAJERUS** Mill *[Coordinateur de la Commission nationale d'éthique]*
Avenue de la Gare, 14 ◊ 2919 Luxembourg (Luxembourg)

[LC 110 L] **MOLITOR** (Dr.) Jules
14-16, rue du Canal ◊ 4050 Esch-5/Alzette (Luxembourg)

[LC 111 L] **WAGNER** Sylvain
Ministère de la Justice, BP 9 ◊ 2934 Luxembourg (Luxembourg)

Netherlands

[LC 112 NL] **AARTSEN** J.M.G.
Health Council, P.O. Box 90517 ◊ 2509 LM La Hague (Netherlands)

[LC 113 NL] **BEERNINK** (Dr.) J. F.
Dignaland 70 ◊ 2591 CD La Hague (Netherlands)

[LC 114 NL] **BENNEBROEK GRAVENHURST** J. *[Chairman]*
Association of Medical Ethics Committees, Dept.

Obstetrics, Univ. Hospital,, Rijnsburgewg, 10 ◊ 2333 AA Leiden (Netherlands)

[LC 115 NL] **BUIS** S.
Erasmus University, Medical Ethics, Klein-Coolstraat, 43 B ◊ 3033 XR Amsterdam (Netherlands)

[LC 116 NL] **CROUGHS** Wim
Wilhenina Kinderziekenhuis, Overslingeland 9, ◊ 422 SNJ Nooderloos (Netherlands)

[LC 117 NL] **DE HILSTER** Ellen
Ministry of Health, P.O. Box 3008 ◊ 2280 MK Rijswijk (Netherlands)

[LC 118 NL] **HARTGERINK** J.W.
Ministry of Health, P.O. Box 3008 ◊ 2280 MK Rijswijk (Netherlands)

[LC 119 NL] **KASTELEIN** W. R.
Koninklijke Nederlandse Maatschappij voor Genees Kunst Commissie Medische Ethiek, Postbus 20051 ◊ 3502 LB Utrecht (Netherlands)

[LC 120 NL] **KITS-NIEUWENKAMP** Johana *[Membre CDBI]*
Ministry of Welfare, Health and Culture, Postbus 5406 ◊ 2280 HK Rijswijk (Netherlands)

[LC 121 NL] **NIERMEIJER** (Prof.) Martinus
Clinical Genetics Erasmus University, Academic Hospital Dijkziq, Dr. Molewaterplein, 40 ◊ 3015 GD Rotterdam (Netherlands)

[LC 122 NL] **RIGTER** (Prof. Dr.) Heuk
Gezondheidsraad, Postbus 90517 ◊ 2509 LM Sgravenhage (Netherlands)

[LC 123 NL] **ROSCAM-ABBIN** (Prof. Dr.) H. *[Legal adviser]*
Ministry of Health, Oosterpark 46 ◊ 1092 AN Amsterdam (Netherlands)

[LC 124 NL] **STEGEMAN** Tineke
Gezondheidsraad, P.O. Box 90517 ◊ 2509 LH La Hague (Netherlands)

[LC 125 NL] **STEMERDING** Dirk
Universiteit Twente, Vakgroep Filosofie van Wetenschap & Techniek, Postbus 217, ◊ 7500 AE Enschede (Netherlands)

[LC 126 NL] **VAN DELDEN** Johanes
Health Council of the Netherlands, P.O. Box 90517 ◊ 2509 LM La Hague (Netherlands)

[LC 127 NL] **ZEILMAKER** (Prof.) G.H.
Université Erasmus, Faculté de médecine, P.O. Box 1738 ◊ 3000 DR Rotterdam (Netherlands)

Portugal

[LC 128 P] **LOURENÇO DOS REYS** (Prof.) J. A.
Istituto de Medicina Legal, Rua Manuel Bento de Sousa, 3 ◊ 1100 Lisbon (Portugal)

United Kingdom

[LC 129 GB] **BERRY** (Prof.) Robert J.
University College, Dept. of Biology ◊ London (United Kingdom)

[LC 130 GB] **BOYD** (Reverend Dr.) Kenneth M. *[Director]*
Institute of Medical Ethics, 1 Donne Terrace ◊ EM3 6DY Edimbourg (United Kingdom)

[LC 131 GB] **BRAMS** Michael *[Medical Officer, membre CDBI]*
Dept. of Health, Richmond House, 79 Whitehall ◊ SW1A 2N8 London (United Kingdom)

[LC 132 GB] **CAMERON** (Dr.) Nigel
Centre for Bioethics and Public, 58 Hanover Gardens ◊ London SE 11 5TN (United Kingdom)

[LC 133 GB] **CICA** Natasha
Medical Law Review, King' s College, Strand ◊ WC ZR 2LS London (United Kingdom)

[LC 134 GB] **DAVIES** J.
World Federation of Right-to-Die Societies, 56 Malborough Road ◊ OX14LR Oxford (United Kingdom)

[LC 135 GB] **MAC FARLANE** Peter
Institute of Law and Ethics in Medicine, Stair Building ◊ G12 8QQ Glasgow (United Kingdom)

[LC 136] **MAC LAREN** Anne
MRC Mammalian Development Unit, 4 Stephenson Way ◊ NW1 2HC London (United Kingdom)

[LC 137 GB] **MORGAN** Derek
Centre for Philosophy and Health Care, Department of Philosophy, Evans University of College ◊ SA2 8PP Swansea (United Kingdom)

[LC 138 GB] **NAIRNE** (Sir) Patrick
Nuffield Council on Bioethics, 28 Bedford Square ◊ WC 1B 3EG London (United Kingdom)

[LC 139 GB] **NATHANSON** Viviane
British Medical Association, Ethical Committee BMA House, Tavistock Square ◊ WC 1N 3AF London (United Kingdom)

[LC 140 GB] **NGWENA** (Dr.) C.
Cardiff Law School, P.O. Box 47 ◊ CF1 1XD Cardiff (United Kingdom)

[LC 141 GB] **SELLER** (Dr.) Mary
Guy's Hospital Department of Medical and Molecular Genetics, 7th Floor, Guy's Tower ◊ SE1 9RD London (United Kingdom)

[LC 142 GB] **TURNER-WARWICK** *[President of the Royal College of Physician' s Committee on Ethical Issues in Medicine]*
11 St. Andrew' s Place ◊ NW1 4LE London (United Kingdom)

Index

Thesaurus

This English thesaurus, containing 879 key words, was developed by the Ethics Documentation and Data Centre (CDIE) of the French National Institute for Health and Medical Research (INSERM) as part of a project supported by the French Research and Space Ministry. It includes all the key words of the Bioethics Thesaurus published by the Kennedy Institute of Ethics (United States).

A1 INFORMATION SOURCES, DATA BASES
A2 — REVIEW
A3 — DATA BASES
A4 — BIBLIOGRAPHY
A5 — INFORMATION CENTERS
A6 — BOOK REVIEW
A7 — LITERATURE
A8 — DOCUMENTATION
A9 — PUBLICATIONS
A10 — EDITORIAL POLICIES

B1 RESPECT OF HUMAN DIGNITY AND HUMAN RIGHTS

B2 — FUNDAMENTAL RIGHTS OF THE INDIVIDUALS
B3 —— human rights
B4 ——— self determination
B5 ——— dignity
B6 ——— women's rights
B7 ——— human equality
B8 ——— integrity
B9 ——— justice
B10 ——— freedom
B11 ——— privacy
B12 —— human person
B13 ——— human body commercialization
B14 ——— reification
B15 ———— traffic of organs

B16 — PROTECTION OF RIGHTS — INVOLVED INSTITUTIONS
B17 —— institutions involved in the protection of rights
B18 ——— Amnesty International
B19 ——— CNIL — Commission Nationale Informatique et Libertés (France)
B20 ——— World Medical Assembly
B21 ——— animal care committees
B22 ——— ethics committees
B23 ———— CCNE — Comité Consultatif National d'Ethique (France)
B24 ——— Council for International Organization of Medical Sciences
B25 ——— Council of Europe

B26 ——— professional deontology organs
B27 ———— College of Physicians
B28 ——— office of science and technology assessment
B29 ——— United Nations
B30 ——— International Court
B31 ——— US Military Court
B32 ——— CCPPRB — French committees for the protection of human subjects of biomedical research
B33 —— manifests and declarations concerning human rights
B34 ——— International Charter of Human Rights
B35 ——— Nuremberg Code
B36 ——— European Convention on Human Rights
B37 ——— Helsinki Declaration
B38 ——— Declaration on Human and Citizen Rights
B39 ——— Universal Declaration of Human Rights
B40 —— due process
B41 ——— public advocacy
B42 ——— patient advocacy
B43 ——— equal protection
B44 ——— state interest
B45 ———— common good
B46 —— humanity heritage

B47 — ETHICAL RULES AND PRINCIPLES
B48 —— ethics
B49 —— codes of ethics
B50 ——— moral obligations
B51 ———— competence
B52 ————— professional competence
B53 ———— obligations of society
B54 ———— obligations to society
B55 ——— moral policy
B56 ———— altruism
B57 ————— beneficence
B58 ———— authoritarianism
B59 ———— deontology
B60 ————— code of deontology
B61 ———— morality
B62 ———— paternalism
B63 ———— respect
B64 ———— virtues
B65 ——— solidarity
B66 ———— humanitarian organizations
B67 ————— Red Cross

B68 — INFORMATION — COMMUNICATION — MEDIA
B69 —— communication
B70 ——— interdisciplinary communication
B71 ——— information dissemination
B72 ——— mass media
B73 ———— press conference
B74 ——— audiovisual aids
B75 ——— public debates
B76 —— education
B77 ——— health education
B78 ——— biomedical ethics education
B79 ——— nursing education
B80 ——— medical education
B81 ———— internship and residency
B82 ——— students
B83 ———— curriculum
B84 ——— universities
B85 —— health and biology mediatisation

B86 ——— famous persons
B87 ——— advertising
B88 —— uncontrolled information

B89 — BIOETHICS
B90 —— codes of biomedical ethics
B91 —— CCNE advices (France)
B92 —— ethicists
B93 —— ethical review
B94 —— history of biomedical ethics
B95 ——— bioethics movement
B96 —— bioethical issues
B97 —— European Convention of Bioethics
B98 —— Bioethics bill, 1992 (France)

C1 SPECIFIC APPROACHES TO ETHICS

C2 — PHILOSOPHICAL ETHICS
C3 —— ethical analysis
C4 ——— wedge argument
C5 —— normative ethics
C6 ——— deontological ethics
C7 ——— teleological ethics
C8 ———— utilitarianism
C9 —— situational ethics
C10 ——— ethical relativism
C11 —— metaethics
C12 —— morals
C13 ——— conscience
C14 ——— moral development
C15 ——— values
C16 ———— quality of life
C17 ———— social worth
C18 ———— value of life

C19 — RELIGION AND RELIGIOUS ETHICS
C20 —— clergy
C21 —— religious ethics
C22 ——— Christian ethics
C23 ———— Roman catholic ethics
C24 ————— natural law
C25 ————— totality
C26 ———— eastern orthodox ethics
C27 ———— Protestant ethics
C28 ———— Islamic ethics
C29 ———— Jewish ethics
C30 —— religion and sects
C31 ——— Roman catholicism
C32 ——— religious beliefs
C33 ——— Islam
C34 ——— Judaism
C35 ——— Protestantism
C36 ——— religion
C37 ——— Christian science
C38 ——— Jehovah's witnesses
C39 —— religious sciences
C40 ——— scriptural interpretation
C41 ——— theology

C42 — PHILOSOPHY OF BIOLOGY
C43 —— biology and human future
C44 ——— future generations
C45 —— consequences

C46 ——— double effect
C47 —— evolution
C48 ——— speciesism
C49 —— biological life
C50 —— primates

C51 — PROFESSIONAL ETHICS
C52 —— medical ethics
C53 —— nursing ethics

C54 — ETHICS AND HUMANITIES
C55 —— humanities
C56 —— history
C57 ——— historical aspects
C58 ——— history before the 20th century
C59 ———— ancient history
C60 ——— history of the 20th century
C61 —— philosophy
C62 ——— determinism
C63 ———— sociobiology
C64 ——— existentialism
C65 ——— hedonism
C66 ——— humanism
C67 ——— Marxism
C68 —— psychology
C69 ——— mental processes
C70 ———— comprehension
C71 ———— judgement
C72 ———— recall
C73 ——— behavior
C74 ———— attitudes
C75 ———— intention
C76 ———— motivation
C77 ————— incentives
C78 ———— normality
C79 ——— emotions
C80 ———— love
C81 ———— trust
C82 ——— personality
C83 ———— self concept
C84 ———— egoism
C85 ———— intelligence
C86 ——— applied psychology
C87 ———— counseling
C88 ————— pastoral care
C89 —— social sciences

D1 SOCIAL AND POLITICAL ISSUES

D2 — SOCIAL ISSUES
D3 —— social control
D4 ——— informal social control
D5 ———— coercion
D6 ———— social dominance
D7 ———— public opinion
D8 ———— punishment
D9 ———— exclusion
D10 ——— formal social control
D11 ———— imprisonment
D12 ————— prisoners
D13 ———— legislation
D14 ————— model legislation
D15 ———— capital punishment

D16 ——— regulation
D17 ———— government regulation
D18 —— socioeconomic factors
D19 ——— employment
D20 ——— strikes
D21 ——— scarcity
D22 —— social groups
D23 ——— aliens
D24 ——— family members
D25 ———— single persons
D26 ———— couple
D27 ————— cohabitation
D28 ————— married persons
D29 —————— marital relationship
D30 ———— sibling
D31 ———— family relationship
D32 ———— parents
D33 ————— mothers
D34 ————— fathers
D35 ——— females
D36 ——— ethnic groups
D37 ———— whites
D38 ———— Hispanic Americans
D39 ———— American Indians
D40 ———— Jews
D41 ———— Blacks
D42 ——— minority groups
D43 ——— handicapped
D44 ———— physically handicapped
D45 ——— males
D46 ——— disadvantaged
D47 ——— associations
D48 ———— indigents
D49 —— social impact
D50 —— social interaction
D51 —— cultural pluralism
D52 —— social problems
D53 ——— social discrimination
D54 ———— stigmatization
D55 —— sociology of medicine

D56 — POLITICAL ISSUES
D57 —— political activity
D58 ——— dissent
D59 —— international aspects
D60 ——— North-South relationships
D61 ——— international organizations
D62 ——— EC — European Communities
D63 ———— ECC — European Community Commission
D64 —— government and political systems
D65 ——— army
D66 ———— military personnel
D67 ——— state government
D68 ———— French National Assembly
D69 ———— advisory committees
D70 ———— French Council of State
D71 ———— federal government
D72 ———— municipal government
D73 ———— French Senate
D74 ———— government agencies
D75 ———— state responsibility
D76 ——— politics
D77 ———— public policy

D78 ——— community services
D79 ——— political systems
D80 ———— capitalism
D81 ———— communism
D82 ———— democracy
D83 ———— national socialism
D84 ———— socialism

D85 — VIOLENCE
D86 —— dehumanization
D87 —— war
D88 ——— biological warfare
D89 ——— nuclear warfare
D90 —— killing
D91 ——— justifiable killing
D92 —— sex offenses
D93 ——— rape
D94 —— torture
D95 ——— force feeding

E1 SCIENCE, TECHNOLOGY, METHODS

E2 — SCIENCE
E3 —— history of science

E4 — LIFE SCIENCES
E5 —— biology
E6 ——— genetics
E7 ——— microbiology
E8 ——— molecular biology
E9 —— epidemiology
E10 —— statistics
E11 ——— incidence
E12 ——— morbidity
E13 ——— mortality

E14 — RESEARCH
E15 —— investigators
E16 —— research team
E17 —— research institutes
E18 —— research policy
E19 —— research design
E20 —— pilot projects
E21 —— behavioral research
E22 —— industrial research
E23 —— international research
E24 —— military research
E25 —— scientific misconduct
E26 —— progress

E27 — TECHNOLOGY
E28 —— biomedical technologies
E29 ——— freezing
E30 ——— preservation
E31 ———— cryonic suspension
E32 ——— diagnostic imaging
E33 —— computers
E34 —— records
E35 —— computerized file

E36 — INDUSTRY
E37 —— drug industry

E38 — CONTAINMENT
E39 —— biological containment
E40 —— physical containment
E41 —— quality of environment
E42 ——— review committees
E43 ——— ecology
E44 ——— nuclear energy
E45 ——— fluoridation
E46 ——— occupational medicine
E47 ——— pollution
E48 ——— radiation
E49 ——— health hazards
E50 ——— toxicity
E51 —— safety
E52 ——— data protection

E53 — METHODS
E54 —— decision making
E55 ——— decision analysis
E57 —— goals
E58 —— survey
E59 —— teaching methods
E60 —— standards

E61 — EVALUATION
E62 —— self regulation
E63 ——— peer review
E64 —— social control of science
E65 —— technology assessment

F1 BIOMEDICAL RESEARCH AND EXPERIMENTATION

F2 — BIOMEDICAL RESEARCH
F3 —— clinical trials
F4 —— health services research
F5 —— nontherapeutic research
F6 —— therapeutic research
F7 —— cognitive research
F8 —— human genome project

F9 — EXPERIMENTATION
F10 —— animal experimentation
F11 —— human experimentation
F12 ——— research subjects
F13 ——— healthy volunteers
F14 —— animal testing alternatives

F15 — METHODOLOGY OF RESEARCH AND EXPERIMENTATION
F16 —— debriefing
F17 —— control groups
F18 —— random selection
F19 —— selection of subjects
F20 ——— group of vulnerable subjects
F21 —— medical informatics

G1 ORGANIZATION OF HEALTH CARE ; FACILITIES, MANPOWER AND SERVICES; HEALTH OCCUPATIONS

G2 — HEALTH
G3 —— maternal health
G4 —— mental health

G5 — PUBLIC HEALTH
G6 —— accidents

G7 ——— traffic accidents
G8 —— family planning
G9 —— state medicine
G10 —— maternal welfare
G11 ——— maternal life
G12 —— contact tracing

G13 — HEALTH POLICY
G14 —— health legislation
G15 —— institutional policies
G16 ——— institutional obligations
G17 —— organizational policies
G18 ——— program descriptions
G19 ——— mandatory programs
G20 ———— quarantine
G21 ——— voluntary programs

G22 — HEALTH FACILITIES
G23 —— medical devices
G24 ——— artificial organs
G25 —— health services
G26 ——— residential facilities
G27 ———— nursing home
G28 ——— hospital
G29 ———— mental institutions
G30 ———— public hospitals
G31 ———— private hospitals
G32 ———— religious hospitals
G33 ———— intensive care units
G34 ——— hospices

G35 — ORGANIZATION OF HEALTH CARE
G36 —— WHO — World Health Organization
G37 —— health personnel
G38 ——— patient care team
G39 ——— nurses
G40 ——— physicians
G41 ——— medical staff
G42 ——— pharmacists
G43 ——— nurse midwives
G44 ——— social workers

G45 — HEALTH CARE
G46 —— hospitalization
G47 ——— patient admission
G48 —— health care delivery
G49 —— patient compliance
G50 —— medical records
G51 —— medical evaluation
G52 —— selection for treatment
G53 —— patient care
G54 —— emergency care
G55 —— home care
G56 —— patient transfer

G57 — CONVENIENCE MEDICINE

G58 — PREVENTIVE MEDICINE
G59 —— immunization

G60 — FAMILY PRACTICE

G61 — HEALTH OCCUPATIONS
G62 —— dentistry

G63 —— medicine
G64 ——— sports medicine
G65 ——— occupational medicine
G66 ——— obstetrics and gynecology
G67 ——— ophtalmology
G68 ——— pediatrics
G69 ——— psychiatry
G70 ——— radiology

H1 HEALTH ECONOMICS, POPULATION CHARACTERISTICS

H2 — ECONOMICS
H3 —— administrators
H4 —— resource allocation
H5 —— insurance
H6 ——— health insurance
H7 ———— national health insurance
H8 ——— life insurance
H9 —— health care costs
H10 ——— costs and benefits
H11 ——— risks and benefices
H12 —— compensation
H13 —— public participation
H14 —— remuneration
H15 ——— medical fees
H16 —— price
H17 —— profit
H18 —— financial support

H19 — POPULATION
H20 —— population growth
H21 —— developing countries
H22 —— population control
H23 —— elderly

I1 HEALTH CARE, MEDICAL ACTS

I2 — DIAGNOSIS
I3 —— prenatal diagnosis
I4 ——— amniocentesis
I5 ——— chorionic villus sampling
I6 —— mass screening
I7 —— mandatory screening

I8 — BIOLOGICAL SPECIMENS PROCUREMENT, BLOOD TRANSFUSION, ORGAN TRANSPLANTATION
I9 —— biological substances contamination
I10 —— required request
I11 —— sperm banks
I12 —— tissue banking
I13 —— tissue donation
I14 —— donors
I15 ——— organ donors
I16 ——— donor cards
I17 ——— anonymous donation
I18 ——— directed donation
I19 —— blood transfusions
I20 ——— blood donation
I21 ——— blood substitutes
I22 —— transplantation
I23 ——— transplant recipients
I24 ——— organ transplantation
I25 ———— organ donation

I26 ——— organ procurement
I27 ——— fetal tissue transplantation
I28 ———— intracerebral fetal tissue transplantation
I29 ——— bone marrow transplantation

I30 — TREATMENT
I31 —— anesthesia
I32 —— renal dialysis
I33 —— alternative therapies
I34 —— drugs
I35 ——— insulin
I36 ——— placebos
I37 —— surgery
I38 ——— cosmetic surgery
I39 —— extraordinary treatment
I40 —— physical restraint
I41 —— immunotherapy
I42 —— rehabilitation
I43 —— vaccination

J1 HEALTH CARE PROFESSIONALS, RESEARCHERS AND PATIENT RELATIONSHIP

J2 — PATIENTS' RIGHTS
J3 —— patient association
J4 —— patient information
J5 ——— patient access
J6 ——— parental notification
J7 ——— disclosure
J8 —— consent to treatment
J9 ——— informed consent
J10 ——— presumed consent
J11 ——— third party consent
J12 ———— spousal consent
J13 ———— parental consent
J14 ——— consent forms
J15 ——— treatment refusal
J16 —— patient participation
J17 —— right to treatment
J18 —— advance directives
J19 ——— living wills

J20 — CONFIDENTIALITY
J21 —— privileged communication
J22 —— medical secrecy

J23 — PHYSICIANS AND RESEARCHERS ACCOUNTABILITY
J24 —— referral and consultation
J25 —— professional deontology
J26 —— physician's role

J27 — INTER-PERSONAL RELATIONSHIP
J28 —— medical etiquette
J29 —— investigator subject relationship
J30 —— nurse patient relationship
J31 —— physician nurse relationship
J32 —— physician patient relationship
J33 —— professional patient relationship
J34 —— parent child relationship

K1 LAW, LEGISLATION AND JURISPRUDENCE

K2 — LEGAL PERSONALITY
K3 —— personhood

K4 —— legally incompetent person
K5 —— potentiality of personhood
K6 —— minors
K7 —— legal guardians

K8 — JURISPRUDENCE — ACCOUNTABILITY
K9 —— conflict of interest
K10 —— battery
K11 —— torts
K12 —— wrongful death
K13 —— fraud
K14 —— deception
K15 —— expert evaluation
K16 ——— technical expertise
K17 ——— generalization of expertise
K18 ——— expert testimony
K19 —— malpractice
K20 ——— negligence
K21 —— legitimacy
K22 —— child donation
K23 —— misconduct
K24 —— child's interest
K25 —— moratory
K26 —— injuries
K27 ——— preconception injuries
K28 —— economic value of life
K29 —— accountability
K30 ——— legal liability
K31 —— ghost surgery
K32 —— therapeutic risk
K33 —— wrongful life

K34 — LEGISLATION AND LAW
K35 —— judicial action
K36 —— constitutional amendments
K37 —— authorization
K38 —— prohibition
K39 —— contracts
K40 —— court decision
K41 ——— Supreme Court decisions
K42 —— rights
K43 ——— civil laws
K44 ———— civil code
K45 ——— legal rights
K46 ——— property rights
K47 ———— patents
K48 ——— constitutional law
K49 ——— European law
K50 ——— international law
K51 ——— medical law
K52 ———— health legislation
K53 ——— criminal law
K54 ———— criminal code
K55 —— guidelines
K56 —— law
K57 ——— law enforcement
K58 ——— legal aspects
K59 ——— legal obligation
K60 ——— bill
K61 —— trade secrets
K62 —— state courts
K63 —— conscience clause
K64 —— therapeutic injunction

L1 STAGES OF LIFE — PROBLEMS PROPER TO CHILDHOOD AND ELDERLY

L2 — ANATOMY, PHYSIOLOGY, DEVELOPMENT
L3 —— human body
L4 ——— body parts and fluids
L5 ———— hearts
L6 ———— brain
L7 ———— livers
L8 ———— bone marrow
L9 ———— kidneys
L10 ———— hormones
L11 ————— somatotropin
L12 ——— human development
L13 ——— age
L14 ———— adolescents
L15 ———— adults
L16 ——— nutrition
L17 ———— food
L18 ——— animal organs
L19 ——— human characteristics
L20 ——— mind

L21 — CHILDHOOD DIFFICULTIES, DISEASES, PROTECTION
L22 —— breast feeding
L23 —— child development
L24 ——— beginning of life
L25 —— children's rights
L26 ——— childhood defense
L27 —— children
L28 ——— infants
L29 ———— newborns
L30 ————— prematurity
L31 —— child and family
L32 ——— adoption
L33 ——— abandoned (child)
L34 ——— aid children
L35 ——— unwanted children
L36 —— children and media
L37 —— child diseases and problems
L38 ——— handicapped children
L39 ——— ill-treat children
L40 ———— exploited children
L41 ———— child abuse
L42 ———— infanticide
L43 ——— child diseases
L44 ———— growth disorders

L45 — AGED — PROBLEMS RELATED WITH AGING
L46 —— disease and aged
L47 —— life extension
L48 —— health care services and aged
L49 —— aged
L50 —— aging
L51 —— old person abuse

L52 — HUMAN ACTIVITIES
L53 —— sports

M1 SEXUALITY AND PROCREATION — UNBORN CHILD

M2 — SEXUALITY AND PROCREATION
M3 —— contraception
M4 ——— condom

M5 ——— immunologic contraception
M6 —— sterilization
M7 ——— involuntary sterilization
M8 ——— volontary sterilization
M9 —— reproductive organs and embryonic structures
M10 ——— embryos
M11 ——— fetuses
M12 ——— ovum
M13 ——— placentas
M14 ——— sperm
M15 —— procreation
M16 ——— wish of children
M17 ——— reproduction
M18 ———— fertility
M19 ———— infertility
M20 —— sexuality
M21 ——— sexual behavior
M22 ——— homosexuality
M23 ———— homosexuals
M24 ——— transsexualism
M25 ——— prostitution

M26 — REPRODUCTIVE TECHNOLOGIES
M27 —— ovum donors
M28 —— semen donors
M29 —— in vitro fertilization
M30 ——— FIV center
M31 —— artificial insemination
M32 ——— AID
M33 ——— AIH
M34 —— excess embryos
M35 —— host mothers
M36 —— embryo transfer
M37 —— GIFT (Gamete Intrafallopian Transfer)
M38 —— embryo donation

M39 — PREGNANCY AND CHILDBIRTH
M40 —— childbirth
M41 ——— caesarean section
M42 —— pregnant women
M43 ——— mother fetus relationship
M44 —— pregnancy
M45 ——— pregnancy in adolescence
M46 ——— multiple pregnancy

M47 — ABORTION
M48 —— illegal abortion
M49 —— selective abortion
M50 —— therapeutic abortion
M51 —— aborted fetuses
M52 —— abortion on demand
M53 —— mifepristone

M54 — FETAL DEVELOPMENT
M55 —— twinning
M56 —— viability

M57 — FETAL THERAPY

N1 DISEASE

N2 — COMMON SYMPTOMATOLOGY
N3 —— iatrogenic disease

N4 —— case studies
N5 —— patients
N6 ——— critically ill
N7 ——— chronically ill
N8 —— physically
N9 ——— physically handicapped
N10 —— self induced illness
N11 —— suffering
N12 ——— pain
N13 —— follow-up studies
N14 —— prognosis
N15 —— transmission
N16 ——— mother-child transmission

N17 — CANCER
N18 —— leukemia

N19 — CARDIOVASCULAR DISEASES
N20 —— heart diseases
N21 —— hypertension

N22 — DIABETES

N23 — CONGENITAL DEFECTS
N24 —— neural tube defects
N25 ——— anencephaly
N26 ——— spina bifida

N27 — COMMUNICABLE DISEASES
N28 —— whooping cough
N29 —— influenza
N30 —— hepatitis
N31 —— venereal diseases
N32 ——— syphilis
N33 —— poliomyelitis
N34 —— rubella
N35 —— acquired immunodeficiency syndrome
N36 ——— HIV seropositivity
N37 —— virus diseases

N38 — EYE DISEASES

N39 — KIDNEY DISEASES

N40 — NERVOUS SYSTEM DISEASES
N41 —— coma
N42 ——— persistent vegetative state
N43 —— central nervous system diseases
N44 ——— brain diseases
N45 ———— temporal lobe epilepsy

N46 — INJURIES, OCCUPATIONAL DISEASES, INTOXICATIONS

N47 —— burns patients
N48 —— occupational diseases
N49 —— injuries
N50 —— prenatal injuries

O1 **DEATH AND RESUSCITATION**

O2 — DETERMINATION OF DEATH
O3 —— autopsies
O4 —— cadavers

O5 —— death
O6 —— brain death

O7 — TERMINAL CARE
O8 —— palliative care
O9 —— life-sustaining treatment
O10 —— withholding treatment
O11 —— allowing to die
O12 —— terminally ill
O13 —— prolongation of life
O14 —— quality adjusted life years

O15 — EUTHANASIA
O16 —— active euthanasia
O17 —— involuntary euthanasia
O18 —— voluntary euthanasia

O19 — SUICIDE

O20 — ATTITUDES TO DEATH

O21 — RESUSCITATION

P1 NEUROSCIENCES — PSYCHIATRY

P2 — BEHAVIORAL AND MENTAL DISORDERS
P3 —— substance dependence
P4 ——— alcohol abuse
P5 ——— smoking
P6 ——— drug abuse
P7 —— mentally handicapped
P8 ——— mentally retarded
P9 ——— mentally ill
P10 —— psychotic disorders
P11 ——— dementia
P12 ——— schizophrenia
P13 —— behavior disorders
P14 ——— aggression
P15 ———— dangerousness
P16 ——— hyperkinesis
P17 ——— psychological stress

P18 — OUTPATIENT COMMITMENT
P19 —— volontary admission
P20 —— duration of commitment
P21 —— involontary commitment
P22 —— deinstitutionalized persons
P23 —— institutionalized persons

P24 — PSYCHIATRIC TECHNICS
P25 —— behavior control
P26 ——— operant conditioning
P27 ———— negative reinforcement
P28 ———— positive reinforcement
P29 —— psychiatric diagnosis
P30 —— psychotherapy
P31 ——— hypnosis
P32 ——— group therapy
P33 —— psychoactive drugs
P34 ——— heroin
P35 ——— LSD
P36 —— electrical stimulation of the brain
P37 —— physical psychiatric methods

P38 ——— psychosurgery
P39 ——— electroconvulsive therapy

Q1 GENETICS AND APPLICATIONS — BIOTECHNOLOGY

Q2 — GENETIC DEFECTS AND HEREDITARY DISEASES
Q3 —— genetic defects
Q4 ——— chromosomal disorders
Q5 ———— XYY karyotype
Q6 ———— down's syndrome
Q7 ——— single gene defects
Q8 ———— dominant genetic conditions
Q9 ————— Huntington's chorea
Q10 ———— recessive genetic conditions
Q11 ————— sickle cell anemia
Q12 ————— Duchenne muscular dystrophy
Q13 ————— hemophilia
Q14 ————— Tay Sachs disease
Q15 ————— cystic fibrosis
Q16 ————— phenylketonuria
Q17 ————— thalassemia
Q18 —— hereditary diseases
Q19 —— sex linked defects

Q20 — BIOTECHNOLOGY — GENETIC ENGINEERING
Q21 —— biotechnology
Q22 —— genetic intervention
Q23 ——— genome mapping
Q24 ——— cloning
Q25 ———— clones
Q26 ——— sex determination
Q27 ——— hybrids
Q28 ——— genetically modified organisms
Q29 ———— transgenic animals
Q30 ——— sex preselection
Q31 ——— recombinant DNA research
Q32 ———— DNA fingerprinting
Q33 ———— DNA linkage

Q34 — BEHAVIORAL GENETICS

Q35 — MEDICAL GENETICS
Q36 —— genetic counseling
Q37 —— genetic screening
Q38 —— preimplantation genetic diagnosis
Q39 —— eugenics
Q40 ——— negative eugenics
Q41 ——— positive eugenics
Q42 —— gene therapy

Q43 — POPULATION GENETICS

Q44 —— genetic identity
Q45 —— gene pool
Q46 —— carriers
Q47 —— human genome

Thematic Index

Smith Kline [O I 33], Fulford [GB 20], Garanis [GR 7], Gillen [L 3], Gillon [GB 21], Gormally [GB 22], Gress-Heister [D 16], Grimaldi [F 98], Heeger [NL 14], Hennaux [B 28], Honnefelder [D 22], Hübner [D 23], Istituto Internazionale di Teologia Pastorale Sanitaria Camillianum [O I 37], Istituto di Bioetica, Università Cattolica del S. Cuore [O I 38], Jacquemin [B 31], Kahlke [D 28], Katsas [GR 9], Lamau [F 116], Le Sommer-Péré [F 123], Lopez Azpitarte [E 23], Lunshof [D 32], Mattei [F 142], Michaud [F 145], Montagut [F 150], Mordini [I 35], Noach [NL 21], Pinto Machado [P 22], Riis [DK 42], Roessler [D 43], Sann [F 176], Schaerer [F 178], Scocozza [DK 43], Serrao [P 31], Skou [DK 45], The Linacre Centre for the Health Care Ethics [O GB 33], Unité de philosophie des sciences biomédicales [O B 10], Van Cutsem [B 47], Vedrinne [F 197], Veldhuis [NL 42], Verspieren [F 198], Viafora [I 51], Vidal [F 201], Von Schubert [D 56], Widdershoven [NL 47], Zorrilla [B 52], Zwart [NL 49].

biomedical research [F2]

Airaudo [F 2], André [B 2], Association internationale droit, éthique et science (Groupe de Milazzo) [O F 8], Barret [F 17], Boeri [I 7], Brazier [GB 3], Breteau [F 35], Centro di Bioetica dell'Istituto Gramsci di Roma [O I 10], Chermann [F 44], Coen [F 49], Comité d'éthique du CHU de Nancy [O F 27], Commissione etica della facoltà di medicina e chirurgia [O I 24], Conseil d'éthique en psychiatrie [O F 33], Conseil National de l'Ordre des Médecins [O F 35], d'Hautefeuille [F 52], Dalla-Vorgia [GR 3], Dauphin [F 56], Département d'éthique biomédicale — Centre Sevres [O F 37], Di mola [I 20], Doppelfeld [D 7], Doray [F 70], Doyal [GB 13], Eggermont [B 17], Fedida [F 86], Feltz [B 20], Giraud [F 95], Godfraind [B 24], Goetze-Claren [D 15], Grimaldi [F 98], Guenther [D 17], Houtepen [NL 16], Huber [F 108], Institute of Medical Ethics [O GB 18], Kahn [F 112], Lachaux [F 115], Langlois [F 117], Larsen [DK 28], Lehn [F 125], Leroy, O.G. [B 34], Lévy [F 132], Lopes [F 136], Maio [D 33], Marc-Vergnes [F 139], Moisselin [F 149], Novaes [F 155], Palacios [E 27], Prof. dr. G.A. Lindeboom Instituut [O NL 7], Rappaport [F 166], Roessler [D 43], Romeo Casabona [E 30], Scocozza [DK 43], Serusclat [F 182], Siksou [F 184], Taillemite [F 186], Tempe [F 188], ten Have [NL 29], Toellner [D 52], Unité de philosophie des sciences biomédicales [O B 10], Vedrinne [F 197], Vidal [F 201], Wells [GB 48].

biomedical research and experimentation [F1]

Airaudo [F 2], Anthony [GR 2], Assenmacher [F 7], Association internationale droit, éthique et science (Groupe de Milazzo) [O F 8], Bernardo, O.P. [P 4], Blanc [F 30], Boucaud [F 32], Breteau [F 35], Centro di Deontologia e di etica medica dell'Università [O I 11], Chiarelli [I 12], Coen [F 49], Comitato Etico dell' Istituto Nazionale per la ricerca sul cancro [O I 18], Comitato Etico Fondazione Floriani [O I 19], Comité d'éthique du CHU de Nancy [O F 27], Di mola [I 20], Duroux [F 78], Emile [F 81], Eser [D 10], Fagot-Largeault [F 83], Fondation Marcel-Mérieux [O F 44], Gros [F 99], Groupe franco-québécois de Recherche en Bioéthique — Association du B O F [O F 51], Guenther [D 17], Hennau-Hublet [B 27], Hirsch [F 106], Iglesias [IRL 4], Jasmin [F 111], Kemp [DK 23], Lachaux [F 115], Linneman [DK 31], Lluis [F 134], Löw [D 31], Maio [D 33], Marc-Vergnes [F 139], Mattei [F 142], Nys [B 40], Reiter [D 41], Reiter-Theil [D 42], Romeo Casabona [E 30], Rouffy [F 172], Santé, Ethique et Libertés [O F 61], Seminar für Moraltheologie [O D 19], Sgreccia [I 44], Siksou [F 184], Sinaccio [I 46], Sinet [F 185], Spagnolo [I 47], Tempe [F 188], The Medical Research Council [O GB 34], Vakgroep Ethiek Filosofie en Geschiedenis der Geneeskunde [O NL 13], Veldhuis [NL 42], Viafora [I 51], Walsh [GB 47].

biomedical technologies [E28]

Agazzi [I 1], Artioli [I 4], Assenmacher [F 7], Association internationale droit, éthique et science (Groupe de Milazzo) [O F 8], Audhuy [F 12], Bayertz [D 3], Byk [F 39], CECOS Italia [O I 8], Center for Ethics [O NL 2], Centre Interfacultaire de droit, éthique et sciences de la santé [O B 5], Chermann [F 44], Comitato Etico dell' Istituto Nazionale per la ricerca sul cancro [O I 18], Conseil National de l'Ordre des Médecins [O F 35], Conselho Nacional de Etica para as Ciencias da vida [O P 4], Debry [B 12], Dekkers [NL 8], Egozcue [E 13], Englert [B 18], Flamigni [I 24], Fondation Marcel-Mérieux [O F 44], Garde [DK 13], Goossens [F 96], Holme Hansen [DK 18], Iglesias [IRL 4], Institute of Medical Ethics — Medical Postgraduate Department [O GB 19], Jacquemin [B 31], Kahlke [D 28], Montagut [F 150], Nunez Cubero [E 26], Poortman [NL 24], Quintana Trias [E 28B], Reinders [NL 26], Romeo Casabona [E 30], Rusticheli [I 42], Salvino Leone [I 43], Scocozza [DK 43], Sinaccio [I 46], The Medical Research Council [O GB 34], Toscano Rico [P 34], Unité de philosophie des sciences biomédicales [O B 10].

biotechnology [Q21]

Achterberg [NL 1], Andersen [DK 2], Assenmacher [F 7], Association internationale droit, éthique et science (Groupe de Milazzo) [O F 8], Binamé [B 4], Boné [L 2], Byk [F 39], Chermann [F 44], Comitato Etico dell' Istituto Nazionale per la ricerca sul cancro [O I 18], Conselho Nacional de Etica para as Ciencias da vida [O P 4], Danchin [F 54], De Carli [I 18], Dupret [F 75], Duyme [F 79], Egozcue [E 13], Farsides [GB 17], Honnefelder [D 22], Jäger [D 26], Lahaye-Bekaert [B 33], Lombardi Vallauri [I 30], Lopes [F 136], Salvino Leone [I 43], Sinaccio [I 46], Sutton [GB 46], Taillemite [F 186], Thomsen [DK 47], van der Burg [NL 34].

biotechnology — genetic engineering [Q20]

Association internationale droit, éthique et science (Groupe de Milazzo) [O F 8], Beaufils [F 21], Belgische Vereniging voor medische moraal en ethiek [O B 2], Binamé [B 4], Centre for Business and Professional Ethics [O GB 4], Changeux [F 43], Chermann [F 44], Coen [F 49], Commissione etica della facoltà di medicina e chirurgia [O I 24], Conselho Nacional de Etica para as Ciencias da vida [O P 4], Cortina [E 10], Costa [P 9], Danchin [F 54], Dausset [F 57], De Carli [I 18], Demaison [F 66], Dupret [F 75], Folscheid [F 88], Fondation Marcel-Mérieux [O F 44], Gafo [E 17], Gunning [GB 23], Institute of Medical Ethics [O GB 18], Kahlke [D 28], Kahn [F 112], Kemp [DK 23], Lacadena [E 22], Lehn [F 125], Linneman [DK 31], Lopez Azpitarte [E 23], Mahoney [GB 34], Müller [D 35], Nivelon-Chevalier [F 154], Norby [DK 36], Palacios [E 27], Poortman [NL 24], Renard [F 167], Romeo Casabona [E 30], Salvino Leone [I 43], Schockweiler [L 6], Susanne [B 44], Taillemite [F 186], Toellner [D 52], van den Beld [NL 31], van Tongeren [NL 40], Vial [F 199], Von Schubert [D 56], Vorstenbosch [NL 45], Walsh [GB 47].

Blacks [D41]

Audras de la Bastie [F 13], Boucaud [F 32].

blood donation [I20]

Association internationale droit, éthique et science (Groupe de Milazzo) [O F 8], Berghmans [NL 2], Binamé [B 4], Centre de sociologie de l'éthique [O F 20], Chermann [F 44], Commission consultative nationale d'éthique pour les sciences de la vie et de la santé [O L 1], Conselho Nacional de Etica para as Ciencias da vida [O P 4], De Wachter [NL 6], Rubellin-Devichi [F 174], Skou [DK 45], Von Schubert [D 56], Zwart [NL 49].

blood substitutes [I21]

Berghmans [NL 2], Binamé [B 4], Chermann [F 44], Commission consultative nationale d'éthique pour les sciences de la vie et de la santé [O L 1], Conselho Nacional de Etica para as Ciencias da vida [O P 4], Skou [DK 45].

Index of names

C

D

N

O

P

QR

S

T

U

V

W-Z

EUROPEAN DIRECTORY OF BIOETHICS

Under the direction of Gérard HUBER

Reading committee

Editorial coordination :
Christian Legrand (Association Descartes)

Conception, editorial, computer and survey engineering under the direction of
Philippe Amiel (Médialexis) *and* Christine Clerc (Onotaxis)

Typing and correction : Dominique Atalay, Muriel Bordeau, Katia Boucebci, Sonia Codhant, Dany Misse.
Translation : Kathleen Broman, Nicole Candelier-Bourrier, Duncan Davidson, Tilly Gaillard, Magali Guenette, Clara Mazzi, Leonor Michelsen-Koutoulakis, Alejandra Pinto, Mark W. Suits.
Correction: André Amiel, Sylvie Duverger, Bernadette Dremière, Yvette Fantin, Harold Hyman
Maps : Patrick Vallée
Economic and demographic Information on countries : with the authorisation of *Courrier International* and *The Economist.*

John Libbey Eurotext / Association Descartes

Editions de l'Association Descartes 1994

Patrimoine génétique et droit de l'humanité, sous la direction de Gérard Huber, Osiris, Paris, 1990.
Individus sous influence : drogues, alcools, médicaments psychotropes, sous la direction de Alain Ehrenberg, Éditions Esprit, Paris, 1991.
Drogues, politique et société, sous la direction de Alain Ehrenberg et Patrick Mignon, Le Monde éditions / Éditions Descartes, Paris, 1992.
Terre, patrimoine commun, sous la direction de Martine Barrère, Éditions Descartes / La découverte, Paris.
Figures de l'altérité, Jean Baudrillard et Marc Guillaume, Éditions Descartes, Paris 1992.
L'Allemagne en puissance, textes réunis par François Bafoil et Ivan Samson, Éditions Descartes, Paris, 1992.
Modes de consommation : mesure et démesure, textes réunis par Pierre Chambat, Éditions Descartes, Paris, 1992.
Les nouveaux outils du savoir, textes réunis par Pierre Chambat et Pierre Lévy, Éditions Descartes, Paris, 1992.
Vers un anti-destin ? Patrimoine génétique et droits de l'humanité, sous la direction de François Gros et Gérard Huber, Éditions Odile Jacob, Paris, 1992.
Atelier "L'éducation face à la crise des valeurs", Unesco, Association Descartes, Paris, 1992.
Technologies : l'ouverture des frontières, Bengt-Arne Vedin, Éditions Descartes, Paris, 1992.
Les mots pour le faire. Conception des modes d'emploi, sous la direction de Dominique Boullier et Marc Legrand, Éditions Descartes, Paris, 1992.
Penser la drogue/penser les drogues (3 volumes), décembre 1992 :

- État des lieux, vol. I, sous la direction de Alain Ehrenberg ;
- Les marchés interdits de la drogue (Evaluation européenne des connaissances), vol. II, sous la direction de Michel Schiray ;
- Bibliographies, vol. III.

Communication et lien social, sous la direction de Pierre Chambat, Éditions Descartes, Paris, novembre 1992.
Annuaire européen de bioéthique, sous la direction de Gérard Huber, Éditions Descartes, Paris, janvier 1993.
La nature en politique, sous la direction de Dominique Bourg, Association Descartes / Éditions l'Harmattan, Paris, 1993.
Les sentiments de la nature, sous la direction de Dominique Bourg, Éditions Descartes / La découverte, Paris, 1993.
Agricultures et société: pistes pour la recherche, sous la direction de Catherine Courtet, Martine Berlan et Yves Demarne, Association Descartes / INRA Editions, Paris, 1993.
L'identité philosophique européenne, sous la direction de Jacques Poulain et Patrice Vermeren, Association Descartes/ Éditions l'Harmattan, Paris, 1993.

Achevé d'imprimer par  Corlet, Imprimeur, S.A.
14110 Condé-sur-Noireau (France)
N° d'Imprimeur : 1882 - Dépôt légal : décembre 1993

Imprimé en C.E.E.